COMMONLY ASKED QUESTIONS IN THERMODYNAMICS

Commonly Asked Questions in
THERMODYNAMICS

Marc J. Assael
Aristotle University, Thessaloniki, Greece

Anthony R. H. Goodwin
Schlumberger Technology Corporation, Sugar Land, Texas, USA

Michael Stamatoudis
Aristotle University, Thessaloniki, Greece

William A. Wakeham
University of Southampton, United Kingdom

Stefan Will
Universitat Bremen, Bremen, Germany

CRC Press
Taylor & Francis Group
Boca Raton London New York

CRC Press is an imprint of the
Taylor & Francis Group, an **informa** business

CRC Press
Taylor & Francis Group
6000 Broken Sound Parkway NW, Suite 300
Boca Raton, FL 33487-2742

Printed in the United States of America on acid-free paper
10 9 8 7 6 5 4 3 2 1

International Standard Book Number: 978-1-4200-8695-9 (Paperback)

Library of Congress Cataloging-in-Publication Data

Commonly asked questions in thermodynamics / Marc J. Assael ... [et al.].
p. cm.
Summary: "Accurate and clearly explained answers to common questions. Every scientist and engineer encounters problems that may be solved at least in part using the principles of thermodynamics. The importance of thermodynamics is often so fundamental to life that we should all have a fairly detailed understanding of this core field. This clearly written, easy-to-follow guide allows even nonscientists considering use of alternative fuel sources to achieve a solid grounding in thermodynamics. The authors cover topics spanning from energy sources to the environment to climate change. A broad audience of general readers, students, industry professionals, and academic researchers will appreciate the answers found in this book"-- Provided by publisher.
Includes bibliographical references and index.
ISBN 978-1-4200-8695-9 (pbk.)
1. Thermodynamics--Miscellanea. I. Assael, Marc J.

QC319.C66 2011
536'.7--dc22 2010050698

Visit the Taylor & Francis Web site at
http://www.taylorandfrancis.com

and the CRC Press Web site at
http://www.crcpress.com

The authors are indebted individually and collectively to a large body of students whom they have taught in many universities in different countries of the world. It is the continually renewed inquisitiveness of students that provides both the greatest challenge and reward from teaching in a university. It is not possible for us to single out individual students who have asked stimulating and interesting questions over a career of teaching in universities.

Contents

Preface

The concept of a series of books entitled *Commonly Asked Questions in...* is inherently attractive in an educational context, an industrial context, or even a research context. This is, of course, at least in part because the idea of a tutorial on a topic to be studied and understood provides a means of seeking personal advice and tuition on special elements of the topic that cannot be understood through the primary medium of education. The primary means can be a lecture, a text book, or a practical demonstration. Equally the motivation for the study can be acquisition of an undergraduate degree, professional enhancement, or the development of a knowledge base beyond one's initial field to advance a technical project or a research activity. Thus, the spectrum of motivations and the potential readership is rather large and at very different levels of experience. As the authors have developed this book, they have become acutely aware that this is especially the case for thermodynamics and thermophysics. The subjects of thermodynamics and thermophysics play a moderate role in every other discipline of science from the nanoscale to the cosmos and astrophysics with biology and life sciences in between. Furthermore, while some aspects of thermodynamics underpin the very fundamentals of these subjects, others aspects of thermodynamics have an impact on almost every application in engineering. In consequence, the individuals who may have questions about thermodynamics and its applications encompass most of the world's scientists and engineers at different levels of activity ranging from the undergraduate to the research frontier.

The task of writing a single text that attempts to answer all questions that might arise from this group of people and this range of disciplines is evidently impossible, partly because only one section of the text is likely to be of use to most people, and partly because the sheer extent of the knowledge available in this subject would be beyond the scope of the book.

We have therefore not attempted to write such a comprehensive text. We have instead been selective about the areas and disciplines we have decided to concentrate on: thermodynamics as opposed to thermophysics, chemical thermodynamics in particular, with a focus on chemists, chemical engineers, and mechanical engineers. Of course, this focus represents the bias of the authors' own backgrounds but this also covers the content required by a large number of those who will wish to make use of the material. In addition, the nature of the subject is such that even within the limited scope we have set, we have not always been able to be deductive and take a rigorous pedagogical approach. Thus in some sections the reader will find references to substantive texts devoted entirely to topics that we merely sketch.

It is our hope that this book will be useful to some of the wide audience who might benefit from answers to common questions in thermodynamics. It is often true in this subject that the most common questions are also rather profound and have engendered substantial debate both in the past and sometimes even today. We indicate a pragmatic way forward with these topics in this text, but we would not suggest that such a pragmatic approach should stifle further debate.

Accordingly, the first chapter answers questions about the fundamentals of the subject and provides some simple examples of applications. The second chapter briefly expounds the basis of statistical mechanics, which links the macroscopic observable properties of materials in equilibrium with the properties and interactions of the molecules they are composed of. Chapter 3 deals with the applications of the second law of thermodynamics and a range of thermodynamic functions. In Chapter 4 we consider the topic of phase equilibrium and the thermodynamics of fluid mixtures, which is vital for both chemists and chemical engineers. Chapter 5 deals with the topic of chemical reactions and systems that are not in equilibrium. This leads to Chapter 6 where we illustrate the principles associated with heat engines and refrigeration. In both cases our emphasis is on using examples to illustrate the earlier material.

Finally, we focus on the sources of data that a scientist or engineer can access to find values for the properties of a variety of materials that allow design and construction of process machinery for various industrial (manufacturing) or research purposes. Even here it is not possible to be comprehensive with respect to the wide range of data sources now available electronically, but we hope that the data sources we have listed will provide a route toward the end point, which will continue to extend as the electronic availability of information continues to expand. Here we are at pains to point out that each values obtained from a particular data source has an uncertainity associated with it. It is generally true that the uncertainty is at least as valuable as the data point itself because it expresses the faith that a design engineer should place in the data point and thus, in the end, on the final design.

Authors

Marc J. Assael, BSc, ACGI, MSc, DIC, PhD, CEng, CSci, MIChemE, is a professor in thermophysical properties. He is also the vice-chairman of the Faculty of Chemical Engineering at the Aristotle University of Thessaloniki in Greece.

Marc J. Assael received his PhD from Imperial College in 1980 (under the supervision of Professor Sir William A. Wakeham) for the thesis "Measurement of the Thermal Conductivity of Gases." In 1982 he was elected lecturer in heat transfer in the Faculty of Chemical Engineering at the Aristotle University of Thessaloniki, where he founded the Thermophysical Properties Laboratory. In 1986 he was elected assistant professor, in 1991 associate professor, and in 2001 professor of thermophysical properties at the same faculty. During the years 1991–1994 he served as the vice-chairman of the faculty and during 1995–1997 he served as the chairman of the Faculty of Chemical Engineering. In 2005, the laboratory was renamed Laboratory of Thermophysical Properties and Environmental Processes, to take into account the corresponding expansion of its activities.

In 1998, Marc J. Assael was TEPCO Chair Visiting Professor in Keio University, Tokyo, Japan, and from 2007 he has also been holding the position of adjunct professor in Jiaotong University, Xi'an, China. He has published more than 250 papers in international journals and conference proceedings, 20 chapters in books, and six books. In 1996, his book *Thermophysical Properties of Fluids: An Introduction to their Prediction* (coauthored by J. P. M. Trusler and T. F. Tsolakis) was published by Imperial College Press (a Greek edition was published by A. Tziola E.), while in 2009, his latest book, *Risk Assessment: A Handbook for the Calculation of Consequences from Fires, Explosions and*

Toxic Gases Dispersion (coauthored by K. Kakosimos), was published by CRC Press (a Greek edition is also published by A. Tziolas E.). He is acting as a referee for most journals in the area of thermophysical properties, while he is also a member of the editorial board of the following scientific journals: *International Journal of Thermophysics, High Temperatures – High Pressures, IChemE Transactions Part D: Education for Chemical Engineers,* and *International Review of Chemical Engineering.*

Marc J. Assael is a national delegate in many committees in the European Union, in the European Federation of Chemical Engineering, as well as in many international scientific organizations.

He is married to Dora Kyriafini and has a son named John-Alexander.

Dr. Anthony R. H. Goodwin is a scientific advisor with Schlumberger and is currently located in Sugar Land, Texas. Dr. Goodwin obtained his PhD from the laboratory of Professor M. L. McGlashan at University College, London, under the supervision of Dr. M. B. Ewing.

After graduation, Dr. Goodwin worked at BP Research Centre, Sunbury, United Kingdom, and then moved to the Physical and Chemical Properties Division of the National Institute of Standards and Technology, Gaithersburg, Maryland. He then took a post at the Department of Chemical Engineering and Centre for Applied Thermodynamic Studies at the University of Idaho from where he joined Schlumberger, first in Cambridge, United Kingdom, then Ridgefield, Connecticut, and now Texas.

Dr. Goodwin's interests include experimental methods for the determination of the thermodynamic and transport properties of fluids and the correlation of these properties. Previously, Dr. Goodwin was an editor of the *Journal of Chemical Thermodynamics* and is now an associate editor of the *Journal of Chemical and Engineering Data*. At Schlumberger he focuses on the measurement of the properties of petroleum reservoir fluids, especially the development of methods to determine these properties down hole in adverse environments. In particular, Dr. Goodwin has extended his research to the use of instruments developed using micro-electromechanical systems (MEMS), which combines the process of integrated circuits with bulk micromachining, for the determination of the thermophysical properties of fluids.

Dr. Goodwin has over 148 publications. This includes 81 refereed journals, 25 granted patents, 16 published patents, 3 edited books and 6 chapters contributed to multiauthor reviews, and 17 publications in conference proceedings. These articles report both state-of-the-art experimental methods and experimental data on the thermophysical properties of alternative refrigerants and hydrocarbon fluids, as well as the measurement of thermophysical properties related to oil field technologies.

He is an active member of several professional organizations, including chairman and former treasurer of the International Association of Chemical Thermodynamics, Fellow of the Royal Society of Chemistry, and member of the American Chemical Society. Dr. Goodwin is an associate member of the Physical and Biophysical Chemistry Division of the International Union of Pure and Applied Chemistry. He has edited two books for the International Union of Pure and Applied Chemistry entitled *Experimental Thermodynamic Volume VI, Measurement of the Thermodynamic Properties of Single Phases* and *Applied Thermodynamics* with Professors J.V. Sengers and C. J. Peters.

Michael Stamatoudis received his bachelor of science in chemical engineering from Rutgers University in 1971 and his master of science in chemical engineering from Illinois Institute of Technology in 1973. Michael Stamatoudis also received his PhD in chemical engineering from Illinois Institute of Technology in 1977 under the supervision of Professor L.L. Tavlarides. In 1982 he was elected lecturer in the Faculty of Chemical Engineering at the Aristotle University of Thessaloniki, Greece. In 1986 he was elected assistant professor and in 1992 associate professor. Currently he serves as a professor of unit operations. During the years 1995–1997 and 2003 he served as the vice-chairman of the Faculty of Chemical Engineering. He has published several papers on applied thermodynamics and on two-phase systems. He is married and has four children.

Professor Sir William A. Wakeham retired as vice-chancellor of the University of Southampton in September 2009 after eight years in that position. He began his career with training in physics at Exeter University at both undergraduate and doctoral level. In 1971, after a postdoctoral period in the United States at Brown University, he took up a lectureship in the Chemical Engineering Department at Imperial College London and became a professor in 1983 and head of department in 1988. His academic publications include six books and about 400 peer-reviewed papers.

From 1996 to 2001 he was pro-rector (research), deputy rector, and pro-rector (resources) at Imperial College. Among other activities he oversaw the college's merger with a series of medical schools and stimulated its entrepreneurial activities.

He is a Fellow of the Royal Academy of Engineering, a vice-president and its International Secretary, and a Fellow of the Institution of Chemical Engineers, the Institution of Engineering and Technology, and the Institute of Physics. He holds a higher doctorate from Exeter University and honorary degrees from Lisbon, Exeter, Loughborough and Southampton Solent Universities and is a Fellow of Imperial College London and holds a number of international awards for his contributions to research in transport processes.

He has, until this year, been chair of the University and Colleges Employers Association and the Employers Pensions Forum and a member of the Board of South East of England Development Agency. In 2008 he chaired a Review of Physics as a discipline in the United Kingdom for Research Councils UK and completed a review of the effectiveness of Full Economic Costing of Research for RCUK/UUK in 2010.

He is a council member of the Engineering and Physical Sciences Research Council and chair of its Audit Committee. He is also currently a visiting professor at Imperial College London; Instituto Superior Técnico, Lisbon; and University of Exeter, as well as chair of the Exeter Science Park Company, Non-Executive Director of Ilika plc, chair of the South East Physics Network, trustee of Royal Anniversary Trust, and the Rank Prizes Fund. He was made a knight bachelor in the 2009 Queen's Birthday Honours list for services to chemical engineering and higher education.

Stefan Will is a professor in Engineering Thermodynamics in the Faculty of Production Engineering of the University of Bremen in Germany.

After graduation in physics Stefan Will received a doctoral degree in engineering from the Technical Faculty of the University Erlangen-Nuremberg in 1995 for a thesis on "Viscosity Measurement by Dynamic Light Scattering." After holding several academic positions at different universities he is a full professor at the University of Bremen since 2002. During the years 2003–2009 he served as deputy dean and dean, respectively, of the Faculty of Production Engineering.

Stefan Will's research interests include optical techniques in engineering, particle and combustion diagnostics, thermophysical properties, heat and mass transfer, and desalination. In these fields he has authored and coauthored more than 100 publications in international journals, conference proceedings, and books. He is an active member and delegate in several national and international organizations in thermodynamics and mechanical/process engineering. He is married and has two children.

Chapter 1

Definitions and the 1st Law of Thermodynamics

1.1 INTRODUCTION

The subjects of thermodynamics, statistical mechanics, kinetic theory, and transport phenomena are almost universal within university courses in physical and biological sciences, and engineering. The intensity with which these topics are studied as well as the balance between them varies considerably by discipline. However, to some extent the development and, indeed, ultimate practice of these disciplines requires thermodynamics as a foundation. It is, therefore, rather more than unfortunate that for many studying courses in one or more of these topics thermodynamics present a very great challenge. It is often argued by students that the topics are particularly difficult and abstract with a large amount of complicated mathematics and rather few practical examples that arise in everyday life. Probably for this reason surveys of students reveal that most strive simply to learn enough to pass the requisite examination but do not attempt serious understanding. However, our lives use and require energy, its conversion in a variety of forms, and understanding these processes is intimately connected to thermodynamics and transport phenomena; the latter is not the main subject of this work. For example, whether a particular proposed new source of energy or a new product is genuinely renewable and/or carbon neutral depends greatly on a global energy balance, on the processes of its production, and its interaction with the environment. This analysis is necessarily based on the laws of thermodynamics, which makes it even more important now for all scientists and engineers to have a full appreciation of these subjects as they seek to grapple with increasingly complex and interconnected problems.

This book sets out to provide answers to some of the questions that undergraduate students and new researchers raise about thermodynamics and statistical mechanics. The list of topics is therefore rather eclectic and, perhaps in some sense, not entirely coherent. It is certainly true that the reader of any level should not expect to "learn" any of these subjects from this book alone. It is, instead, intended to complement existing texts, dealing in greater detail and in a different way with "some" of the topics deemed least straightforward by our own students over many years. If you do not find the question that you have treated in this text, then we apologize. Alternative sources of information include Cengel and Boles (2006), Sonntag et al. (2004), and Smith et al. (2004).

This chapter provides definitions that are required in all chapters of this book along with the definition of intermolecular forces and standard states.

1.2 WHAT IS THERMODYNAMICS?

Thermodynamics provides a rigorous mathematical formulation of the interrelationships among measurable physical quantities that are used to describe the energy and equilibria of macroscopic systems, as well as the experimental methods used to determine those quantities. These formulations include contributions from pressure, volume, chemical potential, and electrical work, but there can also be significant energy contributions arising from electromagnetic sources, gravitation, and relativity. The contributions that are important change with the discipline in which the problem arises. For example, for the majority of chemists the inclusion of gravitational and relativistic contributions is unimportant because of their dominant requirement to understand chemical reactions and equilibrium, whereas for physicists the same contributions may be dominant and chemical and mechanical engineers may need to include electromagnetic forces but will also need to account for phenomena associated with nonequilibrium states such as the processes that describe the movement of energy, momentum, and matter.

The fact that thermodynamics relates measurable physical quantities implies that measurements of those properties must be carried out for useful work to be done in the field. Generally speaking, the properties of interest are called *thermophysical properties*, a subset that pertains to equilibrium states being referred to as thermodynamic properties and a further subset that refers to dynamic processes in nonequilibrium states being called *transport properties*. Thermodynamics is an exacting experimental science because it has turned out to be quite difficult and time consuming to make very accurate measurements of properties over a range of conditions (temperature, pressure, and composition) for the wide range of materials of interest in the modern world. Given the exact relationship between properties that follows from thermodynamics the lack of accuracy has proved problematic. Thus, very considerable efforts have been made over many decades to refine experimental measurements, using methods for which complete working equations are available in the series *Experimental Thermodynamics* (Vol. I 1968, Vol. II 1975, Vol. III 1991, Vol. IV 1994, Vol. V 2000, Vol. VI 2003, Vol. VII 2005, Vol. VIII 2010). It has been important that any such measurements have a quantifiable uncertainty because of properties derived from them, for example, are required to design an effective and efficient air conditioning system. In this paragraph itself, several terms have been used, such as "system," which, in the field of thermodynamics, have a particular meaning and require definition; we have provided these definitions in the following text.

1.3 WHAT VOCABULARY IS NEEDED TO UNDERSTAND THERMODYNAMICS?

The A–Z of thermodynamics has been prepared by Perrot in 1998; hence we do not provide a comprehensive dictionary of thermodynamics here, but instead give some clear definitions of commonly encountered terms.

1.3.1 What Is a System?

A system is the part of the world chosen for study, while everything else is part of the surroundings. The system must be defined in order that one can analyze a particular problem but can be chosen for convenience to make the analysis simpler. Typically, in practical applications, the system is macroscopic and of tangible dimensions, such as a bucket of water; however, a single molecule is a perfectly acceptable microscopic system. A system is characterized both by its contents and the system boundary; the latter in the end is always virtual. For example, if one considers a container with a rigid enclosure, the boundary of the system is set in a way to include all the material inside but to exclude the walls. Especially in engineering applications, a careful and advantageous choice of the system boundary is of enormous importance; defining the right system boundary may considerably ease setting up energy and mass balances, for example.

1.3.2 What Is a State?

The state of a system is defined by specifying a number of thermodynamic variables for the system under study. In principle, these could be any or all of the measurable physical properties of a system. Fortunately, not all of the variables or properties need to be specified to define the state of the system because only a few can be varied independently; the exact number of independent variables depends on the system but rarely exceeds five. The exact choice of the independent variables for a system is a matter of convenience, but pressure and temperature are often included within them. As an illustration of this point, if the temperature and pressure of a pure gas are specified then the density of the gas takes a value (dependent variable) that is determined. The general rule for calculating the number of independent variables for a system at equilibrium is given by the phase rule that will be introduced and discussed in Question 4.1.1.

1.3.3 What Are the Types of Property: Extensive and Intensive?

For a system that can be divided into parts any property of the system that is the sum of the property of the parts is extensive. For example, the mass of the system is the sum of the mass of all parts into which it is divided. Volume and amount of substance (see Question 1.3.11) are all extensive properties as are energy, enthalpy, Gibbs function, Helmholtz function, and entropy, all of which are discussed later. A system property that can have the same value for each of the parts is an intensive property. The most familiar intensive properties are temperature and pressure. It is also worth remembering that the quotient of two extensive properties gives an intensive property. For example, the mass of a system (extensive) divided by its volume (extensive) yields its density, which is intensive.

1.3.4 What Is a Phase?

If a system has the same temperature and pressure, and so on throughout, and if none of these variables change with time, the system is said to be in equilibrium. If, in addition, the system has the same composition and density throughout, it is said to be homogeneous and is defined as a *phase*. When the system contains one or more phases so that the density and composition may vary but the system is still at equilibrium it is termed *heterogeneous*. Water contained in a closed metallic vessel near ambient conditions will have a layer of liquid water at the lowest level (liquid phase) and a vapor phase above it consisting of a mixture of air and water vapor. Necessarily, this picture implies that an interface exists between the liquid and the vapor. The properties of the system are therefore discontinuous at this interface, and, generally, interfacial forces that are not present in the two phases on either side will be present at the interface.

A phase that can exchange material with other phases or surroundings, depending on how the system boundary is defined, is termed *open*, while a closed phase is one that does not exchange material with other phases or surroundings. Consequently, an open system exchanges material with its surroundings and a closed one cannot. In the example given above, the closed metallic vessel contained liquid water and water vapor. If we define the system to include the two phases then the system is closed, but it contains two open phases exchanging material within it.

1.3.5 What Is a Thermodynamic Process?

A thermodynamic process has taken place when at two different times there is a difference in any macroscopic property of the system. A change in the macroscopic property is infinitesimal if it has occurred through an infinitesimal process. Processes can be categorized as follows: (1) natural, which proceed toward equilibrium, (2) unnatural, which occurs when the process proceeds away from equilibrium, and (3) reversible, which is between items 1 and 2 and proceeds either toward or away from equilibrium and which will be discussed further in Section 1.3.8. To illustrate items 1 and 3 we consider a system of substance B in both liquid and gaseous phases of vapor pressure p_B^{sat}, where the phases are at a pressure p. For the case that $p < p_B^{sat}$ the liquid will evaporate in a process that occurs naturally and is categorized by item 1. When $p > p_B^{sat}$ evaporation will not occur and the process is unnatural according to item 2.

The term process can have a variety of other implications for mechanical and chemical engineers, and while some are discussed in this chapter and briefly for irreversible thermodynamics in Chapter 6 others are not.

1.3.6 What Is Adiabatic?

As we have seen, a system is characterized as open or closed, depending on whether mass can cross the system boundary or not. Provided that any chemical reactions in the system have ceased, the state of a closed system is unchanging unless work or heat are transferred across the system boundary. When the system is thermally insulated, so that heat cannot cross the system boundary, it is called *adiabatically enclosed.* A Dewar flask with a stopper approximates an adiabatic enclosure. A system with thermally conducting walls, such as those made of a metal, is called *diathermic*. When a closed system is adiabatic and when no work can be done on it the system is termed *isolated.*

1.3.7 What Is Work?

When a system has electrical or mechanical effort expended within it or upon it, it is termed as *work done* on the system. The work can, and most often does, flow into the system from the surroundings. For example, an electric resistive heater mounted within a fluid, which is defined as the *system*, has work done on it from the surroundings when an electric current I flows through the resistor at a potential difference E, and both E and I are constant from the time the circuit is turned on t_1 to the time it is turned off t_2; the work done W is given by

$$W = EI(t_2 - t_1). \tag{1.1}$$

Work can also be done by changing the volume occupied by the system and by the energy dissipated by a stirrer. For an electrically driven stirrer, the energy dissipated is the energy consumed by the electric motor of the stirrer held in the surroundings, minus the energy used to increase the temperature of the motor, and to overcome the frictional losses within the mechanism used to transmit the power from the motor to the stirrer. There are other forms of work, including that done when the area of an interfacial layer separating two phases increases. Work is also done when a solid is stretched and when a substance is exposed to an electromagnetic field. In general, the work done is the sum of the terms of form $X\,\mathrm{d}y$, where X is an intensive variable, such as a force, and $\mathrm{d}y$ is an extensive quantity, for example, a displacement.

1.3.8 What Is a Reversible Process or Reversible Change?

In Section 1.3.5 an example was used to illustrate natural and unnatural processes, and this will be used for the topic of reversibility; we again define a system of substance B in both liquid and gaseous phases of vapor pressure $p_\mathrm{B}^\mathrm{sat}$, where the phases are at a pressure p. If $p = p_\mathrm{B}^\mathrm{sat}$ both evaporation and condensation can occur for any infinitesimal decrease or increase in p respectively, and the process is reversible, that is, for $p = p_\mathrm{B}^\mathrm{sat} - \delta p$, when $\delta p > 0$ the process conforms to item 1 of Section 1.3.5, and when $p = \lim_{\delta p \to 0} p_\mathrm{B}^\mathrm{sat}$ the process is reversible, it can be considered to be a passage through a continuous series of equilibrium states between the system and the surroundings.

Another, albeit difficult to comprehend but more important example of a reversible process concerns the work done on a phase α by the surroundings. In this case, if the work on α is restricted to an external pressure p_e^α, acting on the phase α, which is at a pressure p^α, then the change in volume of α is $\mathrm{d}V^\alpha$ and in the absence of friction given by

$$W = -p_\mathrm{e}^\alpha\,\mathrm{d}V^\alpha. \tag{1.2}$$

When $p_\mathrm{e}^\alpha = p^\alpha$ the change in volume is said to be reversible. That is, if $p_\mathrm{e}^\alpha = p^\alpha + \delta p$, where δp is an infinitesimal change in pressure, then $\mathrm{d}V^\alpha < 0$ and the phase α contracts. When $p_\mathrm{e}^\alpha = p^\alpha - \delta p$ then $\mathrm{d}V^\alpha > 0$ and the phase α expands. In both cases the change can be reversed by a change in p_e^α equal in magnitude to δp but of opposite sign: $\delta p_\mathrm{e}^\alpha = -\delta p$. When the pressure of the phase $p_\mathrm{e}^\alpha \neq p^\alpha$ the change in volume is not reversible.

However, when we refer to the passage of the system through a sequence of internal equilibrium states without the establishment of equilibrium with the surroundings this is referred to as a reversible change. An example that combines the concept of reversible change and reversible process will now be considered.

For this example, we define a system as a liquid and a vapor of a substance in equilibrium contained within a cylinder that on one circular end has a rigid immovable wall and on the other end has a piston exerting a pressure equal to the vapor pressure of the fluid at the system temperature. Energy in the form of heat is now applied to the outer surface of the metallic cylinder and the heat flows through the cylinder (owing to the relatively high thermal conductivity), increasing the liquid temperature. This results in further evaporation of the liquid and an increase in the vapor pressure. Work must be done on the piston at constant temperature to maintain the pressure. This change in the system is termed a *reversible change*. It can only be called a *reversible process* if the temperature of the substance surrounding the cylinder is at the same temperature as that of the liquid and vapor within the cylinder. This requirement arises because if the temperatures were not equal the heat flow through the walls would not be reversible, and thus, the whole process would not be reversible. If the system is only the liquid and the gas within the cylinder the process is reversible. Another example is provided by considering two systems both in complete equilibrium and in which the heat flows from one to the other. Each system undergoes a reversible change provided each remains at equilibrium. The heat flow is not reversible process unless the temperature of both systems is equal.

The importance of reversible processes and changes along with the content of Section 1.3.5 will first become apparent in Sections 1.7.4, 1.7.5, and 1.7.6, as well as in Chapter 6.

1.3.9 What Are Thermal Equilibrium and the Zeroth Law of Thermodynamics?

If an adiabatically enclosed system is separated into two parts by a diathermic wall then the two parts will be in thermal equilibrium with each other. This implies that the states of the two subsystems that are at thermal equilibrium are dependent on each other. In other words, there is a relationship between the independent variables that define the states of the two subsystems. Mathematically, for a system consisting of two parts A and B with independent variables Γ_A and Γ_B at thermal equilibrium there is a function f that relates the two sets of variables:

$$f(\Gamma_A, \Gamma_B) = 0. \tag{1.3}$$

For three systems A, B, and C that are all adiabatically enclosed, if A is in thermal equilibrium with B, which is also in equilibrium with C, then A must be in thermal equilibrium with C. This is often referred to as the zeroth law of thermodynamics. This of course assumes that sufficient time has elapsed to permit attainment of internal thermal equilibrium. This will be important when we consider temperature and its measurement.

1.3.10 What Is Chemical Composition?

The properties of a system consisting of a mixture of chemical components depend on the composition of the phase, which is specified by a measure of the amount of each chemical component present. The composition of a phase can change by virtue of the extent of a chemical reaction or by the gain or loss of one or more components. To study the variation of the properties of a mixture it is convenient to define other, nonthermodynamic quantities. The purpose of the following sections is to introduce these parameters.

1.3.11 What Is the Amount of Substance?

The amount of substance n_B of a chemical entity B in a system is a physical quantity defined by its proportionality to the number of entities N_B in the system that is given by $N_B = L \cdot n_B$, where L is the Avogadro constant (Mohr et al. 2008). For example, if the chemical entity B is an atom of argon then N_B is the number of atoms of argon in the system. The SI unit for the amount of substance is the mole defined currently by *Le Système international d'unités (SI)* (2006):

> The mole is the amount of substance of a system which contains as many elementary entities as there are atoms in 0.012 kilogram of carbon 12. When the mole is used, the elementary entities must be specified and may be atoms, molecules, ions, electrons, other particles, or specified groups of such particles.

The SI symbol for mole is mol. The specified groups need not be confined to independent entities or groups containing integral numbers of atoms. For example, it is quite correct to state an amount of substance of $0.5H_2O$ or of $(H_2 + 0.5O_2)$ or of $0.2MnO_4^-$.

Proposals to revise the definitions of the kilogram, ampere, Kelvin, and mol to link these units to exact values of the Planck constant h, the electron charge e, the Boltzmann constant k, and the Avogadro constant L, respectively, have been reported (Mills et al. 2006). One proposed definition for the mole is

> The mole is the amount of substance of a system that contains exactly $6.022\ 141\ 5 \cdot 10^{23}$ specified elementary entities, which may be atoms, molecules, ions, electrons, other particles or specified groups of such particles. (Mills et al. 2006)

We digress briefly here to consider, in the same context, the definition of the kilogram, which is currently as follows: The kilogram is the unit of mass; it is equal to the mass of the international prototype of the kilogram sanctioned by the 1st General Conference on Weights and Measures in 1889.

One proposed definition for the kilogram that removes the requirement for an arbitrary artifact whose mass is known to drift is

> The kilogram is the mass of a body whose equivalent energy is equal to that of a number of photons whose frequencies sum to exactly $[(299\ 792\ 458)^2/662\ 606\ 93] \cdot 10^{41}$ hertz. (Mills et al. 2006)

With similar redefinitions of the ampere and the Kelvin it would be possible to define six of the seven base units of the SI system in terms of true invariants of nature, fundamental physical constants. The current weakness of the definitions of the ampere, the mole, and the candela is derived in large measure from their dependence on the definition of the kilogram and its representational artifact.

1.3.12 What Are Molar and Mass or Specific Quantities?

The molar volume of a phase is the quotient of the volume and the total amount of substance of the phase. Generally, any extensive quantity X divided by the total amount of substance $\Sigma_B n_B$ is, by definition, an intensive quantity called the *molar quantity* X_m:

$$X_m = \frac{X}{\sum_B n_B}. \tag{1.4}$$

In Equation 1.4, the subscript m designates a molar quantity and can be replaced by the chemical symbol for the substance in this example, subscript B; when no ambiguity can result the subscripts m and B may be omitted entirely.

In engineering applications quantities are very often related to the mass instead of the amount of substance. The specific volume of a phase is the quotient of the volume and the total mass of substance of the phase. By analogy with molar quantities, any extensive quantity X divided by the total substance mass $\Sigma_B m_B$ is an intensive variable called the *specific quantity* x:

$$x = \frac{X}{\sum_B m_B}. \tag{1.5}$$

Specific quantities are normally designated by lowercase letters.

To elucidate the differences between molar and mass quantities a few examples are provided. The volume of a phase is given the symbol V, and when this refers to a molar volume the symbol V_m is used; the quantity is given by $V_\mathrm{m} = M/\rho = \rho_n^{-1}$, where M is the molar mass and ρ is the mass density, which is given by $\rho = m/V$, where m is the mass and ρ_n is the amount-of-substance density, which is related to the mass density by $\rho_n = \rho/M$. The specific volume v is given by $v = V/m = \rho^{-1}$ and defines the volume of a mass of material.

In the remainder of this book we make use of both molar and mass notation. The choice depends on whether the focus of the discussion is on chemistry and the (fundamental) properties of matter, whereas for engineering applications the use of mass or specific quantities is usually adopted. We may occasionally switch between molar and mass quantities without explicit mention. Throughout the text we have defined each symbol when it has either been first introduced or when it is used for a different purpose.

1.3.13 What Is Mole Fraction?

The mole fraction y of a substance B in a phase is given by y_B, which is an intensive quantity:

$$y_\mathrm{B} = \frac{n_\mathrm{B}}{\sum_\mathrm{B} n_\mathrm{B}}, \tag{1.6}$$

the sum of the mole fractions in a phase must then equal unity. An analogous definition is, of course, possible for mass fraction.

1.3.14 What Are Partial Molar Quantities?

The partial molar quantity X_B (which is an intensive quantity) of substance B in a mixture is defined by

$$X_\mathrm{B} = \left(\frac{\partial X}{\partial n_\mathrm{B}}\right)_{T,p,n_{\mathrm{A}\neq\mathrm{B}}}, \tag{1.7}$$

where $n_\mathrm{A} \neq n_\mathrm{B}$ means all the n's except n_B are held constant; for a pure substance B $X_\mathrm{B} = X/n_\mathrm{B} = X_\mathrm{m}$. Thus an extensive quantity X can be written as

$$X = \left(\frac{\partial X}{\partial T}\right)_{p,n_\mathrm{B}} \mathrm{d}T + \left(\frac{\partial X}{\partial p}\right)_{T,n_\mathrm{B}} \mathrm{d}p + \sum_\mathrm{B} X_\mathrm{B}\,\mathrm{d}n_\mathrm{B}, \tag{1.8}$$

and, by the use of Euler's theorem (see Question 1.11.2)

$$X = \sum_\mathrm{B} n_\mathrm{B} X_\mathrm{B}, \tag{1.9}$$

or is recast as

$$X_{\mathrm{m}} = \sum_{\mathrm{B}} x_{\mathrm{B}} X_{\mathrm{B}}, \tag{1.10}$$

on division of both sides by $\Sigma_{\mathrm{B}}\, n_{\mathrm{B}}$. Differentiation of Equation 1.9 and combination with Equation 1.8 gives

$$0 = -\left(\frac{\partial X}{\partial T}\right)_{p,n_{\mathrm{B}}} \mathrm{d}T - \left(\frac{\partial X}{\partial p}\right)_{T,n_{\mathrm{B}}} \mathrm{d}p + \sum_{\mathrm{B}} n_{\mathrm{B}}\,\mathrm{d}X_{\mathrm{B}}, \tag{1.11}$$

so that at constant temperature and pressure we have

$$0 = \sum_{\mathrm{B}} n_{\mathrm{B}}\,\mathrm{d}X_{\mathrm{B}}, \tag{1.12}$$

which, when substituted into the total derivative of Equation 1.9, gives

$$\mathrm{d}X = \sum_{\mathrm{B}} X_{\mathrm{B}}\,\mathrm{d}n_{\mathrm{B}}. \tag{1.13}$$

It can also be shown that

$$\mathrm{d}X_{\mathrm{m}} = \sum_{\mathrm{B}} X_{\mathrm{B}}\,\mathrm{d}x_{\mathrm{B}}. \tag{1.14}$$

Equations 1.10 and 1.14 can be used to determine all partial molar quantities of a mixture as a function of composition.

For a binary mixture of chemical species $\{x\mathrm{A} + (1 - x)\mathrm{B}\}$ Equations 1.10 and 1.14 are

$$X_{\mathrm{m}} = (1-x)X_{\mathrm{A}} + xX_{\mathrm{B}}, \tag{1.15}$$

and

$$\mathrm{d}X_{\mathrm{m}} = (X_{\mathrm{B}} - X_{\mathrm{A}})\,\mathrm{d}x. \tag{1.16}$$

When Equations 1.15 and 1.16 are solved for X_{A} and X_{B} they give

$$X_{\mathrm{A}} = X_{\mathrm{m}} - x\left(\frac{\partial X_{\mathrm{m}}}{\partial x}\right)_{T,p}, \tag{1.17}$$

and

$$X_{\mathrm{B}} = X_{\mathrm{m}} + (1-x)\left(\frac{\partial X_{\mathrm{m}}}{\partial x}\right)_{T,p}. \tag{1.18}$$

The partial molar quantities X_{A} and X_{B} for a particular composition can be obtained from measurements of X_{m} and the variation of X_{m} with x provided that the latter is nearly linear. When this is not so, as is often the case, for example, for the volume, then an alternative approach must be sought and this is provided by the molar quantity of mixing.

1.3.15 What Are Molar Quantities of Mixing?

For a binary mixture $\{(1 - x)\mathrm{A} + x\mathrm{B}\}$ the molar quantity of mixing at a temperature and pressure $\Delta_{\mathrm{mix}}X_{\mathrm{m}}$ is given by

$$\Delta_{\mathrm{mix}}X_{\mathrm{m}} = X_{\mathrm{m}} - (1-x)X_{\mathrm{A}}^{*} - xX_{\mathrm{B}}^{*}, \tag{1.19}$$

where X_{A}^{*} and X_{B}^{*} are the appropriate molar quantities of pure A and B. For example, the molar volume of mixing can be determined from measurements of the density ρ of the mixture, the densities of the pure materials, and a knowledge of the molar masses M of A and B from

$$\Delta_{\mathrm{mix}}V_{\mathrm{m}} = \frac{(1-x)M_{\mathrm{A}} + xM_{\mathrm{B}}}{\rho} - \frac{(1-x)M_{\mathrm{A}}}{\rho_{\mathrm{A}}^{*}} - \frac{xM}{\rho_{\mathrm{B}}^{*}}. \tag{1.20}$$

1.3.16 What Are Mixtures, Solutions, and Molality?

Mixture is the word reserved for systems (whether they be gases, liquids, or solids) containing more than one substance; all components in the mixture are treated equally. On the other hand, the term solution is reserved for liquids or solids containing more than one substance, where one substance is deemed to be a solvent and the others are solutes; these entities are not treated in the same way. If the sum of the mole fractions of the solutes is small compared with unity, the solution is termed *dilute*.

The composition of a solution is usually expressed in terms of the molalities of the solutes. The definition of the molality of a solute B m_{B} in a solvent A of molar mass M_{A} is defined by

$$m_{\mathrm{B}} = \frac{n_{\mathrm{B}}}{n_{\mathrm{A}}M_{\mathrm{A}}}, \tag{1.21}$$

and is related to the mole fraction x_{B} by

$$x_{\mathrm{B}} = \frac{m_{\mathrm{B}}M_{\mathrm{A}}}{1 + M_{\mathrm{A}}\sum_{\mathrm{B}} m_{\mathrm{B}}}, \tag{1.22}$$

or

$$m_{\mathrm{B}} = \frac{x_{\mathrm{B}}}{M_{\mathrm{A}}\left(1 - \sum_{\mathrm{B}} x_{\mathrm{B}}\right)}. \tag{1.23}$$

1.3.17 What Are Dilution and Infinite Dilution?

For a mixture of species A and B containing amounts of substance n_A and n_B, the change in a quantity X on dilution by the addition of an amount of substance Δn_A is $\Delta_{dil}X$, which is given by

$$\Delta_{dil}X = \Delta_{mix}X\{(n_A + \Delta n_A)A + n_B B\} - \Delta_{mix}X(n_A A + n_B B), \tag{1.24}$$

or when divided by n_B

$$\frac{\Delta_{dil}X}{n_B} = \frac{\Delta_{mix}X(x_f)}{x_f} - \frac{\Delta_{mix}X(x_i)}{x_i}, \tag{1.25}$$

where the subscripts f and i indicate the final and initial mole fractions of B. As $x_f \to 0$ one speaks of infinite dilution of species B in solvent A and the quantity is given as a superscript ∞ so that Equation 1.25 becomes

$$\left(\frac{\Delta_{dil}X}{n_B}\right)^{\infty} = \left\{\frac{1-x}{x}\right\}\{X_A(x) - X_A^*\} + \{X_B^{\infty} - X_B(x)\}. \tag{1.26}$$

In Equation 1.26 the subscripts f and i were removed because at infinite dilution $x_f \approx x_i$.

When a solid B dissolves in a liquid solvent A to give a solution, the change in X is denoted by $\Delta_{soln}X$, which is given by

$$\frac{\Delta_{soln}X}{n_B} = \left\{\frac{1-x}{x}\right\}\{X_A(l,x) - X_A^*(l)\} + \{X_B(l,x) - X_B^*(s)\} \tag{1.27}$$

in which l denotes the liquid state and s denotes the solid.

At infinite dilution of the solid in the solvent, Equation 1.27 becomes

$$\left(\frac{\Delta_{soln}X}{n_B}\right)^{\infty} = X_B^{\infty}(l) - X_B^*(s) \tag{1.28}$$

and

$$\left(\frac{\Delta_{soln}X}{n_B}\right)^{\infty} - \left(\frac{\Delta_{dil}X}{n_B}\right)^{\infty} = \frac{\Delta_{soln}X}{n_B}. \tag{1.29}$$

Equation 1.27 is also used when the solute is a gas with an appropriate designation of the phase of the solute.

1.3.18 What Is the Extent of Chemical Reaction?

A chemical reaction from reagents R to products P can be written as

$$\sum_{\mathrm{R}}(-\nu_{\mathrm{R}})\mathrm{R} = \sum_{\mathrm{P}}\nu_{\mathrm{P}}\mathrm{P}, \tag{1.30}$$

where ν is the stoichiometric number and is, by convention, negative for reactants and positive for products. The extent of a chemical reaction ξ (an extensive property) for a substance B that reacts according to Equation 1.30 is defined by

$$n_{\mathrm{B}}(\xi) = n_{\mathrm{B}}(\xi = 0) + \nu_{\mathrm{B}}\xi, \tag{1.31}$$

where $n_{\mathrm{B}}(\xi = 0)$ is the amount of substance present when the extent of reaction is zero; for example, before the reaction commenced.

1.4 WHAT ARE INTERMOLECULAR FORCES AND HOW DO WE KNOW THEY EXIST?

The fact that liquids and solids exist at all means that there must exist forces that bind molecules together under some conditions so that individual molecules do not simply evaporate into the gas phase. On the other hand, we know that it is extremely hard (taking considerable energy) to compress solids and liquids so as to reduce their volume. This implies that as we try to push atoms and molecules even closer together a force acts to keep them apart. Thus, we conceive a model of intermolecular forces between two molecules that are highly repulsive at small intermolecular distances but attractive at longer distances. In this section we develop this concept to explore the origins of these forces, how they are modeled, and some other direct demonstrations of their existence.

1.4.1 What Is the Intermolecular Potential Energy?

Consider first the interaction of two spherical neutral atoms a and b. The total energy $E_{\mathrm{tot}}(r)$ of the pair of atoms at a separation r is written as

$$E_{\mathrm{tot}}(r) = E_{\mathrm{a}} + E_{\mathrm{b}} + \phi(r). \tag{1.32}$$

Here, E_{a} and E_{b} are the energies of the isolated atoms, and $\phi(r)$ is the contribution to the total energy arising from interactions between them. We call $\phi(r)$ the *intermolecular pair-potential energy function* and, in the present example it depends only on the separation of the two atoms. Since this energy is equal to the work done in bringing the two atoms from infinite separation to the separation

r, it is given in terms of the intermolecular force $F(r)$ by

$$\phi(r) = \int_r^\infty F(r)\,\mathrm{d}r. \tag{1.33}$$

By convention, the force F is positive when repulsive and negative when attractive.

The general forms of $\phi(r)$ and $F(r)$ are illustrated in Figure 1.1 (Maitland et al. 1981). We see as foreshadowed above that, at short range, a strong repulsion acts between the molecules while, at longer range, there is an attractive force, which decays to zero as $r \rightarrow \infty$. Consequently, the potential energy $\phi(r)$ is large and positive at small separations but is negative at longer range. It is known that, for neutral atoms at least, there is only one minimum and no maximum in either $F(r)$ or $\phi(r)$. The parameters σ, r_0, and ε usually employed to characterize the intermolecular pair-potential energy are defined in Figure 1.1. σ is the separation at which the potential energy crosses zero, r_0 is the separation at which $\phi(r)$ is minimum, and $-\varepsilon$ is the minimum energy.

For molecules that are not spherically symmetric the situation is more complex because the force between the molecules, or equivalently the

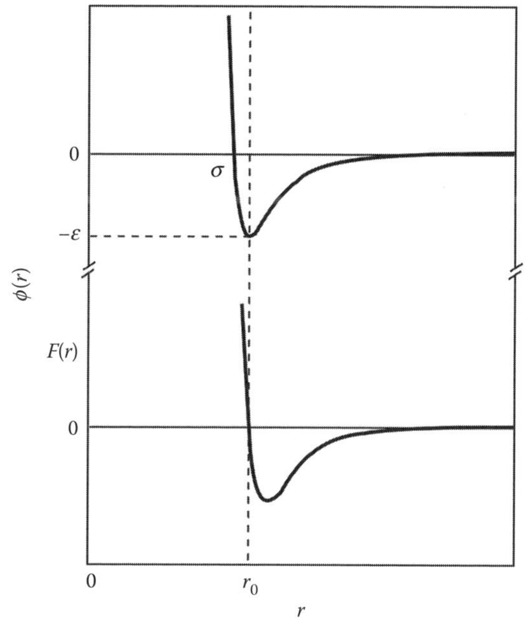

Figure 1.1 The intermolecular pair-potential energy $\phi(r)$ and force $F(r)$ as a function of r about the equilibrium separation r_0.

intermolecular potential energy, depends not just upon the separation of the center of the molecules but also upon the orientation of the two molecules with respect to each other. Thus, the intermolecular potential is not spherically symmetric. We shall consider this in a little more detail later.

In general, the potential energy U of a cluster of molecules is a function of the intermolecular interactions, which in turn depend upon the type and number of molecules under consideration, the separation between each molecule, and their mutual orientation. The term *configuration* is used to define the set of coordinates that describe the relative position and orientation of the molecules in a cluster.

To estimate the potential energy of a configuration it is usual, and often necessary, to make some or all of the following simplifications:

1. The term *intermolecular pair-potential energy* is used to describe the potential energy involved in the interaction of an isolated pair of molecules. It is very convenient to express the total potential energy U of a cluster of molecules in terms of this pair potential ϕ. This leads to a very important assumption, the *pair-additivity approximation*, according to which the total potential energy of a system of molecules is equal to the summation of all possible pair interaction energies. This implies that the interaction between a pair of molecules is unaffected by the proximity of other molecules.
2. The second important assumption is that the pair-potential energy depends only on the separation of the two molecules. As we have argued, this assumption is valid only for monatomic species where, owing to the spherical symmetry, the centers of molecular interaction coincide with the centers of mass.
3. Finally, since the intermolecular potential is known accurately for only a few simple systems, model functions need to be adopted in most cases. Typically, such models give U as a function only of the separation between molecules but nevertheless the main qualitative features of molecular interactions are incorporated.

For a system of N spherical molecules, the general form of the potential energy U may be written as

$$U(\mathbf{r}_1, \mathbf{r}_2, \ldots, \mathbf{r}_N) = \sum_{\mathrm{i}=1}^{N-1} \sum_{\mathrm{j}=\mathrm{i}+1}^{N} \phi_{\mathrm{ij}} + \Delta\phi_N, \tag{1.34}$$

where ϕ_{ij} is the potential energy of the isolated pair of molecules i and j, and $\Delta\phi_N$ is an increment to the potential energy, characteristic of the whole system, over and above the strictly pairwise additive interactions. According to the pair-additivity approximation, this reduces to

$$U(\mathbf{r}_1, \mathbf{r}_2, \ldots, \mathbf{r}_N) = \sum_{i=1}^{N-1} \sum_{j=i+1}^{N} \phi_{ij} = \sum_{i<j} \phi_{ij}. \tag{1.35}$$

The approximation of Equation 1.35 implies that the N-body interactions (with $N > 2$) are negligible compared with the pairwise interactions. In fact, many-body forces are known to make a small but significant contribution to the total potential energy when $N \geq 3$ and, for systems at higher density, the pair-additivity approximation can lead to significant errors. However, it is often possible to employ an effective pair potential that gives satisfactory results for the dense fluid while still providing a reasonable description of dilute-gas properties.

1.4.2 What Is the Origin of Intermolecular Forces?

Intermolecular forces are known to have an electromagnetic origin (Maitland et al. 1981) and the main contributions are well established. The strong repulsion that arises at small separations is associated with overlap of the electron clouds. When this happens, there is a reduction in the electron density in the overlap region leaving the positively charged nuclei incompletely shielded from each other. The resulting electrostatic repulsion is referred to as an *overlap force*. At greater separations, where attractive forces predominate, there is little overlap of electron clouds and the interaction arises in a different manner. Here, the attractive forces are associated with electrostatic interactions between the essentially undistorted charge distributions that exist in the molecules; for a more detailed description the reader is referred to the specialized literature (Maitland et al. 1981).

There are in fact three distinct contributions to the attractive forces that will be discussed here only briefly; for a more detailed description the reader is referred to a specialized literature (Maitland et al. 1981). For polar molecules, such as HCl, the charge distribution in each molecule gives rise to a permanent electric dipole and, when two such molecules are close, there is an *electrostatic force* between them that depends upon both separation and orientation. The force between any two molecules may be either positive or negative, depending upon the mutual orientation of the dipoles, but the averaged net effect on the bulk properties of the fluid is that of an attractive force.

Such electrostatic interactions are not associated exclusively with dipole moments. Molecules such as CO_2, which have no dipole moment but a quadrupole moment, also have electrostatic interactions of a similar nature. These interactions exist in general when both molecules have one or more nonzero multipole moments.

There is a second contribution to the attractive force that exists when at least one of the two molecules possesses a permanent multipole moment. This is known as the *induction force* and it arises from the fact that molecules are polarizable; so that a multipole moment is induced in a molecule when it is placed in any electric field including that of another molecule. Thus, a permanent dipole moment in one molecule will induce a dipole moment in an adjacent molecule. The permanent and induced moments interact to give a force that is always attractive and, at long range, proportional to r^{-6}.

The third contribution to the attractive force, and the only one present when both molecules are nonpolar, is known as the *dispersion force*. This arises from the fact that even nonpolar molecules generate fluctuating electric fields associated with the motion of the electrons. These fluctuating fields around one molecule give rise to an induced dipole moment in a second nearby molecule and a corresponding energy of interaction. Like induction forces, dispersion forces are always attractive and, at long range, vary like r^{-6} to leading order.

1.4.3 What Are Model Pair Potentials and Why Do We Need Them?

The difficulties encountered in the evaluation of the intermolecular pair-potential energy from an *ab initio* basis have led to the adoption of the following heuristic approach. We use the spherically symmetric potential as an example. The evaluation procedure starts with the assumption of an analytical form for the relationship between the potential energy ϕ and the distance r between molecules. Subsequently, macroscopic properties are calculated using the appropriate molecular theory. Comparisons between calculated and experimental values of these macroscopic properties provide a basis for the determination of the parameters in the assumed intermolecular potential-energy function. Finally, predictions may be made of thermodynamic properties of the fluid in regions where experimental information is unavailable.

In the following sections, we present some of the most widely used model potential-energy functions. For a more comprehensive discussion the reader is referred to specialized literature (Maitland et al. 1981).

1.4.3.1 What Is a Hard-Sphere Potential?

In this model, the molecules are assumed to behave as smooth, elastic, hard spheres of diameter σ. It is apparent that the minimum possible distance between the molecules is then equal to σ and that the energy needed to bring

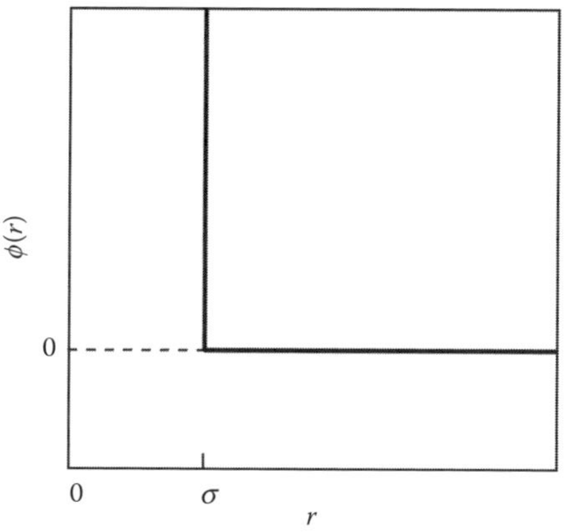

Figure 1.2 Hard-sphere potential $\phi(r)$ as a function of r.

the molecules closer together than $r = \sigma$ is infinite as shown in Figure 1.2. For separation $r > \sigma$, there is no interaction between the molecules. The mathematical form of the potential is given by the following discontinuous function

$$\begin{aligned} \phi(r) &= \infty \quad \text{for } r < \sigma \quad \text{and} \\ \phi(r) &= 0 \quad \text{at } r \geq \sigma. \end{aligned} \tag{1.36}$$

Although this model is not very realistic, it does incorporate the basic idea that the molecules themselves occupy some of the system volume. The hard-sphere model is especially important in the theory of the transport properties of dense fluids.

1.4.3.2 What Is a Square Well Potential?

This potential function is a more realistic one in the sense that it includes an attractive potential field, of depth ε and range $g\sigma$, surrounding the spherical hard core shown in Figure 1.3. Commonly used values of g are between 1.5 and 2.0. The mathematical form of the model is

$$\begin{aligned} \phi(r) &= \infty \quad \text{at } r < \sigma, \\ \phi(r) &= -\varepsilon \quad \text{for } \sigma \leq r < g\sigma, \quad \text{and} \\ \phi(r) &= 0 \quad \text{for } r \geq g\sigma. \end{aligned} \tag{1.37}$$

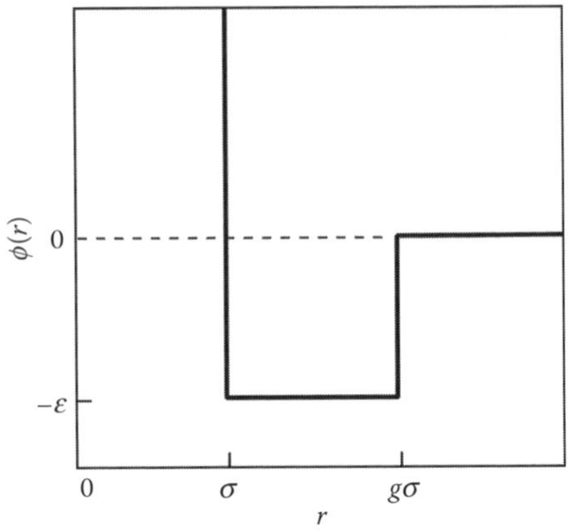

Figure 1.3 Square well potential $\phi(r)$ as a function of r.

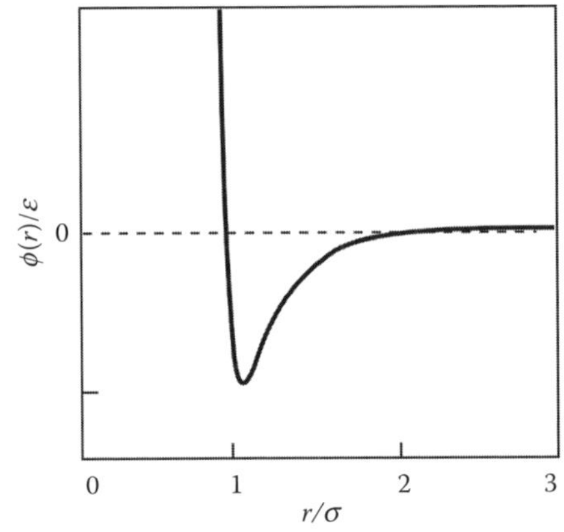

Figure 1.4 Lennard-Jones (12–6) potential $\phi(r)$ as a function of r.

1.4.3.3 What Is a Lennard-Jones (12–6) Potential?

The Lennard-Jones (12–6) potential illustrated in Figure 1.4, accounts for both attractive and repulsive energies, and assumes that the interaction between the molecules occurs along the line joining their centers of mass. It is one of the most commonly used models owing to its mathematical simplicity and the fact that it embodies the most important features of many real interactions, especially because its attractive component conforms to the leading term for the dispersion interaction for real neutral atoms.

The functional form of the Lennard-Jones (12–6) pair-potential is given by

$$\phi(r) = 4\varepsilon\left[\left(\frac{\sigma}{r}\right)^{12} - \left(\frac{\sigma}{r}\right)^{6}\right]. \qquad (1.38)$$

Although the model (given by Equation 1.38) has some realistic characteristics, it is not actually an accurate representation of any of the few intermolecular potentials that are well known. However, despite its approximate nature, the parameters of the potential model can be chosen so as to give a useful representation of the bulk behavior of many real systems and for that reason it is very often used in practical systems.

1.4.3.4 What Is the Potential for Nonspherical Systems?

In the more general case of the interaction of polyatomic molecules, the angular dependence of the potential must be considered as we have illustrated. It may be necessary to include up to five angular variables to describe the relative orientation of a pair of molecules explicitly. However, should we wish to do so, we can still think in terms of a one-dimensional function for any fixed orientation of the molecules. As an example, Figure 1.5 shows two sections through a model potential, which has been proposed for the system Ar + CO_2. In this case, the potential is quite strongly anisotropic and the parameters σ and ε characterizing the interaction along different paths of fixed orientation show marked differences.

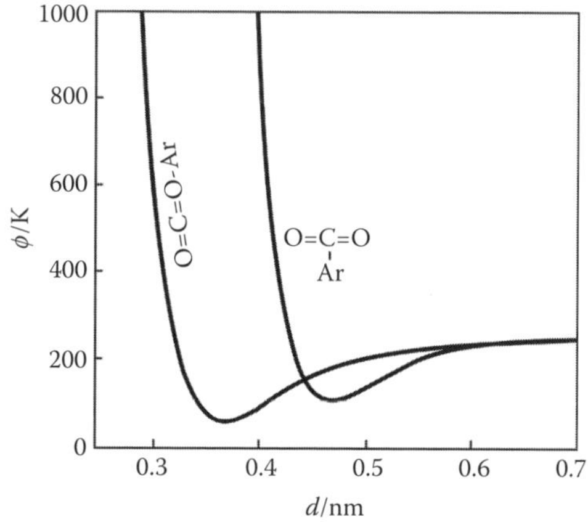

Figure 1.5 Sections through a pair potential for the system Ar + CO_2 for a "T" shaped and a linear configuration as a function of the separation d between Ar and CO_2.

Clearly, the exact mathematical description of such potentials is very complicated. The key features of nonspherical molecules that give rise to anisotropic forces are:

1. The nonspherical "core" geometry that dominates the anisotropy of the repulsive part of the potential
2. The presence of electric multipoles, especially dipole or quadrupole moments, which give rise to anisotropic electrostatic forces that may be dominant at longer range

This last point is of considerable importance and dipolar forces are often included in model intermolecular pair potentials where appropriate. The most common model that includes such forces is the *Stockmayer potential,* which consists of a central Lennard-Jones (12–6) potential plus the energy of interaction of two dipole moments:

$$\phi(r,\theta_1,\theta_2,\psi) = 4\varepsilon\left[\left(\frac{\sigma}{r}\right)^{12} - \left(\frac{\sigma}{r}\right)^{6}\right] - \frac{\mu^2}{4\pi\varepsilon_o r^3}(2\cos\theta_1\cos\theta_2 - \sin\theta_1\sin\theta_2\cos\psi). \quad (1.39)$$

In Equation 1.39 the angles θ_1, θ_2, and ψ define the mutual orientation of the dipole moments. θ_i is the angle made between the dipole moment on molecule i and the intermolecular axis, while ψ is the relative azimuthal angle between the two dipoles about the same axis.

1.4.4 Is There Direct Evidence of the Existence of Intermolecular Forces?

Capillary action, or wicking, is the ability of a substance to draw another substance into it. The standard reference is made to a tube in plants but can also be seen readily with porous paper. Capillary action occurs when the attractive intermolecular forces between the liquid at a surface and (usually) a solid substance are stronger than the cohesive intermolecular forces in the bulk of the liquid. If the solid surface is vertical the liquid "climbs" the wall made by the solid and a concave meniscus forms on the liquid surface.

A common apparatus used to demonstrate capillary action is the *capillary tube.* When the lower end of a vertical glass tube is placed in a liquid such as water, a concave meniscus is formed. Surface tension pulls the liquid column up until there is a sufficient mass of liquid for gravitational forces to overcome the intermolecular forces. The weight of the liquid column is proportional to the square of the tube's diameter, but the contact length (around the edge) between the liquid and the tube is proportional only to the diameter of the tube, so a narrow tube will draw a liquid column higher than a wide tube. For example, a glass capillary tube 0.5 mm in diameter will lift a column of water approximately 2.8 mm high.

With some pairs of materials, such as mercury and glass, a convex meniscus forms, and capillary action works in reverse so that the liquid is depressed in the tube relative to that in the absence of interfacial forces. These forces are known generally as surface tension or more properly as interfacial tension since they may arise at any interface between different materials.

There are many areas where capillary action is important. In hydrology, capillary action describes the attraction of water molecules to soil particles. Capillary action is responsible for moving groundwater from wet areas of the soil to dry areas. Capillary action is also essential for the drainage of constantly produced tear fluid from the eye; two canalicula of tiny diameter are present in the inner corner of the eyelid, also called the *lachrymal ducts*; their openings can be seen with the naked eye within the lachrymal sacs when the eyelids are turned inside-out. Paper towels absorb liquid through capillary action, allowing a fluid to be transferred from a surface to the towel. The small pores of a sponge act as small capillaries, causing it to adsorb a comparatively large amount of fluid. Some modern sport and exercise fabrics use capillary action to "wick" sweat away from the skin. These are often referred to as wicking fabrics, presumably after the capillary properties of a candle wick. Chemists utilize capillary action in thin layer chromatography, in which a solvent moves vertically up a plate via capillary action. Dissolved solutes travel with the solvent at various speeds depending on their polarity.

Maybe it is, finally, worth mentioning that Albert Einstein's first paper submitted to *Annalen der Physik* was on capillarity. It was titled *Conclusions from the capillarity phenomena* and was published in 1901 (Einstein, 1901).

1.5 WHAT IS THERMODYNAMIC ENERGY?

For an adiabatically enclosed system the work needed to change the state of the system from an initial state 1 to a final state 2 in the absence of kinetic energy is given by the change of energy of the system ΔU (often referred to as the change of internal energy)

$$W = \Delta U = U_2 - U_1, \tag{1.40}$$

where U is the thermodynamic energy.

1.6 WHAT IS THE 1ST LAW OF THERMODYNAMICS?

We consider a system that is enclosed by a diathermic wall so that the system can do work on the surroundings and the surroundings can do work on the system. If, for example, the pressure of the system is changed by the surroundings then energy has flowed into the system. Energy can also flow into the system by virtue of a heat flow Q from the surroundings into the system when the system is surrounded by a diathermic wall. In this case, the work done by mechanical or electrical methods to change the state of the system is not equal to the work required for the same change when it is adiabatically enclosed. That is, the change of the energy of the system U depends on the initial and final state of the system but the work done W and the heat flow Q depend on the method used to bring about the change of state often referred to as the path. For example, an increase in the energy U $(\Delta U > 0)$ can be obtained by a path for which both Q and W are positive (Q, W) or by $Q \approx 0$ and $W > 0$ or by $Q > 0$ and $W \approx 0$. Application of the law of energy conservation (i.e., the fact that energy can neither be created nor destroyed) to a thermodynamic system gives

$$\Delta U = W + Q. \tag{1.41}$$

Equation 1.41 is an expression of the 1st law of thermodynamics; roadmaps for this and the other laws of thermodynamics are given elsewhere (Atkins and de Paula 2009).

When $Q = 0$ the enclosure is adiabatic and Equation 1.40 is obtained from Equation 1.41. For a system isolated from all external work so that $W = 0$ and contained within a diathermal enclosure Equation 1.41 reduces to $\Delta U = Q$. Finally, for an adiabatically enclosed isolated system $\Delta U = 0$ and both $Q = 0$ and $W = 0$.

1.7 QUESTIONS THAT SERVE AS EXAMPLES OF WORK AND THE 1ST LAW OF THERMODYNAMICS

In this section we seek to pose and answer a number of practical and realistic problems, using the notions and laws of thermodynamics and thermophysics we have covered so far. As is the case throughout this book the examples are chosen to illustrate particular features of the subjects that are often found difficult by students; the list of topics is not exhaustive but is intended to be illustrative.

1.7.1 How Does a Dewar Flask Work?

A *Dewar* flask, shown schematically in Figure 1.6, is a vessel used for maintaining materials at temperatures other than those of the surroundings for a finite duration. This is accomplished by slowing down the heat transfer between the object in the vessel and the surroundings. Heat can be transferred from

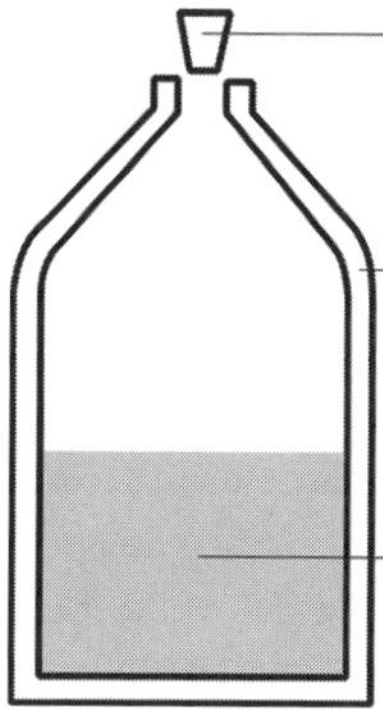

Figure 1.6 Dewar flask.

one region to another or from one body to another by three mechanisms. One mechanism is by conduction, where heat transfer takes place from one part of a body to another part of the same body, or between two bodies in physical contact through the combination of molecular motion that transports the kinetic energy of the molecules or through collisions between the molecules that allow transfer of energy from one molecule to another. A second mechanism is convection, where heat transfer takes place from a point to another within a fluid, or between a fluid and a solid or another fluid, by virtue of the bulk motion of the fluid as a continuum that transports warmer fluid from one location to another. Evidently, convection is not a mechanism of heat transfer that has any meaning for the transfer within solids. A third mechanism of heat transfer is by the exchange of electromagnetic radiation. The radiation can be emitted by one region of a material and absorbed and/or reflected by other regions of the material or by surfaces.

The Dewar flask is constructed so as to inhibit all these three modes of heat transfer to some extent. First, as can be seen in Figure 1.6 it has a double wall and the space between the walls is evacuated to a very low pressure (less than 1 Pa). At such low pressures the mean free path λ of the gas molecules that remain is very long ($\lambda \approx kT/(\pi p\sigma^2)$), where σ is the molecular diameter and, for the pressure quoted, is greater than the distance between the walls of the vessel. As a result the only mechanism for the transport of molecular energy between one wall and another is associated with the kinetic energy of the molecules that collide with the inner wall and then collide at the outer wall. This kinetic energy is very small and the number of molecules making the trip per second is also very small so that the heat conducted between the two walls is very small indeed. The heat transported by convection (bulk fluid motion) is similarly reduced. The magnitude of the heat transported by bulk motion must depend upon the heat capacity of the fluid per unit mass, the mass per unit volume, and the velocity of the motion. The fact that we have a very low pressure in the evacuated space ensures that the density of the gas is very low and in itself this reduces the convective effects to a very small level irrespective of whether the remaining gas has a significant heat capacity per unit mass or there are convective currents.

Finally, the surfaces of the walls of the vessel inside the evacuated space are coated with silver, which is a weak emitter of radiation (it has a low emissivity) and highly reflective. Thus, neither surface emits much radiation according to the Stefan Boltzmann law what it does emit is largely reflected back from the opposing surface. Thus, the amount of heat transported by radiation between the object on the inside of the Dewar flask to the surroundings is very small. This inhibition of all three heat transfer processes results in a long delay of approach to thermal equilibrium between the contents of the flask and the surroundings. Thus, the contents of a Dewar flask will remain either hot or cold for a long time.

In laboratories and industry, vacuum flasks are often used to store liquids, which become gaseous well below ambient temperature, such as O_2, which has a normal boiling temperature of 90.2 K at a pressure of 0.1 MPa and N_2 (normal boiling temperature of 77.3 K). It is possible to maintain such materials in the liquid state for several days without the need for expensive refrigeration equipment.

1.7.2 In a Thermally Isolated Room Why Does the Temperature Go Up When a Refrigerator Powered by a Compressor Is Placed Within?

The reader will recall from the earlier discussion that to answer any thermodynamic question the first thing that must be done is to define the system considered. In this case we define the system to include all of the entities and masses contained within the walls of the isolated room (including the air and the refrigerator itself). Anything outside the walls (the boundary of the system) is defined as the *surroundings*. If no mass enters or leaves the system through the walls (including the doors and the windows), the mass in the system remains fixed (does not change with time) and the system is *closed* as we defined it in Section 1.6. In a closed system only *energy* may be transferred in or out of the system through the boundaries. The 1st law of thermodynamics (or the law of conservation of energy) states that for a time interval Δt the energy accumulated in the system is equal to the energy transfer through the system boundaries.

We assume, in our example, that the room is stationary in some reference frame so that the kinetic energy of the system itself is zero and its potential energy is constant and that there are no magnetic or other external forces. The internal energy of the system U_s is given by

$$U_s = \sum_i m_i u_i, \tag{1.42}$$

where m_i is the mass of one part of the room or refrigerator of specific material and u_i is the specific internal energy of component i. The sum extends over all components within the system.

We remind the reader now that *Heat* is the energy that is transferred between the system and its surroundings and is denoted by Q and *Work* is the energy of interaction between a system and its surroundings as a result of force acting. Thus, a piston compressing a gas, a rotating shaft, and an electric wire heated by a current within the system are all examples of *work*. This gives Equation 1.41.

In the room, there is no heat transfer through the boundaries because it is thermally isolated and so $Q = 0$. The only work crossing the system boundary is the electrical work W^{el} done by the electric current in the wire entering to move the compressor of the refrigerator, which must come from outside in the surroundings. The first law for the isolated room is

$$\Delta U_s = W^{el}. \tag{1.43}$$

The addition of electric work to the system causes the internal energy to increase and, because of the constant mass, the temperature must also increase with time. This description is neither dependent on the position of the refrigerator door nor on the water vapor content of the air within the room that condenses on the cold refrigerator nor anything else that occurs inside the room. Indeed, it has not been necessary to consider these aspects of the problem essentially because of where we placed the system boundary.

1.7.3 What Is the 1st Law for a Steady-State Flow System?

We now consider the application of the 1st law of thermodynamics to the circumstance often faced by engineers and shown in Figure 1.7, where there is a flow of a homogeneous fluid into and out of some process that changes the thermodynamic state in some way. We shall consider the change of state from an initial thermodynamic state designated **1** and characterized by $(p_1, V_1, T_1, \ldots)$ to a new state designated **2** and characterized by $(p_2, V_2, T_2, \ldots)$ by means of a process **A** and then, by means of a process **B**, return to the original state.

This represents a cycle and we make an assumption about the processes that fluid flows in and out of each process, which itself can be reversible or

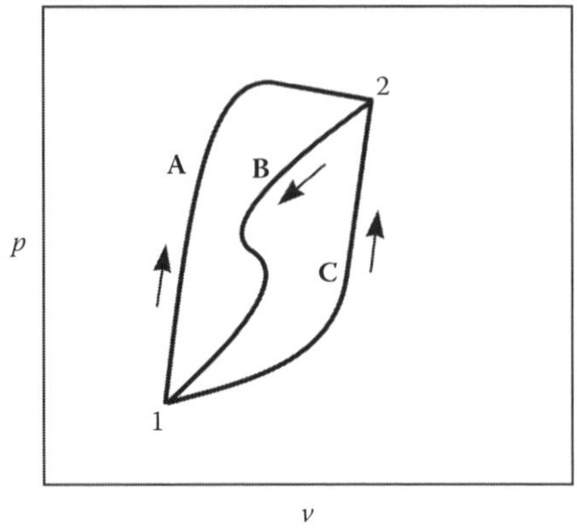

Figure 1.7 Changes of thermodynamic state.

irreversible. We also assume for the present that the kinetic energy of the flow and indeed any potential energy is negligible relative to any work done or heat input. This simplifies our first analysis and this restriction can be removed subsequently. The p, v, T are specific, that is, specified by mass (see Section 1.3.12). Since the properties of the material leaving process **B** are the same as those entering at **A,** the law of conservation of energy requires that no net energy is stored in the flowing material. Thus, after a given amount of material has passed through both processes

$$ {}_1^2Q_A + {}_1^2W_A + {}_2^1Q_B + {}_2^1W_B = 0. \tag{1.44}$$

Equation 1.44 is formally the same as the equation we obtained earlier for a closed system Equation 1.40, but W is now mechanical work. If we were to replace process **A** by process **C** we would obtain a corresponding equation in terms of **B** and **C**,

$$ {}_1^2Q_C + {}_1^2W_C + {}_2^1Q_B + {}_2^1W_B = 0. \tag{1.45}$$

Because the processes are chosen entirely arbitrarily, it follows that for any process $(Q + W)$ must have the same value independent of the process being reversible or irreversible. This combination therefore has the property of a state function; it is called the *enthalpy.* Thus, in a steady-state flow system we use Equation 1.41 and we shall see later that the enthalpy is defined as

$$H = U + pV. \tag{1.46}$$

At the beginning of this argument we restricted ourselves to cases where the kinetic and potential energy of the system were negligible by comparison with Q and W. However, although this is often the case it is not always so and we need to consider these factors separately. First, we recognize that kinetic energy of the bulk material is a mode of energy storage additional to internal energy U and H is the potential energy. Thus, if we use E_k to denote the kinetic energy and E_p to denote potential energy and the subscripts "in" and "out" to denote the amount of energy in a particular mode at input and output of a process, then we can generalize the first law for both closed systems and steady-state flow as

$$\Delta U = Q + W - (E_{k,out} - E_{k,in}) - (E_{p,out} - E_{p,in}), \tag{1.47}$$

and

$$\Delta H = Q + W - (E_{k,out} - E_{k,in}) - (E_{p,out} - E_{p,in}). \tag{1.48}$$

The first of these equations is rarely encountered in thermodynamics, but the second is much more important particularly in the field of fluid mechanics. To see this we consider reversible (frictionless) flow in a nozzle. In this case, Equation 1.48 holds between any two positions 1 and 2 of the nozzle. Thus,

$$\Delta H = \Delta U + \Delta(pV) = Q + W - (E_{k,2} - E_{k,1}) - (E_{p,2} - E_{p,1}). \tag{1.49}$$

If there is no heat transfer and no work done, the internal energy of the fluid remains the same. Thus,

$$p_1V_1 + E_{p,1} + E_{k,1} = p_2V_2 + E_{p,2} + E_{k,2}, \tag{1.50}$$

energy and Equation 1.50 is a constant. Alternatively, Equation 1.50 can be cast as

$$\frac{p_1}{\rho_1} + gz_1 + \frac{1}{2}c_1^2 = \frac{p_2}{\rho_2} + gz_2 + \frac{1}{2}c_2^2, \tag{1.51}$$

where z represents vertical elevation in the direction of gravitational acceleration g and c is the fluid speed. For an incompressible fluid the density ρ is constant so that

$$\frac{p_1}{\rho} + gz_1 + \frac{1}{2}c_1^2 = \frac{p_2}{\rho} + gz_2 + \frac{1}{2}c_2^2 \tag{1.52}$$

and

$$p_1 + \rho gz_1 + \frac{1}{2}\rho c_1^2 = p_2 + \rho gz_2 + \frac{1}{2}\rho c_2^2. \tag{1.53}$$

Equation 1.53 is a constant. This is the classical Bernoulli equation and it holds for incompressible frictionless flow in a conduit.

1.7.4 What Is the Best Mode of Operation for a Gas Compressor?

To answer the question we must, of course, first define what "best" means in this context. Most often in engineering the "best" way to compress gas (in most circumstances used to increase the pressure of a flowing gas) is that which requires minimum work to be done on the system. While there are, of course, many means to compress a gas, all of which could be subject to a similar analysis, we consider here only a reciprocating compressor illustrated in Figure 1.8. This example has the particular advantage that the relationship between the forms of work can be connected with both closed and open systems.

In the reciprocating compressor, shown in Figure 1.8, the gas flows into a compression cylinder through an inlet valve. After closure of the valves the gas is compressed by a piston and then discharged after opening the outlet valve. While the whole process may be regarded as an open system, the actual compression takes place within a closed system. Indeed, this is rather a good illustration of the difference. The total shaft work per unit mass required for the whole process consists of the boundary work exerted to achieve the compression of

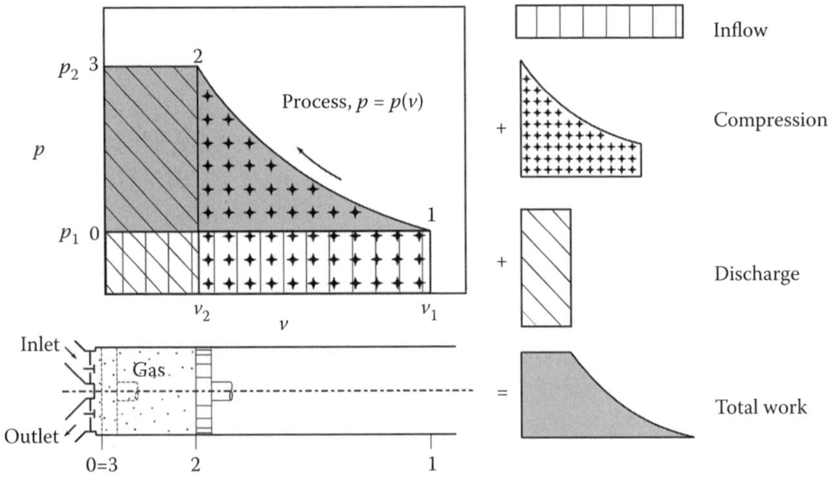

Figure 1.8 Scheme of a reciprocating compressor for the illustration of the total work required for the compression of a gas stream: the gas enters into a compression cylinder ($0 \rightarrow 1$, associated with flow work for the displacement of the piston, negative sign), after closure of the valves the gas is compressed ($1 \rightarrow 2$, boundary work in a closed system) and then discharged after opening of the outlet valve ($2 \rightarrow 3$, flow work). The total shaft work required for the whole process is the sum of these contributions.

the gas and the flow work for moving the piston when the gas enters into the cylinder and again when it is discharged,

$$w_{s12} = w_{b12} + w_{f12}. \tag{1.54}$$

For an ideal (that is reversible) process where

$$(w_{b12})_{rev} = -\int_1^2 p\, dv, \tag{1.55}$$

we may now calculate the total work required:

$$\begin{aligned}(w_{s12})_{rev} &= -\int_1^2 p\, dv + p_2 \cdot v_2 - p_1 \cdot v_1 = \int_1^2 \left[-p\, dv + d(pv)\right] \\ &= \int_1^2 (-p\, dv + v\, dp + p\, dv) = \int_1^2 v\, dp.\end{aligned} \tag{1.56}$$

In a (p, v) diagram the total work may now be illustrated as the area between the ordinate and the line describing the chosen process as shown in Figure 1.8.

We now turn to the question as to what modes of operation are possible for such a compressor, that is, what are reasonable thermodynamic processes for gas compression. It is obvious that the process can neither be performed in an isobaric nor in an isochoric matter. The processes under question are therefore either an isothermal or an adiabatic process.

1.7.5 What Is the Work Required for an Isothermal Compression?

If—for simplicity, but without detriment to the argument—we restrict ourselves to an ideal gas, the ideal gas law provides the simple relation $pv = RT/M = R_s T$ where R_s is called the *specific gas constant*, and M is the molar mass. For an isothermal process (at constant temperature T) the product of pressure and volume is a constant (pv = const.). For the ideal, reversible case the boundary work for compression of the gas in a closed system is given by

$$(w_{b12})_{rev}^{isoth.} = -\int_1^2 p dv = -\int_1^2 \frac{R_s T}{v} dv = -R_s T \ln\left(\frac{v_2}{v_1}\right) = R_s T \ln\left(\frac{p_2}{p_1}\right). \tag{1.57}$$

For an open system we have to add the flow work resulting in

$$(w_{s12})_{rev}^{isoth.} = \int_1^2 v\, dp = \int_1^2 \frac{R_s T}{p} dp = R_s T \ln\left(\frac{p_2}{p_1}\right). \tag{1.58}$$

Equation 1.58 must be identical to that for the closed system in this case because flow work vanishes for the isothermal process (pv = const.).

An isothermal process may be realized by discharging the same amount of heat into cooling water as is supplied by the input of work. This relation directly follows from the 1st law:

$$q_{12} + w_{12} = u_2 - u_1 = c_v\left(T_2 - T_1\right) = 0. \tag{1.59}$$

In Equation 1.59 c_v is the specific heat capacity at constant volume.

1.7.6 What Is the Work Required for an Adiabatic Compression?

Again, we consider the reversible work in connection with the compression of an ideal gas. The process is reversible and adiabatic, and is characterized by a property called the *isentropic (expansion) exponent* κ, defined by $\kappa = -\, vp^{-1}\,(\partial p/\partial v)_s$ (compare Question 3.5.6).

For an ideal gas κ is equal to γ, the ratio of the isobaric and isochoric heat capacities $\gamma = c_p / c_v$. For our derivation we additionally assume that γ is constant with temperature. This statement strictly only holds for monatomic gases, but is a good approximation also for polyatomic gases because of the similar dependence of c_p and c_v on temperature. From the first law in a differential form.*

$$\delta q + \left(\delta w\right)_{\text{rev}} = \mathrm{d}u, \tag{1.60}$$

and with $\delta q = 0$ and $\left(\delta w\right)_{\text{rev}} = -\,p\,\mathrm{d}v$ we obtain:

$$\mathrm{d}u + p\,\mathrm{d}v = 0. \tag{1.61}$$

For the ideal gas, $\mathrm{d}u = c_v\,\mathrm{d}T$, $p = R_s T / v$, $c_p - c_v = R_s$, and $\gamma = c_p / c_v$ so that Equation 1.61 becomes

$$\frac{\mathrm{d}T}{T} + (\gamma - 1)\frac{\mathrm{d}v}{v} = 0. \tag{1.62}$$

Integration from state 1 to state 2 yields

$$\ln\left(\frac{T_2}{T_1}\right) + (\gamma - 1)\ln\left(\frac{v_2}{v_1}\right) = 0, \tag{1.63}$$

or

$$T_2 \cdot v_2^{\gamma - 1} = T_1 \cdot v_1^{\gamma - 1}, \tag{1.64}$$

* In contrast to energy u which is a thermodynamic property, heat q, and work w are path functions that do not have exact differentials; in these two cases differentials are denoted by δ.

and with $T = p \cdot v/R_s$ we obtain

$$p_1 \cdot v_1^{\gamma} = p_2 \cdot v_2^{\gamma} = p \cdot v^{\gamma}. \tag{1.65}$$

This equation generally describes the relation between p and v for a reversible and adiabatic process of an ideal gas. We can finally combine the preceding equations to find a relation between p and T for such a process:

$$T_1 \cdot p_1^{(1-\gamma)/\gamma} = T_2 \cdot p_2^{(1-\gamma)/\gamma}. \tag{1.66}$$

From the first law for an adiabatic process we then obtain the work for the ideal gas in a closed system of

$$(w_{b12})_{rev}^{adiab.} = u_2 - u_1 = c_v(T_2 - T_1) = c_v T_1\left(\frac{T_2}{T_1} - 1\right), \tag{1.67}$$

with $c_v = R_s/(\gamma - 1)$ and the ideal gas law, this expression can be rewritten in the form

$$(w_{b12})_{rev}^{adiab.} = \frac{p_1 v_1}{\gamma - 1}\left(\frac{T_2}{T_1} - 1\right), \tag{1.68}$$

or, using Equation 1.66 into

$$(w_{b12})_{rev}^{adiab.} = \frac{p_1 v_1}{\gamma - 1}\left[\left(\frac{p_2}{p_1}\right)^{(\gamma-1)/\gamma} - 1\right]. \tag{1.69}$$

Similarly, the work for an open system with an ideal gas is

$$(w_{s12})_{rev}^{adiab.} = h_2 - h_1 = c_p(T_2 - T_1) = c_p T_1\left(\frac{T_2}{T_1} - 1\right). \tag{1.70}$$

Equation 1.70 can, with $c_p = R_s\gamma/(\gamma - 1)$, be cast as

$$(w_{s12})_{rev}^{adiab.} = \frac{\gamma\, p_1 v_1}{\gamma - 1}\left(\frac{T_2}{T_1} - 1\right) = \frac{\gamma\, p_1 v_1}{\gamma - 1}\left[\left(\frac{p_2}{p_1}\right)^{(\gamma-1)/\gamma} - 1\right]. \tag{1.71}$$

Having answered the questions for the work required in the most relevant processes (isothermal and adiabatic) we can now generalize it to the case of a polytropic process, which is characterized by the relation $pv^n = \text{const.}$ with the polytropic exponent n, with little extra effort.

A compression is really performed in a manner that lies between the idealized cases of an isothermal (with $n = 1$) and an adiabatic (with $n = \gamma$) process. Rather than comparing the individual equations for the work required, we consider the relations for the boundary and the shaft work with reversible processes,

$$(w_{b12})_{rev} = -\int_1^2 p\, dv, \tag{1.72}$$

and

$$(w_{s12})_{rev} = \int_1^2 v\, dp, \tag{1.73}$$

respectively, and view the (p, v) diagram for the cases of interest.

The most relevant and familiar case is that of the compression of a flowing gas stream from pressure p_1 to pressure p_2. For such an open system the total (shaft) work is the quantity of interest, which for an ideal (i.e., reversible) process may be obtained as the area between the respective process (path) and the ordinate (p axis) from a (p, v) diagram, shown in Figure 1.9a. Because ($\gamma > 1$) the magnitude of $|dp/dv|$ is greater for an adiabatic than for an isothermal process, as shown in Figure 1.9, and more work is required in the adiabatic case. For practical purposes this result implies that a gas compressor should indeed be operated as close as possible to the isothermal case and therefore requires efficient heat exchange.

If the task, however, is to compress a gas in a closed system such as a cylinder, the answer to our initial question for the best mode of operation may be different. The result depends upon whether the gas is to be brought to a defined

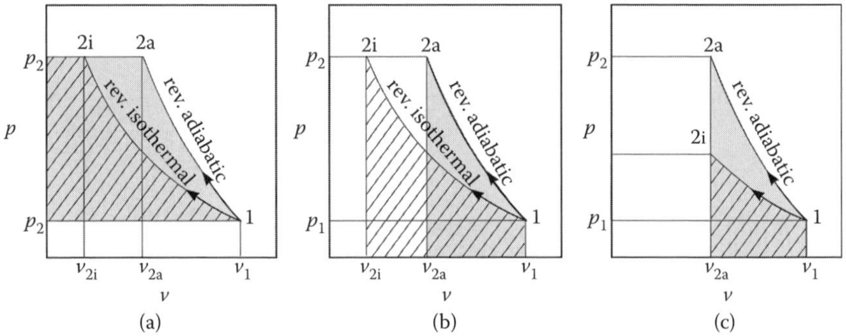

Figure 1.9 Illustration of reversible work required for gas compression as a function of boundary conditions for reversible adiabatic and isothermal compression: (a) for an open system, where the total work is defined by the area between the p axis and the respective process; (b) and (c) for a closed system, where the relevant quantity is boundary work obtained from the area between the process and the v axis; for a closed system the preferred compression method, be it adiabatic or isothermal, depends on if a defined pressure is to be achieved as shown in (b) or a defined volume is to be reached defined as shown in (c).

pressure or to a defined specific volume (the latter case does not make sense for an open system). In this case we have to consider the boundary work for the two processes, which may be identified in a (p,v) diagram as the area between the respective process and the abscissa (v axis, Figure 1.9b and 1.9c). Whereas for the compression to a defined pressure the adiabatic process is the right choice, things reverse when it comes to reaching a defined smaller specific volume, where the isothermal process is to be preferred.

1.8 HOW ARE THERMOPHYSICAL PROPERTIES MEASURED?

Thermodynamics is an experimental science, almost all of the properties that we have discussed so far and that occur in thermodynamics and transport phenomena must be measured by experimental means. Molecular simulation or theoretical calculation based on molecular physics can provide estimates of thermophysical properties that are usually significantly less precise than the measured values. For a very few systems such as helium, quantum mechanical calculations and the fundamental constants have been used, *ab initio*, to determine the pair interaction potential energy. When combined with the methods of statistical mechanics (Chapter 2) this potential provided estimates of the thermophysical properties of helium in the low-density gas phase with an estimated uncertainty less than that obtained from measurements. Thermophysical properties calculated by this approach have been used to provide data for instruments used for measurements on other materials and to form a standard for pressure. However, for most other molecules and atoms even at low density these calculations have yet to be done with sufficient precision to be able to replace measurement; at higher densities in the gas and in the liquid phase there is little likelihood that such calculations will be performed in the near future. For that reason the current reliance on experimental determination of properties will persist for a considerable time.

Question 7.4 addresses some of the techniques used to measure thermophysical properties. Those interested in additional information should consult the series *Experimental Thermodynamics* (Vol. I 1968, Vol. II 1975, Vol. III 1991, Vol. IV 1994, Vol. V 2000, Vol. VI 2003, Vol. VII 2005, Vol. VIII 2010), which also includes two volumes concerned with equations of state for fluids and fluid mixtures that are discussed in Chapter 4. The first of these two volumes (Vol. V 2000) has been updated in 2010 and Volume VIII 2010 places a greater emphasis on the application of theory. The latter volume specifically includes theoretical and practical information regarding equations of state for chemically reacting fluids and methods applicable to nonequilibrium thermodynamics than hitherto provided in Volume V. However, computer simulations for the calculation of thermodynamic properties was omitted from Volume VIII because the subject requires an in-depth coverage such as that given in a special issue of *Fluid Phase Equilibria* (Case et al. 2008; Eckl et al. 2008; Ketko et al. 2008; Li et al. 2008; Müller et al. 2008; Olson and Wilson 2008). The problem of evaluating the thermodynamic properties for industrial use by means of calculation and simulation is treated in other publications (Case et al. 2004; 2005; 2007).

The monographs in the series *Experimental Thermodynamics* (Vol. I 1968; Vol. II 1975; Vol. III 1991; Vol. IV 1994; Vol. V 2000; Vol. VI 2003; Vol. VII 2005; Vol. VIII 2010) were published under the auspices of the International Union of Pure and Applied Chemistry (IUPAC) and since 2004 in association with the International Association of Chemical Thermodynamics (IACT) that is an affiliate of IUPAC. Throughout this text we have adopted the quantities, units, and symbols of physical chemistry defined by IUPAC in the text commonly known as the *Green Book*. We have also adopted, where possible, the ISO guidelines for the expression of uncertainty (*Guide to the Expression of Uncertainty in Measurement* 1995), and vocabulary in metrology (*International Vocabulary of Basic and General Terms in Metrology* 1993). Values of the fundamental constants and atomic masses of the elements have been obtained from literature (Wieser 2006; Mohr 2008).

The series *Experimental Thermodynamics* is complemented by other recent publications associated with IUPAC and IACT that have covered a range of diverse issues reporting applications of solubility data (*Developments and Applications of Solubility* 2007) to the topical issue of alternate sources of energy (*Future Energy: Improved Sustainable and Clean Options for our Planet* 2008) and the application of chemical thermodynamics to other matters of current industrial and scientific research, including separation technology, biology, medicine, and petroleum in one (*Chemical Thermodynamics* 2000) of eleven monographs of an IUPAC series entitled *Chemistry for the 21st Century* and heat capacity measurements (*Heat Capacity* 2010).

1.8.1 How Is Temperature Measured?

The temperature T is the thermodynamic temperature; it can only be measured by means of a primary thermometer such as a gas thermometer through the relationship

$$\frac{T_2}{T_1} = \lim_{p \to 0} \left\{ \frac{p_2 V_2}{p_1 V_1} \right\}, \tag{1.74}$$

or with acoustic thermometers, which make use of measurements of the speed of sound u through the relationship

$$\frac{T_2}{T_1} = \lim_{p \to 0} \left(\frac{u_2^2}{u_1^2} \right). \tag{1.75}$$

When T_1 of Equations 1.74 and 1.75 is chosen to be the triple point of water for which

$$T(\mathrm{H_2O,s{+}l{+}g}) = 273.15\,\mathrm{K}, \tag{1.76}$$

then T_2 can be determined. These few lines do not convey to the uninitiated the effort required to perform the measurements by either method set out above. It is sufficient to state here that either method is an impractical method of determining thermodynamic temperature for routine scientific work. Instead, use is made of one or more types of empirical or secondary thermometers that can reproduce, to the required uncertainty, the temperature that would be obtained with a primary thermometer with the aid of a calibration either against a primary thermometer or a set of accepted reference points. Practical thermometers include liquid-in-glass, the resistivity of platinum, semiconductors or thermistors, and thermocouples. The last three require the measurement of a resistance or of a voltage rather than visual observation. The choice of instrument is determined by the precision required in the temperature and the range in which it is required. The interpretation of the resistivity of platinum in terms of thermodynamic temperature is achieved by the use of the International Temperature Scale of 1990 (ITS-90) (Preston 1990; Nicholas and White 2003).

1.8.2 How Is Pressure Measured?

Pressure can be measured with a piston or "dead-weight" gauge. In this instrument, the pressure to be measured is applied to the base of a piston of known effective cross-sectional area contained within a close-fitting cylinder. Masses are added to a carrier connected to the piston so as to balance the pressure. The force exerted by the masses is determined from the local acceleration due to gravity and the pressure is determined from the known cross-sectional area of the piston. This experiment is far from routine and is very time consuming and delicate. Thus, the majority of pressure measurements are obtained from transducers that have been calibrated against dead-weight gauges. These transducers usually determine the mechanical strain induced by the applied pressure with an appropriately located resistive strain gauge and a Wheatstone bridge or from variations in the resonance frequency of a quartz object. All methods of pressure measurement have been extensively reviewed elsewhere (Suski et al. 2003).

1.8.3 How Are Energy and Enthalpy Differences Measured?

Unfortunately, the absolute value of the energy U cannot be measured directly; only the difference between two states $\Delta U = U_2 - U_1$ can be determined with a calorimeter. A calorimeter is also the name given to the instrument used to measure enthalpy differences. A calorimeter is an adiabatically enclosed container in which work is done to change the state of a material, and from Equation 1.41 $\Delta U = W$. The work is usually obtained by passing a constant current through an electrical resistance within the system for a measured time. The resistance, current, and time determine the electrical work, $W^{\mathrm{el}} = I^2 Rt$. The total work done on the calorimeter must also include any pressure work ($\int p\, \mathrm{d}V$) done by the change of volume at an external pressure p from the calorimeter to the surroundings and by, for example, stirring the contents of the calorimeter or by initiating a chemical reaction W^{o} so that the working equation for the calorimeter becomes

$$\Delta U = W^{\mathrm{el}} - \int p\, \mathrm{d}V + W^{\mathrm{o}} + Q. \tag{1.77}$$

If the calorimeter volume is held constant by rigid walls then $\int p\, \mathrm{d}V = 0$ and if the system is adiabatically enclosed then $Q = 0$ so that Equation 1.77 becomes

$$\Delta U = W^{\mathrm{el}} + W^{\mathrm{o}}, \tag{1.78}$$

which shows how the calorimeter can be used to measure the internal energy difference of two states.

If the pressure in the calorimeter is maintained equal to that of the surroundings and the calorimeter walls are still adiabatic but not rigid (so that the volume of the system changes from V_1 to V_2) then Equation 1.77 becomes

$$\Delta U = U_2 - U_1 = W^{\mathrm{el}} - pV_2 + pV_1 + W^{\mathrm{o}}. \tag{1.79}$$

On rearrangement, Equation 1.79 is

$$\Delta(U + pV) = (U_2 + pV_2) - (U_1 + pV_1) = W^{\mathrm{el}} + W^{\mathrm{o}}. \tag{1.80}$$

The combination $U + pV$ occurs so often in many practical problems that it has been given a special symbol H and is called the *enthalpy*:

$$H = U + pV. \tag{1.81}$$

In terms of enthalpy H, Equation 1.79 becomes

$$\Delta H = W^{\mathrm{el}} + W^{\mathrm{o}}, \tag{1.82}$$

and we see that enthalpy difference between two states can also be determined with a calorimeter.

1.8.4 How Is the Energy or Enthalpy Change of a Chemical Reaction Measured?

For an adiabatically enclosed calorimeter of constant volume that contains reactants at initial temperature T_i and extent of reaction ξ_i, that continues to temperature T_f and extent of reaction ξ_f, the energy change can be written as

$$\Delta U = U(T_\mathrm{f}, V, \xi_\mathrm{f}) - U(T_\mathrm{i}, V, \xi_\mathrm{i}) = Q_1 \approx 0. \tag{1.83}$$

If the temperature is returned to T_i and if the calorimeter is then electrically heated to T_f so that

$$\Delta U = U(T_\mathrm{f}, V, \xi_\mathrm{f}) - U(T_\mathrm{i}, V, \xi_\mathrm{f}) = Q_2 + W^\mathrm{el} \approx W^\mathrm{el}, \tag{1.84}$$

then subtracting Equation 1.84 from Equation 1.83 gives

$$\Delta U = U(T_\mathrm{i}, V, \xi_\mathrm{f}) - U(T_\mathrm{i}, V, \xi_\mathrm{i}) = -W^\mathrm{el} + (Q_1 - Q_2) \approx -W^\mathrm{el}, \tag{1.85}$$

and the right hand side can be achieved if either the calorimeter is adiabatic, so that Q_1 and Q_2 equal 0, or if it arranged experimentally so that $Q_1 = Q_2$. If the pressure of the calorimeter is maintained constant the appropriate function is enthalpy and then

$$\Delta H = H(T_\mathrm{i}, p, \xi_\mathrm{f}) - H(T_\mathrm{i}, p, \xi_\mathrm{i}) \approx -W^\mathrm{el}. \tag{1.86}$$

1.8.5 How Is Heat Capacity Measured?

Consider a system comprised of a substance contained within an adiabatic calorimeter (so that $Q = 0$) when the temperature is changed from T_1 to T_2, while the sample volume is held constant (so that $\int p\,\mathrm{d}V = 0$) and only electrical work W^el is done so that no other external work is done ($W^\mathrm{o} = 0$). The experiment is performed first with the substance in the calorimeter to determine the electrical work necessary to achieve the prescribed change of temperature and then again to determine the electrical work required to change the temperature of solely the calorimeter from T_1 to T_2.

The electrical work required to increase the temperature of the substance from T_1 to T_2 is therefore

$$\Delta U = W^\mathrm{el}(\text{sample + calorimeter}) - W^\mathrm{el}(\text{calorimeter}). \tag{1.87}$$

Because ΔU can be written as

$$\Delta U = \int_{T_1}^{T_2} \left(\frac{\partial U}{\partial T}\right)_V \mathrm{d}T, \tag{1.88}$$

then with the definition

$$C_V = \left(\frac{\partial U}{\partial T}\right)_V, \tag{1.89}$$

for the heat capacity at constant volume, Equation 1.87 becomes

$$\int_{T_1}^{T_2} C_V \, \mathrm{d}T = \Delta U = W^{\mathrm{el}}(\text{sample} + \text{calorimeter}) - W^{\mathrm{el}}(\text{calorimeter}). \tag{1.90}$$

If the heat capacity is independent of temperature then

$$C_V = \frac{W^{\mathrm{el}}(\text{sample} + \text{calorimeter}) - W^{\mathrm{el}}(\text{calorimeter})}{T_2 - T_1}, \tag{1.91}$$

which shows how the heat capacity can be measured.

However, in practice, the pressure required to maintain the volume of a sample constant when it is either a solid or a liquid sample under a temperature change requires a container constructed from a material that has a volume independent of temperature and one that is also rigid. The container would require a linear thermal expansion of zero (which is impractical if not impossible except over a limited temperature range) and, for its construction, either a material of unrealizable elastic properties or with very thick walls. The latter implies a mass much greater than the sample so that the majority of the heat capacity and work done would be that of the container and the effect of the sample would be rather lost in the experiment.

To see the problem clearly we give a comparison. The pressure increase at constant volume is given by $(\partial p/\partial T)_V$ and for a liquid hydrocarbon it is about 1 MPa · K^{-1} so that a 10 K temperature increase gives rise to a pressure increase of 10 MPa. For a gas $(\partial p/\partial T)_V$ is about 0.001 MPa · K^{-1} and a 10 K temperature increase results in a pressure change of only 0.01 MPa, which is more easily contained. Thus, the mass of container required to maintain a zero volume change is much less for a gas; however, the heat capacity of a gas is correspondingly lower than that for a liquid and so the mass of the container is still about 100 times greater than that of the sample. These experimental difficulties, which result in an unacceptable uncertainty, require the measurements to be done at constant pressure rather than at constant volume and, therefore, to be of enthalpy differences rather than energy differences. Thus we have

$$\Delta H = \int_{T_1}^{T_2} \left(\frac{\partial H}{\partial T}\right)_p \mathrm{d}T = W^{\mathrm{el}}(\text{sample} + \text{calorimeter}) - W^{\mathrm{el}}(\text{calorimeter}), \tag{1.92}$$

which, in view of the definition of the heat capacity at constant pressure C_p, of

$$C_p = \left(\frac{\partial H}{\partial T}\right)_p, \quad (1.93)$$

and can be written as

$$C_p = \frac{W^{\mathrm{el}}(\text{sample} + \text{calorimeter}) - W^{\mathrm{el}}(\text{calorimeter})}{T_2 - T_1}. \quad (1.94)$$

If it is assumed that C_p is independent of temperature over the range, the heat capacity at constant volume C_V can then be obtained from C_p with Equation 1.150, which is equivalent to Equation 4.93 of Chapter 4. However, for a gas the experiment defined by Equation 1.94 is a very complex and demanding one and yields the heat capacity only with a high uncertainty. Thus, for gases an alternative method is required and one is described in Question 1.8.7 as a means of demonstrating an application of the 1st law of thermodynamics.

The heat capacity of a gas can also be determined from measurements of the speed of sound as alluded to in Chapter 3. This method has the special advantage that is independent of the amount of substance in the sample, but it is outside the scope of this text and the reader is referred to Goodwin and Trusler (2003 and 2010) for the details.

1.8.6 How Do I Measure the Energy in a Food Substance?

Here we describe the methods used to determine the energy in a food substance and this value is reported on the container of processed food. This energy can be measured by completely burning the substance in the presence of an excess of oxygen. The evolved heat is measured in an adiabatic bomb calorimeter. In this reactor the heat evolved during the combustion is absorbed by a known mass of water that surrounds the calorimeter, resulting in an increase of the water temperature that can be measured. A description of this apparatus is given below for the emotive example of a Snickers bar.

Snickers is a chocolate bar consisting of peanut butter, nougat topped with roasted peanuts and caramel covered with milk chocolate. It is well known that a Snickers contains substantial food energy. *Food energy* is the energy in the food available through digestion. The values for food energy are found on all commercially available processed food. The material within the food comprises large organic molecules that are broken down into smaller molecules by digestion. Some of these molecules are used by the body to build complicated molecules necessary for the body's function. Others are metabolized (burned) with the oxygen we breathe in from air. The products of complete combustion are CO_2 and H_2O and the energy of combustion $\Delta_c U$. It is the $\Delta_c U$ that is used to power the body, including both physical mental activity. The

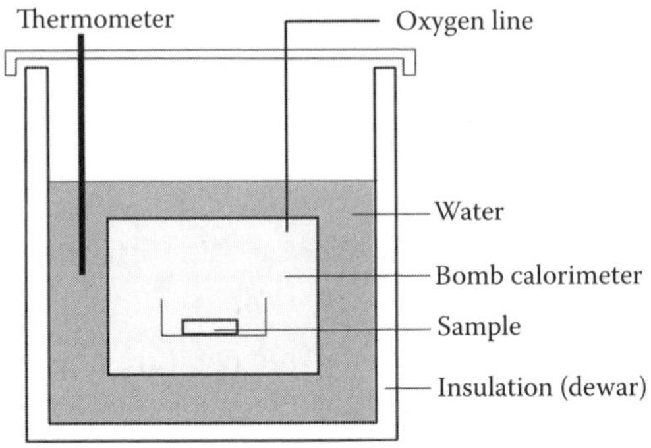

Figure 1.10 Bomb calorimeter.

amount of energy in a substance (including food) can be measured by completely burning the substance in the presence of excess oxygen within a bomb calorimeter shown in Figure 1.10. It is nothing more than a plausible assumption that the heat of complete combustion of a substance can be equated to the *Food* energy.

The "bomb" is actually a high-pressure vessel, usually made of steel, immersed in a water bath. The bomb is designed to change its volume by a negligible amount when the pressure inside it changes. The temperature of this water is continuously monitored with a high-precision thermometer. The water is itself contained in a Dewar flask (Question 1.7.1) that prevents heat flow from the water to the surroundings.

The sample of the Snickers bar is dried and then ground into a powder and placed on a sample container inside the constant volume "bomb." The bomb is then charged with a supply of oxygen up to a pressure of about 2.5 MPa so that there is adequate oxygen for complete combustion of the sample. The sample is then ignited electrically. The heat evolved is transferred to the bomb and the water surrounding it, leading to a temperature increase of both. Owing to the thermal isolation provided by the Dewar flask the food sample and the oxygen can be taken as a closed system and the bomb and the water as the surroundings, and assuming that there is no thermal exchange out of the Dewar flask so that

$$\Delta U_{\text{tot}} = \Delta U_{\text{sys}} + \Delta U_{\text{sur}} = 0, \qquad (1.95)$$

and

$$\Delta U_{\text{sys}} = -\Delta U_{\text{sur}}. \qquad (1.96)$$

The 1st law of thermodynamics for closed systems is given by Equation 1.41 as a process conducted at constant volume in the closed system has $dV = 0$ and the work done, $W = \int p\,dV = 0$ so that

$$Q = \Delta U_{sys}. \tag{1.97}$$

The heat transferred from the system to the surroundings, consisting of the bomb container and the water, is given by

$$\Delta U_{sur} = (mc)_{bomb}\,\Delta T + (mc)_{water}\,\Delta T, \tag{1.98}$$

where c is the specific heat of each component and m is its mass and ΔT is the temperature increase of the bomb wall and that of the water. The energy change of the system is obtained from

$$\Delta U_{sys} = (mc)_{bomb}\,\Delta T + (mc)_{water}\,\Delta T, \tag{1.99}$$

so that the change of the internal energy of the closed system can be determined from the temperature change measured in the water, and knowledge of the masses and heat capacities of the bomb and the water. The heat capacity of the calorimeter can be determined from measurements with a substance for which the heat capacity is known precisely, for example, benzoic acid. However, we are interested in the change in enthalpy (defined by Equation 1.81) of the food in the process; it is given by

$$\Delta H = \Delta U + \Delta(pV). \tag{1.100}$$

The volumes of the solids and liquids and their changes are small compared to those of the gases in the bomb so we assume $\Delta(pv) \approx 0$ for both, and Equation 1.100 becomes

$$\Delta H = \Delta U + \Delta(pV)_{gases}, \tag{1.101}$$

and assuming the gases are ideal so that $pV = n_{gases}RT$, Equation 1.101 is then

$$\Delta H = \Delta U + RT\Delta n_{gases}, \tag{1.102}$$

where Δn_{gases} is the difference of the number of moles between the reactants and the products in the gas. Thus, ΔH, the energy content of the food (heat flow at constant pressure) can be determined from ΔU (heat flow under constant volume) plus the pV work done under constant pressure conditions.

1.8.7 What Is an Adiabatic Flow Calorimeter?

A schematic of an adiabatic flow calorimeter is shown in Figure 1.11. It consists of a thermally isolated tube with a throttle (a constriction), through which gas flows leading to a pressure drop across it, a resistance heater and a measured

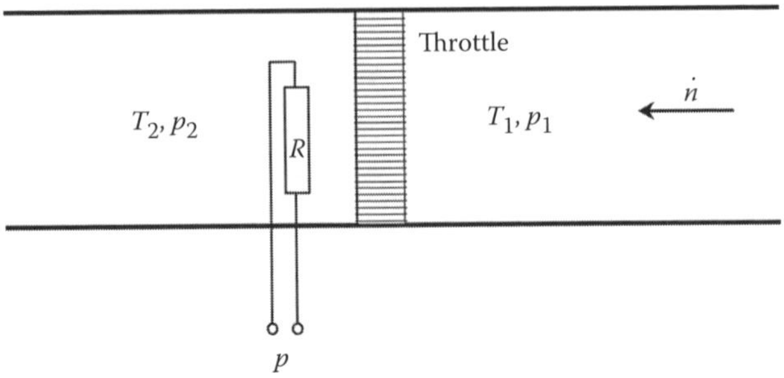

Figure 1.11 Schematic cross-section through an adiabatic flow calorimeter with gas flowing at an amount-of-rate $\dot{n}$ fitted with a throttle and resistor R through which a power P can be dissipated.

source of power P downstream of the throttle, and a means of measuring the temperature and pressure before and after the throttle.

Material present upstream of the throttle at temperature T_1 and pressure p_1 passes at a rate $\dot{n}$ through the throttle where it emerges at temperature T_2 and pressure p_2 in an adiabatic enclosure, where a power P is applied to the resistor R. For an amount of substance n, the 1st law of thermodynamics Equation 1.41 becomes

$$\Delta U = U_1 - U_2 = \frac{Pn}{\dot{n}} + p_1 V(T_1, p_1) - p_2 V(T_2, p_2), \tag{1.103}$$

assuming that the tube is horizontal and that the kinetic energy of the gas is negligible as is often the case. Using the definition $H = U + pV$ we have

$$H(T_2, p_2) - H(T_1, p_1) = \frac{Pn}{\dot{n}}. \tag{1.104}$$

Here we note that when $P = 0$

$$H(T_2, p_2) = H(T_1, p_1), \tag{1.105}$$

the process is enthalpic.

For a gas, if we are able to measure the temperatures of the material on both sides of the throttle and if the process is carried out at low pressures then it is possible to measure the Joule-Thomson coefficient, which is defined as

$$\mu_{\mathrm{JT}} = \left(\frac{\partial T}{\partial p}\right)_H = \lim_{p_2 \to p_1} \left\{ \frac{T(H, p_2) - T(H, p_1)}{p_2 - p_1} \right\}. \tag{1.106}$$

This quantity is of importance in understanding the forces between simple molecules.

Alternatively, it is possible to operate the equipment so as to adjust P and to maintain $T_2 = T = T_1$, and Equation 1.104 becomes

$$\phi_{\mathrm{JT}} = \left(\frac{\partial H}{\partial p}\right)_T = \lim_{p_2 \to p_1}\left\{\frac{H(T,p_2) - H(T,p_1)}{p_2 - p_1}\right\} = \frac{Pn}{\dot{n}(p_2 - p_1)} \qquad (1.107)$$

where ϕ_{JT} is the so-called isothermal Joule-Thomson coefficient.

In the case where $p_1 \approx p_2$ for very slow flow with the throttle removed (since an exact equal rate of fluid flow is not possible) we can also measure the heat capacity at constant pressure

$$C_p = \left(\frac{\partial H}{\partial T}\right)_p = \lim_{T_2 \to T_1}\left\{\frac{H(T_2,p) - H(T_1,p)}{T_2 - T_1}\right\} = \frac{Pn}{\dot{n}(T_2 - T_1)}. \qquad (1.108)$$

The three quantities μ_{JT}, ϕ_{JT}, and C_p are related by Equation 1.142, the −1 rule, as follows:

$$\left(\frac{\partial T}{\partial p}\right)_H \left(\frac{\partial p}{\partial H}\right)_T \left(\frac{\partial H}{\partial T}\right)_p = -1, \qquad (1.109)$$

that can be written as

$$\mu_{\mathrm{JT}} = -\frac{\phi_{\mathrm{JT}}}{C_p}. \qquad (1.110)$$

1.9 WHAT IS THE DIFFERENCE BETWEEN UNCERTAINTY AND ACCURACY?

It is common in the literature for the words accuracy and uncertainty to be used interchangeably but there is a difference between them that is significant and vital. The term “accuracy of measurement” has the internationally agreed definition that paraphrased states; it is the difference between the measured and the true values and is a hypothetical term because in most circumstances the true value is not known. The phrase “uncertainty of measurement” defines the range of values of the result within which it is reasonable with a cited statistical confidence the value will lie. This is achieved without recourse to the assumption of a true value. On the basis of these definitions, the vast majority of measurements are uncertain and not accurate.* Those interested in this topic should also refer to the NIST Technical Note 1297 (Taylor and Kuyatt 1994) and the *Guide to the Expression of Uncertainty in Measurement* (1995).

* http://www.npl.co.uk/server.php?show=ConWebDoc.493

1.10 WHAT ARE STANDARD QUANTITIES AND HOW ARE THEY USED?

The definition of height above the earth relative to mean sea level leads to one entry for each location, while the tabulation of the difference in height between two locations leads to N points for $N(N-1)/2$ entries. By analogy, the same efficiency can be given to the tabulation of thermodynamic quantities by reference to a standard state that is independent of pressure and composition. The tables make use of the general definitions that will be provided for gas, liquids (and solids), solutes, and solvents. For now it is sufficient to note the relationship between the standard chemical potential $\mu_B^\ominus$ and standard absolute activity $\lambda_B^\ominus$ through:

$$\mu_B^\ominus = RT \ln \lambda_B^\ominus. \tag{1.111}$$

Once the relevant standard chemical potential $\mu_B^\ominus$ or standard absolute activity $\lambda_B^\ominus$ have been defined all other standard thermodynamic functions are obtained by differentiation with respect to temperature to give the standard molar entropy

$$S_B^\ominus = -\frac{d\mu_B^\ominus}{dT} = -R \ln \lambda_B^\ominus - RT\frac{d\ln\lambda_B^\ominus}{dT}, \tag{1.112}$$

the standard molar enthalpy

$$H_B^\ominus = \mu_B^\ominus - T\frac{d\mu_B^\ominus}{dT} = -RT^2\frac{d\ln\lambda_B^\ominus}{dT}, \tag{1.113}$$

the standard partial molar Gibbs function

$$G_B^\ominus = \mu_B^\ominus = RT\ln\lambda_B^\ominus, \tag{1.114}$$

and the standard molar heat capacity at constant pressure

$$C_{p,B}^\ominus = -T\frac{d^2\mu_B^\ominus}{dT^2} = -2RT\frac{d\ln\lambda_B^\ominus}{dT} - RT^2\frac{d^2\ln\lambda_B^\ominus}{dT^2}. \tag{1.115}$$

The standard equilibrium constant for a chemical reaction $0 = \Sigma_B \nu_B B$ (Equation 1.30) is defined by

$$K^\ominus(T) = \exp\left\{-\frac{\sum_B \nu_B \mu_B^\ominus(T)}{RT}\right\} = \prod_B \left\{\lambda_B^\ominus(T)\right\}^{-\nu_B}. \tag{1.116}$$

The $K^{\ominus}(T)$ depends only on temperature and not on pressure or composition. From Equations 1.112, 1.113, and 1.114 with Equation 1.116 we arrive at the standard molar entropy

$$\Delta_r S_m^{\ominus} = \sum_B \nu_B S_B^{\ominus} = -R \ln K^{\ominus} - RT \frac{d \ln K^{\ominus}}{dT}, \tag{1.117}$$

the standard molar enthalpy of the reaction

$$\Delta_r H_m^{\ominus} = \sum_B \nu_B H_B^{\ominus} = RT^2 \frac{d \ln K^{\ominus}}{dT}, \tag{1.118}$$

the standard molar Gibbs function

$$\Delta_r G_m^{\ominus} = \sum_B \nu_B \mu_B^{\ominus} = \sum_B \nu_B G_B^{\ominus} = RT \ln K^{\ominus}. \tag{1.119}$$

In Equations 1.117, 1.118, and 1.119 we have introduced the change of standard molar entropy for the reaction $\Delta_r S_m^{\ominus}$, the change of standard molar enthalpy for the reaction, $\Delta_r H_m^{\ominus}$, and the standard molar change in Gibbs function for the reaction $\Delta G_m^{\ominus}$.

Equation 1.118 is usually written as

$$\frac{d \ln K^{\ominus}}{dT} = \frac{\Delta_r H_m^{\ominus}}{RT^2}, \tag{1.120}$$

and called *Van't Hoff's equation* and is important because if a value of $K^{\ominus}$ for a reaction is available at one temperature T_1 the value at another temperature T_2 can be determined from

$$\ln\{K^{\ominus}(T_2)\} = \ln\{K^{\ominus}(T_1)\} + \int_{T_1}^{T_2} \left\{ \frac{\Delta_r H_m^{\ominus}(T)}{RT} \right\} dT. \tag{1.121}$$

Provided $\Delta_r H_m^{\ominus}$ is known over the required temperature range T_1 to T_2 and in the absence of this information $\Delta_r H_m^{\ominus}$ can be assumed independent of temperature. It is more typical to find values of the standard molar heat capacity at constant pressure $C_{p,m}^{\ominus}$ and the $\Delta_r H_m^{\ominus}(T)$ obtained from $\Delta_r H_m^{\ominus}$ at a temperature where it is known (T_3) with Equation 1.93 cast as

$$\Delta_r H_m^{\ominus}(T) = \Delta_r H_m^{\ominus}(T_3) + \int_{T_3}^{T} \sum_B \nu_B C_{p,B}^{\ominus}(T)\, dT, \tag{1.122}$$

and when substituted into Equation 1.121 gives

$$\ln\{K^{\ominus}(T_2)\} = \ln\{K^{\ominus}(T_1)\} + \frac{\Delta_r H_m^{\ominus}(T_3)(T_2 - T_1)}{RT_1T_2}$$

$$+\int_{T_1}^{T_2}\left\{\int_{T_3}^{T}\sum_B \nu_B C_{p,B}^{\ominus}(T)\,dT\right\}\left(\frac{1}{RT^2}\right)dT. \qquad (1.123)$$

If the $C_{p,B}^{\ominus}$ is not known over a temperature range then it can be assumed independent of T.

Primary chemical thermodynamic tables contain values of the standard molar thermodynamic functions related to the standard equilibrium constant. For these tables, careful measurements are made on the pure substance: these might be the enthalpy of combustion at one temperature and the heat capacity as a function of temperature using Equation 1.123. The tabulated values are combined to calculate for any reaction or system the $K^{\ominus}$ at another temperature in the range of those given.

So-called secondary tables list fugacities, virial coefficients, activity coefficients, and osmotic coefficients; the latter are needed to calculate the extent of reaction under pressure and initial composition from $K^{\ominus}$. For these tables, measurements are required for each mixture. The enormous amount of work means that theories of mixtures are vital. A chemical engineer might also need values of the enthalpy of formation $\Delta_f H_m^{\ominus}$ to permit calculation of the energy flow in a chemical plant. These can be obtained from

$$\Delta_f H_m^{\ominus} = -R\frac{d\ln\{K_f^{\ominus}(T)\}}{d(1/T)}. \qquad (1.124)$$

Henceforth the term, formation, will mean the elements that constitute the molecule. We will not provide sources of these values here but leave that until Chapter 7. We now turn to defining the standard thermodynamic functions for gas, liquid, or solid and solutions.

The standard chemical potential for a gas B is defined by

$$\mu_B^{\ominus}(g, T) = \mu_B(g, T, p, x) - RT\ln\left(\frac{x_B p}{p^{\ominus}}\right) - \int_0^p \left\{V_B(g, T, p, x) - \frac{RT}{p}\right\} dp, \qquad (1.125)$$

where results obtained from the second law, to be introduced in Chapter 3, have been used. Equation 1.125, with the Roman character defining the state of the phase, with g for gas, l for liquid, and s for solid, is written according to the nomenclature established by the IUPAC *Green Book* (Quack et al. 2007). This form of representation shows, for example, the standard chemical potential $\mu_{\mathrm{B}}^{\ominus}$ (g, T), which is a function of the phase and temperature. However, as we recognize throughout this text, the language of the chemist is not familiar to all, indeed this particular example generated much discussion among the authors. Consequently, for the general audience we have adopted an approach that deviates from the formal IUPAC symbolism and indicates, when significant, the phase as a subscript after identifying the substance B, also a subscript. Thus, with these rules we now write, for example, $\mu_{\mathrm{B}}^{\ominus}$ (g, T) of Equation 1.125 as $\mu_{\mathrm{B,g}}^{\ominus}(T)$ and cast Equation 1.125 in the form

$$\mu_{\mathrm{B,g}}^{\ominus}(T)=\mu_{\mathrm{B,g}}(T,p,x)-RT\ln\left(\frac{x_{\mathrm{B}}p}{p^{\ominus}}\right)-\int_{0}^{p}\left\{V_{\mathrm{B,g}}(T,p,x)-\frac{RT}{p}\right\}\mathrm{d}p, \quad (1.126)$$

In Equation 1.126, $\mu_{\mathrm{B,g}}(T,p,x)$ is the chemical potential and $V_{\mathrm{B,g}}(T,p,x)$ the partial molar volume of species B in a gas mixture of composition given by mole fractions x for which the mole fraction of B is x_{B} at a pressure p and temperature T. The pressure $p^{\ominus}$ is the standard pressure and is usually 0.1 MPa.* The other standard thermodynamic functions follow from Equations 1.112, 1.114, and 1.115 as follows:

$$S_{\mathrm{B,g}}^{\ominus}(T)=S_{\mathrm{B,g}}(T,p,x)+R\ln\left(\frac{x_{\mathrm{B}}p}{p^{\ominus}}\right)+\int_{0}^{p}\left[\left\{\frac{\partial V_{\mathrm{B,g}}(T,p,x)}{\partial T}\right\}_{p}-\frac{R}{p}\right]\mathrm{d}p, \quad (1.127)$$

$$H_{\mathrm{B,g}}^{\ominus}(T)=H_{\mathrm{B,g}}(T,p,x)-\int_{0}^{p}\left[V_{\mathrm{B}}-\left\{\frac{\partial V_{\mathrm{B,g}}(T,p,x)}{\partial T}\right\}_{p}\right]\mathrm{d}p, \text{ and} \quad (1.128)$$

$$C_{p,\mathrm{B,g}}^{\ominus}(T)=C_{p,\mathrm{B,g}}(T,p,x)+\int_{0}^{p}T\left\{\frac{\partial^{2}V_{\mathrm{B,g}}(T,p,x)}{\partial T^{2}}\right\}_{p}\mathrm{d}p. \quad (1.129)$$

Results obtained from the second law were used to obtain Equations 1.127, 1.128, and 1.129. For a perfect gas the integrals in Equations 1.127, 1.128 and 1.129 vanish. For the case when $p \approx p^{\ominus}$ we find $\Delta H_{\mathrm{m}} \approx \Delta H_{\mathrm{m}}^{\ominus}$, and $C_{p,\mathrm{m}} \approx C_{p,\mathrm{m}}^{\ominus}$ because the integrals in Equations 1.128 and 1.129 are small fractions of ΔH_{m} and $C_{p,\mathrm{m}}$.

* The value for $p^{\ominus}$ is 10^5 Pa and has been the IUPAC recommendation since 1982 and should be used to tabulate thermodynamic data. Before 1982 the standard pressure was usually taken to be $p^{\ominus}$ = 101 325 Pa (=1 bar or 1 atm), called the *standard atmosphere*). In any case, the value for $p^{\ominus}$ should be specified.

The standard chemical potential $\mu^{\ominus}_{\text{B,l}}(T)$ for a liquid B is defined by

$$\mu^{\ominus}_{\text{B,l}}(T) = \mu^{*}_{\text{B,l}}(T, p^{\ominus}), \tag{1.130}$$

where the $\mu^{*}_{\text{B,l}}(T, p^{\ominus})$ is the chemical potential of pure B at the same temperature and at the standard pressure $p^{\ominus}$. Similarly, the standard chemical potential $\mu^{\ominus}_{\text{B,s}}(T)$ for a solid B is

$$\mu^{\ominus}_{\text{B,s}}(T) = \mu^{*}_{\text{B,s}}(T, p^{\ominus}), \tag{1.131}$$

where the only change from Equation 1.130 is the state symbol. When $\mu^{*}_{\text{B,l}}(T, p)$ is at a pressure $p \neq p^{\ominus}$ Equation 1.130 can be written as

$$\mu^{\ominus}_{\text{B,l}}(T) = \mu^{*}_{\text{B,l}}(T, p) + \int_{p}^{p^{\ominus}} V^{*}_{\text{B,l}}(T, p)\,\text{d}p, \tag{1.132}$$

where $V^{*}_{\text{B,l}}(T, p)$ is the molar volume of the pure liquid B at temperature T and pressure p. Equation 1.132 can be cast in terms of the absolute activity as

$$\lambda^{*}_{\text{B,l}}(T, p) = \lambda^{\ominus}_{\text{B,l}}(T)\exp\left[\int_{p^{\ominus}}^{p}\left\{\frac{V^{*}_{\text{B,l}}(T, p)}{RT}\right\}\text{d}p\right]. \tag{1.133}$$

For $(p - p^{\ominus}) = 0.1$ MPa and $V^{*}_{\text{B,l}} = 100\ \text{cm}^3 \cdot \text{mol}^{-1}$ the exponential term in Equation 1.133 is 1.004 and the integral in Equations 1.132 and 1.133 is sufficiently small to neglect so that the approximate forms of Equations 1.132 and 1.133 are

$$\mu^{\ominus}_{\text{B,l}}(T) \approx \mu^{*}_{\text{B,l}}(T, p) \tag{1.134}$$

and

$$\lambda^{*}_{\text{B,l}}(T, p) \approx \lambda^{\ominus}_{\text{B,l}}(T). \tag{1.135}$$

For a liquid mixture containing substance B, Equation 1.132 can be cast as

$$\mu^{\ominus}_{\text{B,l}}(T) = \mu_{\text{B,l}}(T, p, x) + \{\mu^{*}_{\text{B,l}}(T, p) - \mu_{\text{B,l}}(T, p, x)\} + \int_{p}^{p^{\ominus}} V^{*}_{\text{B,l}}(T, p)\,\text{d}p. \tag{1.136}$$

In Equation 1.136, $\mu_{\text{B,l}}(T, p, x)$ is the chemical potential of substance B in the liquid mixture that has a composition given by mole fractions x at the temperature T and pressure p. Equation 1.136 reduces to Equation 1.131 for a pure substance.

In a solution the standard chemical potential of the solvent A is defined by

$$\mu^{\ominus}_{\text{A,l}}(T) = \mu^{*}_{\text{A,l}}(T, p^{\ominus}). \tag{1.137}$$

Equation 1.137 is identical to Equation 1.131, and, thus, Equations 1.132 and 1.136 are applicable for a solvent, albeit with change in notation from mole fractions to molality m and use of the standard molality $m^{\ominus} = 1\,\mathrm{mol \cdot kg^{-1}}$. The standard chemical potential of the solute B in the solvent A is defined by

$$\mu^{\ominus}_{\mathrm{B,sol}}(T) = \left\{ \mu_{\mathrm{B,sol}}(T, p^{\ominus}, m_{\mathrm{C}}) - RT\ln\left(\frac{m_{\mathrm{B}}}{m^{\ominus}}\right) \right\}^{\infty}$$

$$= \left\{ \mu_{\mathrm{B,sol}}(T, p, m_{\mathrm{C}}) - RT\ln\left(\frac{m_{\mathrm{B}}}{m^{\ominus}}\right) \right\}^{\infty} + \int_{p}^{p^{\circ}} V^{*}_{\mathrm{B,sol}}(T, p)\mathrm{d}p. \qquad (1.138)$$

In Equation 1.138 m is the molalities of the solutes and the ∞ indicates infinite dilution that is $\Sigma_{\mathrm{i}}\, m_{\mathrm{i}} \rightarrow 0$.

1.11 WHAT MATHEMATICAL RELATIONSHIPS ARE USEFUL IN THERMODYNAMICS?

Much of this book seeks to provide explanations of the fundamental laws of thermodynamics and thermophysics. Many of the derivations of the results we quote in this book are omitted in the interest of brevity and because they are not actually the intent of this book. However, we recognize that some readers will want to attempt the derivations themselves for further understanding and for that reason we provide here a few useful mathematical relationships. In any case, some of the difficulty in understanding thermodynamics arises because of the long, nonintuitive manipulation of thermodynamic relationships particularly through partial derivatives. Thus, in the last section of this chapter we provide a statement of several important relationships among partial derivatives that will help students of thermodynamics to keep at their fingertips while attempting to understand other texts that do provide (or expect!) full derivations.

1.11.1 What Is Partial Differentiation?

For a function of $x, y, z, \ldots, u = u(x, y, z, \ldots)$ then the total derivative of u is

$$\mathrm{d}u = \left(\frac{\partial u}{\partial x}\right)_{y,z,\ldots} \mathrm{d}x + \left(\frac{\partial u}{\partial y}\right)_{x,z,\ldots} \mathrm{d}y + \left(\frac{\partial u}{\partial z}\right)_{y,x,\ldots} \mathrm{d}z + \cdots. \qquad (1.139)$$

For a function $u(x, y)$ Equation 1.139 becomes

$$\mathrm{d}u = \left(\frac{\partial u}{\partial x}\right)_{y} \mathrm{d}x + \left(\frac{\partial u}{\partial y}\right)_{x} \mathrm{d}y. \qquad (1.140)$$

Equation 1.139 can be used to illustrate three theorems. The first is for the change of variable held constant

$$\left(\frac{\partial u}{\partial x}\right)_z = \left(\frac{\partial u}{\partial x}\right)_y + \left(\frac{\partial u}{\partial y}\right)_x \left(\frac{\partial y}{\partial x}\right)_z. \qquad (1.141)$$

The second is the −1 rule that is

$$\left(\frac{\partial u}{\partial x}\right)_y \left(\frac{\partial x}{\partial y}\right)_u \left(\frac{\partial y}{\partial u}\right)_x = -1. \qquad (1.142)$$

The third theorem is for cross-differentiation

$$\left\{\frac{\partial(\partial u/\partial y)_y}{\partial y}\right\}_x = \left\{\frac{\partial(\partial u/\partial y)_x}{\partial x}\right\}_y. \qquad (1.143)$$

An expression for the difference $(C_p - C_V)$ can be found by applying the first two rules and by using the definitions of C_p, C_V, and $H = U + pV$. First we note from Equations 1.89 and 1.93

$$C_p - C_V = \left(\frac{\partial H}{\partial T}\right)_p - \left(\frac{\partial U}{\partial T}\right)_V$$

$$= \left(\frac{\partial H}{\partial T}\right)_p - \left(\frac{\partial H}{\partial T}\right)_V + V\left(\frac{\partial p}{\partial T}\right)_V. \qquad (1.144)$$

Use of the rule for change of variable held constant on $(\partial H/\partial T)_V$ gives

$$\left(\frac{\partial H}{\partial T}\right)_V = \left(\frac{\partial H}{\partial T}\right)_p + \left(\frac{\partial H}{\partial p}\right)_T \left(\frac{\partial p}{\partial T}\right)_V \qquad (1.145)$$

so that Equation 1.144 can be written as

$$C_p - C_V = \left\{V - \left(\frac{\partial H}{\partial p}\right)_T\right\}\left(\frac{\partial p}{\partial T}\right)_V. \qquad (1.146)$$

From $H = G + TS$ it can be shown that

$$\left(\frac{\partial H}{\partial p}\right)_T = V - T\left(\frac{\partial V}{\partial T}\right)_p \qquad (1.147)$$

when combined with Equation 1.146 this gives

$$C_p - C_V = T\left\{\left(\frac{\partial V}{\partial T}\right)_p\right\}\left(\frac{\partial p}{\partial T}\right)_V. \qquad (1.148)$$

When the −1 rule is used on $(\partial p/\partial T)_V$ we find

$$\left(\frac{\partial p}{\partial T}\right)_V\left(\frac{\partial T}{\partial V}\right)_p\left(\frac{\partial V}{\partial p}\right)_T = -1, \qquad (1.149)$$

and rearrangement gives

$$\left(\frac{\partial p}{\partial T}\right)_V = -\frac{(\partial V/\partial T)_p}{(\partial V/\partial p)_T}. \qquad (1.150)$$

Substitution of Equation 1.150 into Equation 1.148 yields

$$C_p - C_V = -T\frac{\{(\partial V/\partial T)_p\}^2}{(\partial V/\partial p)_T}. \qquad (1.151)$$

Equation 1.51 is an important result because the derivatives $(\partial V/\partial T)_p$ and $(\partial V/\partial p)_T$ can be measured directly. An example of the application of the cross-differentiation rule can be found in a method of evaluating the change in entropy of a material arising from a pressure change at constant temperature. For a phase of fixed composition when the variables are T and p the axiom 2a (Equation 3.14) is

$$\mathrm{d}G = -S\,\mathrm{d}T + V\,\mathrm{d}p, \qquad (1.152)$$

and directly from Equation 1.152 it follows that

$$\left(\frac{\partial G}{\partial T}\right)_p = -S, \qquad (1.153)$$

and

$$\left(\frac{\partial G}{\partial T}\right)_T = V. \qquad (1.154)$$

Differentiating Equation 1.154 with respect to p at constant T gives

$$\left\{\frac{\partial(\partial G/\partial T)_p}{\partial p}\right\}_T = -\left(\frac{\partial S}{\partial p}\right)_T, \qquad (1.155)$$

and Equation 1.154 with respect to T at constant p

$$\left\{\frac{\partial(\partial G/\partial p)_T}{\partial T}\right\}_p = -\left(\frac{\partial V}{\partial T}\right)_p. \tag{1.156}$$

Comparison of Equations 1.153 and 1.156 gives

$$\left(\frac{\partial S}{\partial p}\right)_T = -\left(\frac{\partial V}{\partial T}\right)_p, \tag{1.157}$$

that is called a *Maxwell equation* and by integration provides a means of determining the entropy difference

$$\{S(T,p_2)-S(T,p_1)\} = -\int_{p_1}^{p_2}\left(\frac{\partial V}{\partial T}\right)_p \mathrm{d}p. \tag{1.158}$$

1.11.2 What Is Euler's Theorem?

According to Euler's theorem, when u is a homogeneous function of the n th degree in the variables $x, y, z, \ldots$, then

$$x\left(\frac{\partial u}{\partial x}\right)_{y,z,\ldots} + y\left(\frac{\partial u}{\partial y}\right)_{x,z,\ldots} + z\left(\frac{\partial u}{\partial z}\right)_{x,y,\ldots} + \cdots = nu. \tag{1.159}$$

For a mixture (A + B) containing amounts of substance n_A and n_B, respectively, at constant temperature and pressure the volume V is a homogeneous function of n_A and n_B of the first degree and from Equations 1.7 and 1.9 it is

$$V = n_A\left(\frac{\partial V}{\partial n_A}\right)_{T,p,n_B} + n_B\left(\frac{\partial V}{\partial n_B}\right)_{T,p,n_A}. \tag{1.160}$$

We have already made use of Euler's theorem in this chapter.

1.11.3 What Is the Taylor's Theorem?

For an analytic function $f(x)$ the Taylor's expansion about $x = a$ is

$$f(x) = f(a) + (x-a)\left\{\frac{\partial f(x)}{\partial x}\right\}_{x=a} + (x-a)^2\frac{\{\partial^2 f(x)/\partial x^2\}_{x=a}}{2!} + (x-a)^3\frac{\{\partial^3 f(x)/\partial x^3\}_{x=a}}{3!} + \cdots. \tag{1.161}$$

1.11.4 What Is the Euler–MacLaurin Theorem?

The sum of a function $f(n)$ over all integral values of n from 0 to ∞ is given by

$$\sum_{n=0}^{\infty} f(n) = \int_0^{\infty} f(n)\,\mathrm{d}n + \frac{1}{2}f(0) - \frac{1}{12}\left(\frac{\partial f}{\partial n}\right)_{n=0} + \frac{1}{720}\left(\frac{\partial^3 f}{\partial n^3}\right)_{n=0} - \cdots. \quad (1.162)$$

1.12 REFERENCES

Atkins P., and de Paula, J., 2009, *Physical Chemistry, Resource Section*, Part 1, Oxford University Press, Oxford, pp. 911–913.

Case F., Chaka A., Friend D.G., Frurip D., Golab J., Johnson R., Moore J., Mountain R.D., Olson J., Schiller M., and Storer J., 2004, "The first industrial fluid properties simulation challenge," *Fluid Phase Equilib.* **217**:1–10.

Case F., Chaka A., Friend D.G., Frurip D., Golab J., Gordon P., Johnson R., Kolar P., Moore J., Mountain R.D., Olson J., Ross R., and Schiller M., 2005, "The second industrial fluid properties simulation challenge," *Fluid Phase Equilib.* **236**:1–14.

Case F., Brennan J., Chaka A., Dobbs K.D., Friend D.G., Frurip D., Gordon P.A., Moore J., Mountain R.D., Olson J., Ross R.B., Schiller M., and Shen V.K., 2007, "The third industrial fluid properties simulation challenge," *Fluid Phase Equilib.* **260**:153–163.

Case F., Brennan J., Chaka A., Dobbs K.D., Friend D.G., Gordon P.A., Moore J.D., Mountain R.D., Olson J.D., Ross D.B., Schiller M., Shen V.K., and Stahlberg E.A., 2008, "The fourth industrial fluid properties simulation challenge," *Fluid Phase Equilib.* **274**:2–9.

Cengel Y.A., and Boles M.A., 2006, *Thermodynamics—an Engineering Approach*, 6th Edition, McGraw-Hill, New York.

Chemical Thermodynamics, 2000, ed. Letcher T.M., for IUPAC, Blackwells Scientific Publications, Oxford.

Developments and Applications of Solubility, 2007, ed. Letcher T.M., for IUPAC, Royal Society of Chemistry, Cambridge.

Eckl B., Vrabec J., and Hasse H., 2008, "On the application of force fields for predicting a wide variety of properties: Ethylene oxide as an example," *Fluid Phase Equilib.* **274**:16–26.

Einstein A., 1901, "Folgerungen aus den Capillaritätserscheinungen," *Ann. der Phys.* **4**:513.

Experimental Thermodynamics, Volume I, Calorimetry of Non-Reacting Systems, 1968, eds. McCullough J.P., and Scott D.W., for IUPAC, Butterworths, London.

Experimental Thermodynamics, Volume II, Experimental Thermodynamics of Non-Reacting Fluids, 1975, eds. Le Neindre B., and Vodar B., for IUPAC, Butterworths, London.

Experimental Thermodynamics, Volume III, Measurement of the Transport Properties of Fluids, 1991, eds. Wakeham W.A., Nagashima A., and Sengers J.V., for IUPAC, Blackwell Scientific Publications, Oxford.

Experimental Thermodynamics, Volume IV, Solution Calorimetry, 1994, eds. Marsh K.N., and O'Hare P.A.G., for IUPAC, Blackwell Scientific Publications, Oxford.

Experimental Thermodynamics, Volume V, Equations of State for Fluids and Fluid Mixtures, Parts I and II, 2000, eds. Sengers J.V., Kayser R.F., Peters C.J., and White Jr. H.J., for IUPAC, Elsevier, Amsterdam.

Experimental Thermodynamics, Volume VI, Measurement of the Thermodynamic Properties of Single Phases, 2003, eds. Goodwin A.R.H., Marsh K.N., and Wakeham W.A., for IUPAC, Elsevier, Amsterdam.

Experimental Thermodynamics, Volume VII, Measurement of the Thermodynamic Properties of Multiple Phases, 2005, eds. Weir R.D., and de Loos T.W., for IUPAC, Elsevier, Amsterdam.

Experimental Thermodynamics, Volume VIII, Applied Thermodynamics of Fluids, 2010, eds. Goodwin A.R.H., Sengers J.V., and Peters C.J., for IUPAC, RSC Publishing, Cambridge.

Future Energy: Improved, Sustainable and Clean Options for our Planet, 2008, ed. Letcher T.M., for IUPAC, Elsevier, Amsterdam.

Goodwin A.R.H., and Trusler, J.P.M., 2003, *Sound Speed*, Chapter 6, in *Experimental Thermodynamics, Volume VI, Measurement of the Thermodynamic Properties of Single Phases*, eds. Goodwin A.R.H., Marsh K.N., and Wakeham W.A., for IUPAC, Elsevier, Amsterdam.

Goodwin A.R.H., and Trusler J.P.M., 2010, *Sound Speed*, Chapter 9, in *Heat Capacity*, eds. Letcher T.M., and Willhelm E., for IUPAC, RSC Publishing, Cambridge.

Guide to the Expression of Uncertainty in Measurement, 1995, International Standards Organization, Geneva, Switzerland.

Heat Capacity, 2010, eds. Letcher T.M., and Willhelm E., for IUPAC, RSC Publishing, Cambridge.

International Vocabulary of Basic and General Terms in Metrology, 1993, International Standards Organization, Geneva, Switzerland.

Ketko M.H., Rafferty J., Siepmann J.L., and Potoff J.J., 2008, "Development of the TraPPE-UA force field for ethylene oxide," *Fluid Phase Equilib.* **274**:44–49.

Le Système international d'unités (*SI*), 2006, 8th ed., Bureau International des Poids et Mesures, Pavillion de Breteuil, F-92312 Sevres Dedex, France.

Li X., Zhao L., Cheng T., Liu L., and Sun H., 2008, "One force field for predicting multiple thermodynamic properties of liquid and vapor ethylene oxide," *Fluid Phase Equilib.* **274**:36–43.

Maitland G.C., Rigby M., Smith E.B., and Wakeham W.A., 1981, *Intermolecular Forces. Their Origin and Determination*, Clarendon Press, Oxford.

Mills I.N., Mohr P.J., Quinn T.J., Taylor B.N., and Williams E.R., 2006, "Redefinition of the kilogram, ampere, kelvin and mole: A proposed approach to implementing CIPM recommendation 1 (CI-2005)," *Metrologia* **43**:227–246.

Mohr P.J., Taylor B.N., and Newell D.B., 2008, "CODATA recommended values of the fundamental physical constants: 2006," *J. Phys. Chem. Ref. Data* **37**:1187–1284.

Müller T.J., Roy S., Ahao W., and Maaß A., 2008, "Economic simplex optimization for broad range property prediction: Strengths and weaknesses of an automated approach for tailoring of parameters," *Fluid Phase Equilib.* **274**:27–35.

Nicholas J.V., and White D.R., 2003 *Temperature*, Chapter 2, in *Experimental Thermodynamics, Volume VI, Measurement of the Thermodynamic Properties of Single Phases*, eds. Goodwin A.R.H., Marsh K.N., and Wakeham W.A., for IUPAC, Elsevier, Amsterdam.

Olson J.D., and Wilson L.C., 2008, "Benchmarks for the fourth industrial fluid properties simulation challenge," *Fluid Phase Equilib.* **274**:10–15.

Perrot P., 1998, *A to Z of Thermodynamics*, Oxford University Press, Oxford.

Preston T.H., 1990, "The International Temperature Scale of 1990 (ITS-90)," *Metrologia* **27**:3–10.

Quack M., Stohner J., Strauss H.L., Takami M., Thor A.J., Cohen E.R., Cvitas T., Frey J.G., Holström B., Kuchitsu K., Marquardt R., Mills I., and Pavese F., 2007, *Quantities, Units and Symbols in Physical Chemistry*, 3rd ed., RSC Publishing, Cambridge.

Smith J.M., van Ness H.C., and Abbott M., 2004, *Introduction to Chemical Engineering Thermodynamics*, McGraw-Hill, New York.

Sonntag R.E., Borgnakke C., and van Wylen G.J., 2004, *Fundamentals of Thermodynamics*, 6th ed., John Wiley & Sons, New York.

Suski J., Puers R., Ehrlich C.D., and Schmidt J.W., 2003, *Pressure*, Chapter 3, in *Experimental Thermodynamics, Volume VI, Measurement of the Thermodynamic Properties of Single Phases*, eds. Goodwin A.R.H., Marsh K.N., and Wakeham W.A., for IUPAC, Elsevier, Amsterdam.

Taylor B.N., and Kuyatt C.E., 1994, *Guidelines for Evaluating and Expressing the Uncertainty of NIST Measurement Results*, NIST Technical Note 1297.

Wieser M.E., 2006, "Atomic Weights of the Elements, 2005," *Pure Appl. Chem.* **78**:2051–2066.

Chapter 2

What Is Statistical Mechanics?

2.1 INTRODUCTION

Chapter 1 dealt with the definitions of many of the quantities required for the macroscopic description of the thermodynamic behavior of systems viewed as continua, including the definition of a system. However, we are familiar with the notion that all matter is made up of atomic or molecular entities, and it is the purpose of statistical mechanics to provide a microscopic description of the behavior of a thermodynamic system in terms of the properties, interactions, and motions of the atoms or molecules that make up the system. Because macroscopic thermodynamic systems contain very large numbers of molecules, the task of statistical mechanics is not to describe exactly what happens to every single molecule, but rather to derive results that pertain to the complete assembly of molecules that comprise the system in a probabilistic manner. The atoms and molecules that comprise the system are best described using quantum mechanics rather than classical mechanics so that is the basis for the development of the theory of statistical mechanics.

The solution of Schrödinger's equation of quantum mechanics is a wave that describes the probable state of the system that includes a description of the quantum states (eigenstates) and energy levels (energy eigenvalues) an individual and the system can attain; in quantum theory the energy levels are discrete. It is very much easier to solve Schrödinger's equation for a single molecule (or realistically for a single atom) than for a system of N molecules or atoms to obtain the quantum states or energy levels, so we begin with that problem. For a single, relatively simple molecule (such as nitrogen) the problem of solving Schrödinger's equation is made tractable by separation of the modes of motion of the molecule (translation of the centre of mass, rotation, and vibration) so that each is handled independently. This is legitimate provided that certain conditions are met and a number of texts on quantum mechanics and/or statistical physics will provide you with the means of deducing the allowed energy states for each of these modes of motion.

The question then arises as to how are the molecules distributed between the energy levels available to a single molecule? It is reasonable to anticipate that a system of N molecules will be arranged so that in each quantum state of discrete energy there will be a number of molecules. When the energy of the molecules is very much higher than the difference between the energies of the various quantum levels, which happens for the translational kinetic energy of the molecules of a gas in a macroscopic system at moderate temperature, then the relationship between the number of molecules with a specified energy and the energy of that state, is given by Boltzmann's distribution. For some other type of system other distributions of energy are possible. For these other cases the spin of the molecular system matters. For a set of entities with an integral spin the system will obey Bose–Einstein statistics, while if it is a half-integral spin system Fermi–Dirac statistics are used. When the energy of the system is sufficiently high, both reduce to Boltzmann's distribution. For the molecules and conditions of interest to chemists and engineers, Boltzmann statistics are likely to be appropriate. On the other hand, for Physicists, particularly at low temperatures the other types of distribution are often appropriate. The distinction between low and high temperatures will be quantified as we proceed.

We now consider a system of N molecules (where $N > 10^{15}$). If we suppose that we have a distribution of the molecules so that N_i of the molecules is in the i'th quantum state each with energy ε_i then the total number of molecules is

$$N = \sum_i N_i, \tag{2.1}$$

and the internal energy U is

$$U = \sum_i N_i \varepsilon_i. \tag{2.2}$$

It can also be shown (McQuarrie 2000) that the thermodynamic pressure may be written as

$$p = \sum_i N_i \left(-\frac{d\varepsilon_i}{dV} \right), \tag{2.3}$$

where V is the volume of the system.

Indeed, it has been shown that, building on these methods for low density helium gas, quantum mechanical calculations and fundamental constants can be used alone, *ab initio*, to determine the sum of the pair interaction potential energies of a group of molecules. The methods of statistical mechanics have then been used to provide the thermophysical properties of helium with an estimated uncertainty less than that obtained from measurements (Hurly and Moldover 2000; Hurly and Mehl 2007). For most other molecules and atoms these calculations are yet to be done with sufficient precision. The general principles of statistical mechanics are used within molecular simulation or computational chemistry to provide estimates of the thermophysical properties of materials, but for most systems the calculations are significantly less precise than those available through direct measurement. In this chapter we try to consider the implications of this fact, which separates in significant ways the interests of scientists from those of engineers.

2.2 WHAT IS BOLTZMANN'S DISTRIBUTION?

According to Boltzmann's distribution the number of molecules N_i in the i'th quantum state of energy ε_i is given by

$$N_i = \lambda \exp\left(-\frac{\varepsilon_i}{kT}\right), \tag{2.4}$$

where λ is the absolute activity of a substance that is defined in terms of the chemical potential μ (discussed in Chapter 1) by

$$\lambda = \exp\left(\frac{\mu}{kLT}\right), \tag{2.5}$$

where k is the Boltzmann's constant with the numerical value $(1.380\,6504 \pm 0.000\,0024) \cdot 10^{-23}\ \mathrm{J \cdot K^{-1}}$ (Mohr et al. 2008) and is the proportionality between statistical and classical thermodynamics, L is the Avogadro's number $(=6.0221 \cdot 10^{23}\ \mathrm{mol^{-1}})$, and T is the thermodynamic temperature. The use of Equation 2.4 in Equation 2.1 gives

$$N = \lambda \sum_i \exp\left(-\frac{\varepsilon_i}{kT}\right), \tag{2.6}$$

where the sum on the right hand side is defined by

$$\sum_i \exp\left(-\frac{\varepsilon_i}{kT}\right) = q, \tag{2.7}$$

and called the *molecular partition function*. The combination Lk is also special

$$R = Lk, \tag{2.8}$$

and is known as the universal gas constant.

From Equations 2.6 and 2.7 we then have

$$N = \lambda q, \tag{2.9}$$

so that

$$\mu = RT \ln N - RT \ln q. \tag{2.10}$$

Equation 2.10 was obtained from the definition $\mu = RT \ln \lambda$ that is Equation 2.5.

The fraction of molecules in particular states N_i/N is then given by

$$\frac{N_i}{N} = \frac{\exp(-\varepsilon_i / kT)}{\sum_i \exp(-\varepsilon_i / kT)} = \exp\frac{(-\varepsilon_i / kT)}{q}, \tag{2.11}$$

and, as ε_i/kT increases N_i/N decreases so that fewer molecules are found at higher energies, in line with intuition. The use of these same results in Equations 2.2 and 2.3 gives

$$U = N \frac{\sum_i \varepsilon_i \exp(-\varepsilon_i / kT)}{\sum_i \exp(-\varepsilon_i / kT)} = NkT^2 \left(\frac{\partial \ln q}{\partial T} \right)_V, \tag{2.12}$$

and

$$p = N \frac{\sum_i (-\mathrm{d}\varepsilon_i / \mathrm{d}V) \exp(-\varepsilon_i / kT)}{\sum_i \exp(-\varepsilon_i / kT) = NkT(\partial \ln q / \partial V)_T}. \tag{2.13}$$

In the case when a number of independent quantum states have the same energy it is a matter of convenience to write Equation 2.7 as

$$q = \sum_i g_i \exp\left(-\frac{\varepsilon_i}{kT} \right), \tag{2.14}$$

where g_i is the degeneracy of the energy level ε_i.

2.3 HOW DO I EVALUATE THE PARTITION FUNCTION q?

The concept of solving Schrödinger's equation and, thus, evaluating q from quantum theory was alluded to in Section 2.1. We separate the modes of motion of the molecule so that the energy ε of a molecule in an eigenstate can be written as

$$\varepsilon = \varepsilon_T + \varepsilon_R + \varepsilon_v + \varepsilon_E + \varepsilon_N + \varepsilon_0, \tag{2.15}$$

that is, as the sum of the energy eigenvalues for translational ε_T, rotational ε_R, vibrational ε_v, electronic ε_E, nuclear ε_N, and the lowest energy state ε_0. As was indicated earlier, this separation is valid only under certain circumstances.

It follows from the existence of these various forms of energy and their summation to the whole that the partition function for a molecule can be written as

$$q = q_T\, q_R\, q_V\, q_E\, q_N\, q_0. \tag{2.16}$$

The translational partition function q_T can be separated into three parts, one for each of the Cartesian coordinates. For a molecule of n atoms that is linear there are two rotational modes corresponding to rotation about the two axes perpendicular to the axis of the linear molecule and $(3n - 5)$ vibrational modes, while for a nonlinear molecule there are three rotational modes and $(3n - 6)$ vibrational modes. This ideal separation fails when the molecule is in a high vibrational state because the rotational energy levels depend on the moment of inertia of the molecule, which can change in practice as molecules come near dissociation; then there is some interaction between the modes. However, for many molecules of interest to chemists, biologists, and engineers, only the lowest vibrational energy levels are accessible so that the intramolecular potential nearly approximates that of a simple harmonic oscillator that does not permit dissociation so that the separation of rotational and vibrational modes is very often valid.

For the likely readership of this text the remaining modes of energy of a molecule are of small interest. For example, electronic modes at room temperatures are unimportant for chemists and engineers because only the lowest and at most the first excited states of atomic or molecular orbitals are populated so that the electronic partition function can be written as

$$q_E = g(\varepsilon_{0,E}) \exp\left\{\frac{-\varepsilon_{0,E}}{kT}\right\} \left[1 + \left\{\frac{g(\varepsilon_{1,E})}{g(\varepsilon_{0,E})}\right\} \exp\left\{\frac{-(\varepsilon_{1,E} - \varepsilon_{0,E})}{kT}\right\}\right]. \tag{2.17}$$

The nuclei are all in the ground state for molecular gases of interest with a mass greater than hydrogen and so $q_N = 1$. For homonuclear molecular gases with a nuclear spin and low mass at low temperature (i.e., < 300 K) the effect of nuclear spin contributes to the thermodynamics properties such as for hydrogen but this very special topic will not be considered further here.

Assuming the preceding approximations of mode separation are valid and the nuclei are in the ground state, the molecular partition function can then be written as

$$q = q_x\, q_y\, q_z\, q_R\, q_v\, g(\varepsilon_0) g(\varepsilon_{0,E}) \exp\left\{\frac{-\varepsilon_{0,E}}{kT}\right\}$$

$$\times\left[1 + \left\{\frac{g(\varepsilon_{1,E})}{g(\varepsilon_{0,E})}\right\} \exp\left\{\frac{-(\varepsilon_{1,E} - \varepsilon_{0,E})}{kT}\right\}\right] \exp\left(\frac{-\varepsilon_0}{kT}\right). \tag{2.18}$$

When, as is often the case, $(\varepsilon_{1,E} - \varepsilon_{0,E}) \gg kT$, Equation 2.17 becomes

$$q_E = g(\varepsilon_{0,E})\exp\left\{\frac{-\varepsilon_{0,E}}{kT}\right\}, \tag{2.19}$$

and is electronically unexcited so that Equation 2.18 reduces to

$$q = q_x\, q_y\, q_z\, q_R\, q_v\, g(\varepsilon_0)\exp\left(\frac{-\varepsilon_0}{kT}\right), \tag{2.20}$$

where the $g(\varepsilon_{0,E})$ and $\exp(-\varepsilon_{0,E}/kT)$ of Equation 2.17 are represented by $g(\varepsilon_0)$ and $\exp(-\varepsilon_0/kT)$, respectively.

For translational motion the solution of the Schrödinger's equation for a particle of mass m moving in the x-direction within a box of length l_x gives the energy

$$\varepsilon_x = \frac{n_x^2 h^2}{8 l_x^2 m}, \tag{2.21}$$

where n_x is the quantum number, of value 1, 2, 3, $\cdots$, and h the Planck constant so that the partition function q_x is

$$q_x = \sum_{n_x=1}^{\infty} \exp\left\{\frac{-n_x^2 h^2}{8 l_x^2 mkT}\right\}. \tag{2.22}$$

Because the separation of the energy levels in translational motion is small (as a simple calculation using Equation 2.21 will illustrate), the sum in Equation 2.22 can be replaced by an integral, which is tantamount to the assumption that the energy is a continuous variable. Upon integration with the Euler–MacLaurin theorem (Question 1.11.4 of Chapter 1) we find that

$$q_x = \left(\frac{2\pi mkT}{h^2}\right)^{1/2} l_x - \frac{1}{2}. \tag{2.23}$$

The first term on the right hand side of Equation 2.23 is much larger than 10^6; the second term (½) can therefore be ignored and the motion termed *classical* (because the energy levels have been assumed continuous) so that for the three independent translational directions, recognizing that the system volume $V = l_x l_y l_z$, q_T is given by

$$q_T = q_x\, q_y\, q_z = \left(\frac{2\pi mkT}{h^2}\right)^{3/2} V. \tag{2.24}$$

The form of the rotational partition function is different depending on whether the molecule is linear or nonlinear: both are considered here. For a linear molecule, the solution of Schrödinger's equation gives the rotational energy ε_R as

$$\varepsilon_R = \frac{j(j+1)h^2}{8\pi^2 I}, \tag{2.25}$$

where I is the moment of inertia and j is the quantum number equal to 0, 1, 2, 3, · · ·. Each of the energy levels is $(2j + 1)$ degenerate so that the partition function for rotation is

$$q_R = \sum_{j=0}^{\infty}(2j+1)\exp\left\{\frac{-j(j+1)h^2}{8\pi^2 IkT}\right\}. \tag{2.26}$$

Again, because the separation of the energy levels in rotation is usually small compared with kT the sum can be replaced by an integral (assuming the energy levels are continuous), and on integration with the Euler–MacLaurin theorem (see Question 1.11.4) we find that

$$q_R = \frac{8\pi^2 IkT}{h^2} + 0.42. \tag{2.27}$$

The numerical value of $(8\pi^2 IkT/h^2)$ is usually, (but not always) large compared to 0.42 as a simple calculation reveals. For the extreme case of hydrogen, $(8\pi^2 Ik/h^2) \approx 0.01\ \text{K}^{-1}$; and at a temperature of 300 K $(8\pi^2 IkT/h^2) \approx 3$ so that ignoring 0.42 gives rise to a fractional uncertainty of 0.1 and the treatment of hydrogen as a classical rotator fails. However, for iodine at a temperature of 300 K the fractional error is $<10^{-4}$ and it is thus an acceptable approximation to assume that rotational motion is classical. For molecules with a molar mass greater than hydrogen and for temperatures on the order of 100 K or greater, Equation 2.27 can be approximated by

$$q_R \approx \frac{8\pi^2 IkT}{sh^2}, \tag{2.28}$$

where s is the symmetry number. The symmetry number is $s = 1$ for linear molecules with no center of symmetry such as hydrogen fluoride and 2 for molecules with a center of symmetry, for example, oxygen. The difference arises because, for symmetrical molecules each distinguishable orientation has been counted twice.

The rotational partition function for a nonlinear molecule is

$$q_R = \frac{\left\{8\pi^2 (I_A I_B I_C)^{1/3} kT/h^2\right\}^{3/2} \pi^{1/2}}{s} \tag{2.29}$$

where the I's are the moments of inertia for each direction of the coordinate system selected and s is the symmetry number, which has a similar interpretation as that given above; $s = 1$ for molecules with no symmetry axis, 2 for water, 3 for ammonia, 4 for ethane, and 12 for methane and ethane.

If the vibrational motion is assumed to be simple harmonic, which means the molecule can never dissociate, but allows the modes to be treated as separable the solution to Schrödinger's equation gives the energy

$$\varepsilon_V = (\nu + 1/2)h\upsilon \tag{2.30}$$

where the quantum number $\nu = 0, 1, 2, 4, \cdots$, and υ is the frequency of vibration so that the partition function is

$$q_V = \sum_{\nu=0}^{\infty} \frac{\exp\{-(\nu + \frac{1}{2})\, h\upsilon\}}{kT}. \tag{2.31}$$

Even at room temperature the molecular vibrations of most molecules do not behave classically and, indeed may not be excited at all, so the sum cannot be replaced by an integral. However, the summation is a geometrical progression; hence we can write the vibrational partition function as

$$q_V = \left[1 - \exp\left\{-\frac{h\upsilon}{kT}\right\}\right]^{-1} \exp\left\{-\frac{h\upsilon}{kT}\right\}. \tag{2.32}$$

2.4 WHAT CAN BE CALCULATED USING THE MOLECULAR PARTITION FUNCTION?

Now that we have the molecular partition function, it is possible to evaluate a number of thermodynamic properties of several systems, and we provide five examples. They are all in some way or other results for idealized systems and are important because they illustrate the power of the methodology of statistical thermodynamics. For many systems encountered in engineering practice the results derived for these idealized systems are useful limiting values for the results of calculations that are generally much more complex and where approximations must be used to obtain meaningful results.

2.4.1 What Is the Heat Capacity of an Ideal Diatomic Gas?

For an electronically unexcited diatomic molecule the molecular partition function is given by

$$q = \left(\frac{2\pi mkT}{h^2}\right)^{3/2} V \left\{\frac{8\pi^2 IkT}{sh^2}\right\} \left\{1 - \exp\left(-\frac{h\upsilon}{kT}\right)\right\}^{-1} g(\varepsilon_0)\exp\left(-\frac{\varepsilon_0}{kT}\right). \tag{2.33}$$

Use of Equation 2.33 in Equation 2.12 gives the internal energy of N molecules of a gas of diatomic molecules as

$$U = NkT^2\left(\frac{\partial \ln q}{\partial T}\right)_V = \frac{3}{2}NkT + NkT + \frac{Nh\upsilon}{\{\exp(h\upsilon/kT)-1\}} + N\varepsilon_0. \quad (2.34)$$

The molar heat capacity at constant volume $C_{V,\mathrm{m}}$, which is the molar form of Equation 1.89, can be obtained from Equation 2.34 cast in terms of the molar internal energy U_m using

$$\frac{(\partial U_\mathrm{m}/\partial T)_V}{R} = \frac{C_{V,\mathrm{m}}}{R} = \frac{5}{2} + \left(\frac{h\upsilon}{kT}\right)^2 \frac{\exp(h\upsilon/kT)}{\{\exp(h\upsilon/kT)-1\}^2}, \quad (2.35)$$

which is the formal result for the heat capacity of a diatomic molecule. Evidently, a knowledge of the single vibrational frequency allows calculation of the heat capacity quite simply. The frequency required can be measured spectroscopically.

For a monatomic gas both the rotational and vibrational partition functions are omitted from Equation 2.33 and so Equation 2.35 becomes

$$\frac{C_{V,\mathrm{m}}}{R} = \frac{3}{2}, \quad (2.36)$$

which is independent of temperature.

For a polyatomic molecule with n atoms the vibrational partition function must be multiplied by a factor of $(3n - 5)$ for a linear molecule and $(3n - 6)$ for a nonlinear molecule. The appropriate rotational partition function should also be used and if electronic excitation is significant it must also be included.

2.4.2 What Is the Heat Capacity of a Crystal?

A perfect crystal formed from N identical atoms has no modes of motion except the N three-dimensional vibrations of each atom about the occupied lattice sites. If this motion is considered simple harmonic the partition function q of an atom is then

$$q = \prod_{i=1}^{3N}\left[\frac{\exp(-0.5h\upsilon_i/kT)}{\{1-\exp(-h\upsilon_i/kT)\}}\right]. \quad (2.37)$$

To evaluate the partition function and, thus, the heat capacity of such a crystal we must, in principle, know the $3N$ vibrational frequencies of the entire crystal from measurement. Alternatively, an approximation can be used for υ_i and it is to examples of alternatives that we now turn.

Einstein assumed all υ_i were equal to υ_E so that Equation 2.37 reduces to

$$q = \frac{\exp(-1.5h\upsilon_E / kT)}{\{1 - \exp(-h\upsilon_E / kT)\}^3}, \tag{2.38}$$

and the molar heat capacity at constant volume is, from Equation 1.89, given by

$$\frac{C_{V,m}}{R} = 3\left(\frac{h\upsilon_E}{kT}\right)^2 \frac{\exp(h\upsilon_E / kT)}{\{\exp(h\upsilon_E / kT) - 1\}^2} \tag{2.39}$$

at high temperatures; it is left as an exercise to show that the result of Equation 2.39 tends to equal $3R$ at high temperatures.

Another approximation for the vibrational frequencies was proposed by Debye, who assumed the number of vibrational frequencies did not exceed a value υ_D was proportional to υ_i^2; υ_D was selected so that there are $3N$ frequencies in total. With these constraints the Debye expression for the molar heat capacity at constant volume is easily shown to be

$$\frac{C_{V,m}}{R} = 9\left(\frac{kT}{h\upsilon_D}\right)^3 \int_0^{h\upsilon_D/kT} y^4 \left(\frac{e^y}{e^y - 1}\right) dy. \tag{2.40}$$

At low temperatures Equation 2.40 becomes

$$\frac{C_{V,m}}{R} \to \left(\frac{12}{5}\right)\pi^4 \left(\frac{kT}{h\upsilon_D}\right)^3. \tag{2.41}$$

Comparison of predictions obtained from both Equation 2.39 and Equation 2.41 with measurements at temperatures less than 60 K show that Equation 2.41 provides the better agreement. Equation 2.41 shows that $C_{p,m} \propto T^3$, and this relationship has been used to determine the enthalpy and entropy of a crystal as $T \to 0$ by extrapolating measurements of heat capacity from the lowest experimentally accessible temperature to $T = 0$.

2.4.3 What Is the Change of Gibbs Function Associated with the Formation of a Mixture of Gases?

In subsequent material the significance of the Gibbs function will become apparent, and so the reader is invited to look forward to Chapter 3 to explore the definition of the Gibbs function $G = U + pV - TS = H - TS$. It is also worth saying that the Gibbs function, which is a measure of energy, is the appropriate thermodynamic variable for conditions where we specify temperature and pressure that are themselves the most easily controlled experimental variables. Here we use the Gibbs function to consider the formation of a perfect-gas mixture.

For any species in a mixture of substances from Equation 2.9 we have

$$\lambda_i = \frac{N_i}{q}. \tag{2.42}$$

We also have from Chapter 3, Question 3.3.1,

$$G = \sum_i N_i \mu_i, \tag{2.43}$$

and in view of the definition

$$\mu_i \overset{\text{def}}{=} kT \ln \lambda_i, \tag{2.44}$$

we can cast Equation 2.43 as

$$G = kT \sum_i N_i \ln\left(\frac{N_i}{q_i}\right). \tag{2.45}$$

Equation 2.24 contains for each q_i a factor V. If the gas mixture is perfect then we can substitute

$$pV = kT \sum_i N_i, \tag{2.46}$$

and write Equation 2.45 as

$$\frac{G}{kT} = \sum_i N_i \ln\left(\frac{N_i V}{q_i}\right) - \left(\sum_i N_i\right) \ln\left(\sum_i N_i\right) - \left(\sum_i N_i\right) \ln\left(\frac{kT}{p}\right). \tag{2.47}$$

In view of the definition of the molar quantity of mixing (Equation 1.19) the Gibbs function for mixing a perfect gas at constant pressure is then given by

$$\frac{\Delta_{\text{mix}} G}{kT} = -\left(\sum_i N_i\right) \ln\left(\sum_i N_i\right) + \sum_i N_i \ln N_i, \tag{2.48}$$

and the change of molar Gibbs function by

$$\Delta_{\text{mix}} G_m = RT \sum_i x_i \ln x_i, \tag{2.49}$$

where the mole fraction defined by Equation 1.5 has been used. The molar entropy of mixing is

$$\Delta_{\text{mix}} S_m = -R \sum_i x_i \ln x_i. \tag{2.50}$$

Both Equations 2.49 and 2.50 are important because they can also be derived from thermodynamic assumptions and are mentioned again in Chapter 3. Perhaps of greater importance is the fact that the values obtained from

Equation 2.50 agree with measurements of the osmotic pressure of a gas mixture at low pressure (see Question 4.3.3).

The Helmholtz function A also defined in Chapter 3 can be written as

$$A = U - TS = G - pV = G - \sum_{\mathrm{i}} N_{\mathrm{i}} kT \tag{2.51}$$

or

$$A = -kT \ln\left\{\prod_{\mathrm{i}} \left(\frac{q_{\mathrm{i}}^{N_{\mathrm{i}}}}{N_{\mathrm{i}}^{N_{\mathrm{i}}} \mathrm{e}^{-N_{\mathrm{i}}}}\right)\right\}. \tag{2.52}$$

Defining a quantity Q called the *canonical partition function* of the system we can cast Equation 2.52 as

$$A = -kT \ln Q. \tag{2.53}$$

In Section 2.5 we will discuss Q for interacting particles and how it may be used to calculate the properties of substances from statistical mechanics. However, for now we are considering a system of independent particles when Q is given by

$$Q = \prod_{\mathrm{i}} \left(\frac{q_{\mathrm{i}}^{N_{\mathrm{i}}}}{N_{\mathrm{i}}^{N_{\mathrm{i}}} \mathrm{e}^{-N_{\mathrm{i}}}}\right). \tag{2.54}$$

Using Stirling's approximation, which states that for large N

$$N! = N^N \mathrm{e}^{-N}, \tag{2.55}$$

Equation 2.54 can be cast as

$$Q = \prod_{\mathrm{i}} \left(\frac{q_{\mathrm{i}}^{N_{\mathrm{i}}}}{N_{\mathrm{i}}!}\right), \tag{2.56}$$

which gives a simple means to calculate A for the perfect-gas mixture. Here we note that Equation 2.55 is nearly exact already for $N > 10^3$, which is still small compared with the number of particles in a mole of $\approx 10^{23}$ in which we really have interest.

2.4.4 What Is the Equilibrium Constant for a Chemical Reaction in a Gas?

For chemists, an important quantity to be determined is the equilibrium constant of a chemical reaction. For a chemical reaction, as discussed already in Chapter 1 (Equation 1.30),

$$0 = \sum_{\mathrm{i}} \nu_{\mathrm{i}} \mathrm{i}, \tag{2.57}$$

where ν is the stoichiometric number, thermodynamic equilibrium is given by

$$\prod_{i}(\lambda_i)^{\nu_i}=1. \tag{2.58}$$

From Equation 2.9, at a pressure called the *standard pressure* $p^{\ominus}$ (so that the property of the system depends only on temperature and not on pressure or composition as discussed in Question 1.10) the absolute activity coefficient of Equation 2.42 can be written as

$$\lambda_i=\frac{N_i}{q_i}=\left(\frac{x_i p}{p^{\ominus}}\right)^{-1}=\left(\frac{p^{\ominus}}{kT}\right)\left(\frac{q_i}{V}\right)^{-1}. \tag{2.59}$$

From Chapter 1 Equation 1.116 the definition of the standard equilibrium constant is given by

$$K^{\ominus}(T)=\prod_{i}\{\lambda_i^{\ominus}(T)\}^{-\nu_i}, \tag{2.60}$$

and the standard absolute activity of substance i is related to the standard chemical potential by the definition

$$\lambda_i^{\ominus}(T)=\exp\left(\frac{\mu_i^{\ominus}}{RT}\right). \tag{2.61}$$

For a gaseous phase with mole fractions x_i Equation 2.60 can, in light of Equation 2.59, be written as

$$K^{\ominus}(T)=\prod_{i}\left\{\frac{x_i p}{p^{\ominus}}\right\}^{\nu_i}=\prod_{i}\left\{\frac{kTq_i}{Vp^{\ominus}}\right\}^{\nu_i}. \tag{2.62}$$

For a reaction between diatomic gases for which the electronic modes are unexcited at a temperature T, substitution of q_i/V from Equation 2.33, yields the following:

$$\begin{aligned}\ln\{K^{\ominus}(T)\}=&\sum_i \nu_i \ln\left\{\left(\frac{kT}{p^{\ominus}}\right)\left(\frac{2\pi M_i RT}{L^2h^2}\right)^{3/2}\right\}+\sum_i \nu_i \ln\left(\frac{T8\pi^2 I_i k}{s_i h^2}\right)\\ &-\sum_i \nu_i \ln\left\{1-\exp\left(-\frac{h\upsilon_i}{kT}\right)\right\}+\sum_i \nu_i \ln\{g(\varepsilon_{0,i})\}\\ &-\sum_i \frac{\nu_i\varepsilon_{0,i}}{kT}.\end{aligned} \tag{2.63}$$

The standard equilibrium constant for the reaction posed can be evaluated from Equation 2.63 at a temperature T if we have values of the molar mass M_i, vibrational frequency υ_i, the moment of inertia I_i, the degeneracy g_i, and $\Sigma_i \nu_i \varepsilon_{0,i}$ for each reacting substance i. Apart from the molar mass and $\Sigma_i \nu_i \varepsilon_{0,i}$ all quantities can be determined spectroscopically. To make Equation 2.63 useful the term $\Sigma_i \nu_i \varepsilon_{0,i}$ in Equation 2.63 must be eliminated. To do so requires the use of van't Hoff's equation (Equation 1.120 of Chapter 1)

$$\Delta H_m^\ominus = RT^2 \frac{dK^\ominus}{dT}. \tag{2.64}$$

Differentiation of Equation 2.63 and the use of Equation 2.64 gives

$$\frac{\Delta H_m^\ominus(T^\ominus)}{RT} = \frac{7T^\ominus}{2T}\sum_i \nu_i + \frac{T^\ominus}{T}\sum_i \nu_i \left[\frac{h\upsilon_i/(kT^\ominus)}{\{\exp(-h\upsilon_i/kT^\ominus) - 1\}}\right] + \sum_i \frac{\nu_i \varepsilon_{0,i}}{kT}, \tag{2.65}$$

where $T^\ominus$ is a temperature for which $\Delta H_m^\ominus(T^\ominus)$ is known; as discussed in Chapter 1, $T^\ominus$ is usually chosen to be 298.15 K. Addition of Equation 2.63 and Equation 2.65 gives

$$\begin{aligned}\ln\{K^\ominus(T)\} = {} & \sum_i \nu_i \ln\left\{\left(\frac{kT}{p^\ominus}\right)\left(\frac{2\pi M_i RT}{L^2h^2}\right)^{3/2}\right\} + \sum_i \nu_i \ln\left(\frac{8\pi^2 I_i kT}{s_i h^2}\right) \\ & - \sum_i \nu_i \ln\left\{1 - \exp\left(-\frac{h\upsilon_i}{kT}\right)\right\} + \sum_i \nu_i \ln\{g(\varepsilon_{0,i})\} \\ & - \frac{\Delta H_m^\ominus(T^\ominus)}{RT} + \frac{7T^\ominus}{2T}\sum_i \nu_i + \frac{T^\ominus}{T}\sum_i \nu_i \left[\frac{h\upsilon_i/kT^\ominus}{\{\exp(-h\upsilon_i/kT^\ominus) - 1\}}\right],\end{aligned} \tag{2.66}$$

and we see that we have replaced the requirement to obtain $\Sigma_i \nu_i \varepsilon_{0,i}$ from spectroscopic measurements with the need for the calorimetric determinations of $\Delta H_m^\ominus(T^\ominus)$. This is one reason for the effort expended in measuring precise values of $\Delta H_m^\ominus(T^\ominus)$ at a temperature of 298.15 K, which the reader will find decorates the literature of chemical thermodynamics.

2.4.5 What Is the Entropy of a Perfect Gas?

From the definition of the standard molar entropy for species i

$$S_i^\ominus = -\frac{d\mu_i^\ominus}{dT} = -R\ln\lambda_i^\ominus - RT\frac{d\lambda_i^\ominus}{dT}, \tag{2.67}$$

for a perfect pure gas (see Equation 2.10) it follows that

$$S_i^\ominus(T) = R\ln\left\{\left(\frac{kT}{p^\ominus}\right)\left(\frac{q_i}{V}\right)\right\} + RT\left(\frac{\partial \ln q_i}{\partial T}\right)_V, \tag{2.68}$$

so that the evaluation of the entropy requires the molecular partition function for the pure gas.

For an electronically unexcited monatomic gas, Equation 2.68 becomes

$$S_i^\ominus(T) - R\ln\{g(\varepsilon_{0,i})\} = R\ln\left\{\frac{(2\pi M_i)^{3/2}(RT)^{5/2}}{p^\ominus L^4 h^3}\right\} + 2.5R. \tag{2.69}$$

For an electronically unexcited diatomic gas in Equation 2.68 we find

$$S_i^\ominus(T) - R\ln\{g(\varepsilon_{0,i})\} = R\ln\left\{\frac{(2\pi M_i)^{3/2}(RT)^{5/2}}{p^\ominus L^4 h^3}\right\} + R\ln\left(\frac{8\pi^2 I_i kT}{s_i h^2}\right)$$
$$- R\ln\left\{1 - \exp\left(-\frac{h\upsilon_i}{kT}\right)\right\} + 3.5R + R\left[\frac{h\upsilon_i / kT^\ominus}{\{\exp(-h\upsilon_i / kT^\ominus) - 1\}}\right]. \tag{2.70}$$

The relationship between $S_i^\ominus(T)$ and $R\ln\{g(\varepsilon_{0,i})\}$ for a solid as the temperature tends to zero will be the topic of discussion in Question 3.7 of Chapter 3 under Nernst's heat theorem.

2.5 CAN STATISTICAL MECHANICS BE USED TO CALCULATE THE PROPERTIES OF REAL FLUIDS?

The idealized systems that have been examined in Question 2.4 are of immense value as limiting cases approached occasionally by real systems. The analysis presented is necessarily simplified in a number of ways compared to that which needs to be applied to real materials. The majority of the differences between real systems and the idealized models we have considered lie in the fact that the noninteracting particles of the idealized system must be replaced by particles that interact. In the case of molecular entities they interact through intermolecular forces which can affect the total energy of the ensemble of molecules because the total internal energy is not simply the sum of that of individual molecules. It is that difference which is the subject of this question where as we illustrate the use of statistical mechanics for the evaluation of the thermodynamic properties of fluids. We are not attempting to be comprehensive in this question, and the reader is referred to specialized texts for greater detail and breadth (e.g., McQuarrie 2000).

2.5.1 What Is the Canonical Partition Function?

As has been explained above, the role of statistical mechanics is that of a bridge between the microscopic and macroscopic descriptions of the system. The statistical mechanics of systems at equilibrium, from which the thermodynamic properties may be obtained, is based upon the two postulates. The first postulate, introduced earlier, has enabled us to evaluate some of the properties of some idealized systems. To try to calculate the properties of more complex systems that are less than ideal in some way, in particular, where the molecules interact with each other, we need to move away from the single molecular partition function discussed earlier to the canonical partition function Q. To introduce this concept we first consider a real system in a thermodynamic state defined by the macroscopic variables of thermodynamics and consisting of N molecules. The individual molecules in this system are in an unknown quantum state, but we know that a very large number of systems must exist in which individual molecules are in different states but the overall thermodynamic state is the same. The collection of all of these possible systems consistent with the real system, each of which is a unique quantum state of the system is called the *canonical ensemble*.

The second postulate of statistical mechanics states that the only dynamic variable upon which the quantum states of the entire canonical ensemble depend is the total ensemble energy. From this postulate we deduce that all states of the ensemble having the same energy are equally probable. It can then be shown (Hill 1960; Reed and Gubbins 1973) that the probability Π_i that a system selected at random from the ensemble will be found in quantum state i varies exponentially with the energy E_i of that state. That is

$$\Pi_i(E_i) \propto \exp\left(-\frac{E_i}{kT}\right), \tag{2.71}$$

since, however, there is unit probability that the system resides in *some* state we have that $\Sigma_i \Pi_i(E_i) = 1$ and

$$\Pi_i(E_i) = \frac{\exp\{-E_i/kT\}}{Q}, \tag{2.72}$$

where

$$Q(N,V,T) = \sum_i \exp\left(-\frac{E_i}{kT}\right). \tag{2.73}$$

Equation 2.73 defines the quantity Q, known as the canonical partition function, which plays a central role in statistical thermodynamics. It does not have a well-defined physical meaning but it serves as a useful statistical device in terms of which all of the thermodynamic properties of a system may be expressed. We now examine the relation between the thermodynamic properties and the canonical partition function for the most general case.

The internal energy of the system is just the ensemble average system energy. Following the first postulate of statistical mechanics the ensemble average of the energy is defined as

$$\langle E \rangle = \sum_{\mathrm{i}} E_{\mathrm{i}} \Pi_{\mathrm{i}}, \tag{2.74}$$

where Π_{i} is the probability that a system chosen at random from the ensemble will be found in the quantum state i with energy E_{i}. According to the first postulate, this ensemble average will approach the thermodynamic internal energy U of the real system as $N \rightarrow \infty$:

$$U = \lim_{N \to \infty} \sum_{\mathrm{i}} E_{\mathrm{i}} \Pi_{\mathrm{i}}. \tag{2.75}$$

Combining Equation 2.73 and Equation 2.75 with Equation 2.72 we obtain the expression

$$U = \frac{1}{Q} \sum_{\mathrm{i}} E_{\mathrm{i}} \exp\left(-\frac{E_{\mathrm{i}}}{kT}\right), \tag{2.76}$$

which, in view of the definition of Q, may be written as

$$U = kT^2 \left(\frac{\partial \ln Q}{\partial T}\right)_{N,V}. \tag{2.77}$$

Equation 2.77 provides a direct relation between the internal energy and the canonical partition function.

To obtain an expression for the entropy in terms of the partition function, we compare the relation between internal energy and the probability function with the 2nd law of thermodynamics (see Question 3.2.1 of Chapter 3). According to macroscopic thermodynamics, the fundamental equation for a change in the state of a system of fixed composition is Equation 3.1

$$\mathrm{d}U = T\,\mathrm{d}S - p\,\mathrm{d}V. \tag{2.78}$$

Now, according to our statistical-mechanical arguments, when N is constant, a change in the internal energy of the system can occur only if either the probability function or the energy levels change. Thus, from Equation 2.77,

$$\mathrm{d}U = \sum_{\mathrm{i}} E_{\mathrm{i}}\,\mathrm{d}\Pi_{\mathrm{i}} + \sum_{\mathrm{i}} \Pi_{\mathrm{i}}\,\mathrm{d}E_{\mathrm{i}}. \tag{2.79}$$

Let us start with the second term of Equation 2.79. With N constant, the energy levels may change only if the volume changes and hence $dE_i = (\partial E_i/\partial V)\cdot dV$. Thus, comparing Equations 2.78 and 2.79, we see that

$$pdV = -\sum_i \Pi_i dE_i \tag{2.80}$$

and

$$pdS = -\sum_i E_i d\Pi_i. \tag{2.81}$$

To obtain the entropy, we eliminate the energy levels E_i from Equation 2.81 in favor of the partition function Q. We do this by obtaining an expression for E_i from the logarithm of Equation 2.73 with the result

$$T dS = -kT\left(\sum_i \ln\Pi_i d\Pi_i + \ln Q\sum_i d\Pi_i\right) = -kT\sum_i \ln\Pi_i d\Pi_i, \tag{2.82}$$

where we have used the fact that $\Sigma_i \Pi_i = 1$ and hence $\Sigma_i d\Pi_i = 0$.

Equation 2.82 can be also written as

$$dS = -kd\left(\sum \Pi_i \ln\Pi_i\right) \tag{2.83}$$

and, since dS is an exact differential, we see that the right hand side of this equation is the product of a constant and an exact differential. We may therefore integrate Equation 2.83 directly with the result

$$S = -k\sum_i \Pi_i \ln\Pi_i. \tag{2.84}$$

Finally, using Equations 2.72 and 2.73 to eliminate Π_i in favor of Q, we obtain

$$S = kT\left(\frac{\partial \ln Q}{\partial T}\right)_{N,V} + k\ln Q, \tag{2.85}$$

which is the desired relation between S and Q.

We now have expressions for both U and S in terms of Q from which the Helmholtz free energy A can readily be obtained through the relation

$$A = U - TS. \tag{2.86}$$

Combining Equations 2.75, 2.76, and 2.86, we find that A is given by the simple relation

$$A = -kT\ln Q. \tag{2.87}$$

Since A is the characteristic state function for the choice of N, V, and T as the independent variables, all of the other thermodynamic properties follow from this quantity. For example, from Equation 2.86 we have

$$dA = dU - T\,dS - S\,dT, \tag{2.88}$$

but we shall see in Chapter 3 the law 2a given by Equation 3.1, which for a closed phase of fixed composition, becomes

$$dU - T\,dS = -p\,dV, \tag{2.89}$$

so that the total differential of Equation 2.88 is

$$dA = -p\,dV - S\,dT, \tag{2.90}$$

and

$$dA = \left(\frac{\partial A}{\partial V}\right)_T dV + \left(\frac{\partial A}{\partial T}\right)_V dT, \tag{2.91}$$

from which we can deduce that

$$p = -\left(\frac{\partial A}{\partial V}\right)_T, \tag{2.92}$$

so that

$$p = kT\left(\frac{\partial \ln Q}{\partial V}\right)_{N,T}. \tag{2.93}$$

2.5.2 Why Is the Calculation so Difficult for Real Systems?

The difficulty of applying statistical mechanics to the evaluation of all the thermodynamic properties of real systems is twofold. First, the fact that the energy of the system of molecules in a real system arises not just from the energies of individual isolated molecules but the energies arising from their interactions with each other in pairs or other many-body configurations. Those interactions as a function of the distance between the atoms or molecules are not, in general, available. It has been pointed out that the potential energy that characterizes the forces between just two atoms or molecules at a time has been evaluated theoretically for only two systems, hydrogen atoms and helium atoms. For other systems the forces have been deduced empirically (Maitland et al. 1981) and are now known for the monatomic gases and for two or three simple polyatomic gases such as nitrogen and water.

Even if the interaction energies were known with great precision, to compute the thermodynamic properties of such a system exactly for the large number of molecular interactions that would be involved is evidently a very large problem that is beyond even the fastest computers today. As a consequence, a means of sampling the ensemble has been introduced followed by various means of averaging through techniques known as equilibrium molecular simulation. This subject is beyond the scope of this book and an interested reader is referred, for example, to McQuarrie (2000).

Because of the difficulties of exact calculation of the properties, while possible in principle, a series of methods have been developed, which rely on models of systems, and they have provided the basis of much of the development of the engineering application of the properties of fluids. To introduce these models we first characterize a number of limiting models. We then sketch the development of statistical mechanics for real systems and quote results derived elsewhere in the interests of brevity. In this section we are more interested in the practical application of the methods than their derivation. A reader wishing to know more than we can include here is invited to consult a number of suitable texts such as van Ness and Abbott (1982); Poling et al. (2001); Prausnitz et al. (1986); Assael et al. (1996).

2.6 WHAT ARE REAL, IDEAL, AND PERFECT GASES AND FLUIDS?

At very low pressures every gas conforms to the very simple, ideal equation of state for n moles

$$pV = nRT, \tag{2.94}$$

which is also the equation of state for the perfect gas, composed of infinitesimal particles that exert no forces on each other.

The behavior of a real material is shown in Figure 2.1, alongside that for the perfect gas in a general (p, V, T) diagram. The diagram reveals the liquid and solid phases as well as the vapor phase of a real substance. The behavior of even the vapor phase of this real system departs considerably from that embodied in Equation 2.94. The very existence of the liquid and solid phases is a result of the attractive forces that hold the molecules together, while their incompressibility reveals the strong repulsive forces that must exist between the same molecules at small separations as has been pointed out in Chapter 1.

The transition between vapor and liquid phases received systematic attention in 1823 from Faraday, but it was not until the work of Andrews on carbon dioxide in 1869 that the volumetric and phase behavior of a pure fluid was established over appreciable ranges of temperature and density. This behavior is illustrated by the three-dimensional phase diagram shown, together with its projections on to the (p, V) and (p, T) planes, in Figure 2.1. The pioneering

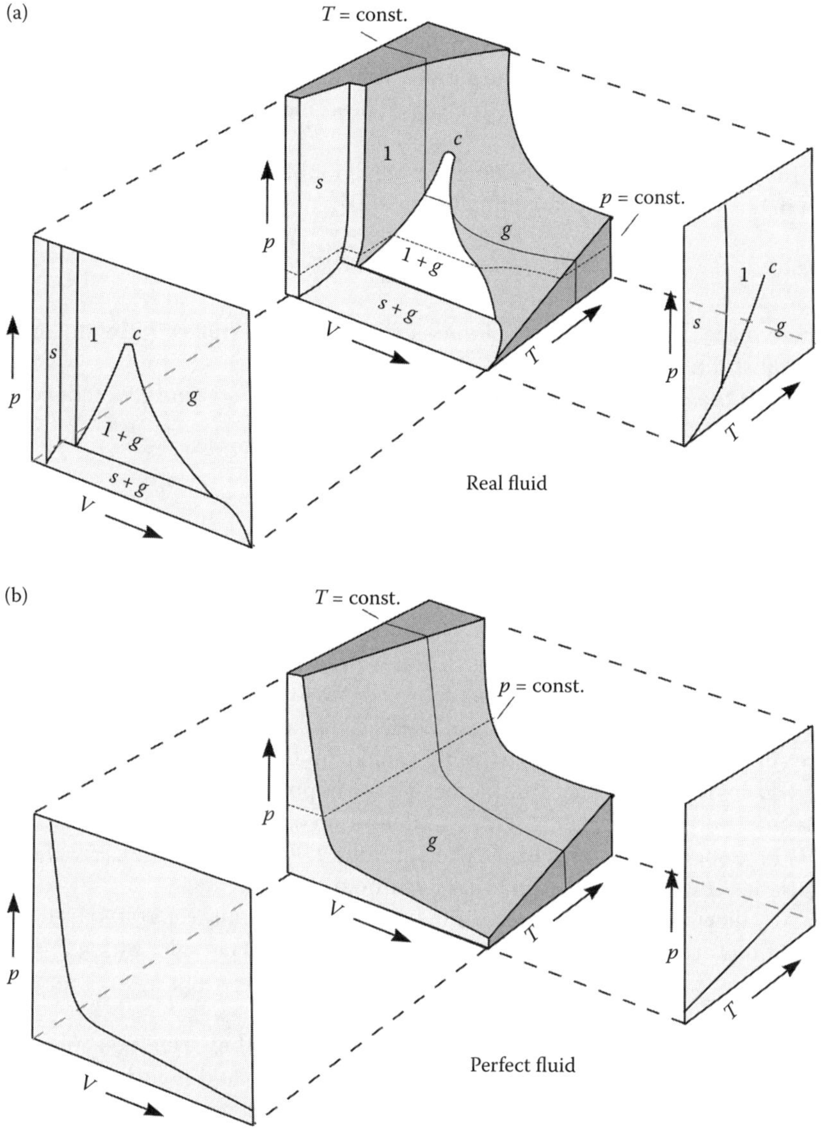

Figure 2.1 (p, V, T) of real fluid (a) and a perfect gas (b).

experimental work of Andrews and others paved the way for the modern view of the equation of state and led van der Waals to postulate in his dissertation in 1873 "On the continuity of the gas and liquid states" the famous equation of state that now bears his name, and which described for the first time gas and liquid phases.

The general equation describing the behavior of a real fluid is usually written in terms of the compressibility factor Z as

$$Z = \frac{pV}{n\,RT}, \tag{2.95}$$

where p the pressure, V the volume, n the number of moles, R the universal gas constant, and T the thermodynamic temperature.

When the compressibility factor equals unity then Equation 2.95 reduces to Equation 2.94, which can be written as

$$p = \rho_n \, RT, \tag{2.96}$$

where ρ_n is the amount-of-substance density.

A real gas that obeys Equation 2.94 under some conditions is then called an "ideal gas," or is said to be acting as an "ideal gas"; Equation 2.94 is referred to as the ideal-gas equation of state. Of course, from a different perspective, one can say that the compressibility factor Z is used to modify the ideal-gas equation so that it can account for the real-gas behavior.

As implied earlier, a "perfect gas" is the model of a material in which there are supposed to be point particles that make up the gas, that have no volume and do not interact. Hence, the perfect gas is a hypothetical substance for which the total potential energy is zero. This definition of the perfect gas implies that $p(V, T)$ properties conform exactly to Equation 2.94, which can be derived from statistical-mechanical methods or from kinetic theory.

Any thermodynamic property X of a real fluid is usually separated into contributions arising from a perfect gas X^{pg} and a residual part X^{res} by

$$X = X^{\text{pg}} + X^{\text{res}}, \tag{2.97}$$

In Equation 2.97, the X^{res} arises from the interactions between molecules. The calculation of the residual part from first principles would require calculation of the canonical partition function, and this is, in general, an impossible task. We consider several techniques to obviate the need for this calculation in a later section. Here we first concentrate on the perfect gas contribution.

The perfect-gas contribution can be obtained in many different ways that follow what has been discussed in Question 2.4. We have already seen in this chapter how some of the properties of a gas, treated as a perfect gas containing molecules with translational, rotation, and vibrational degrees, can be calculated. It was made clear earlier that to perform these calculations it is essential to know some properties of the molecules so that the energy levels of its quantum states (or at least molecular constants) that relate to them such as the molecular moment of inertia or the vibrational constant are known. In practice, the various molecular constants required for the calculation of perfect-gas properties are obtained from spectroscopic measurements of rotational, vibrational, and electronic energy levels. Such data are readily available for a wide variety of molecules (Herzberg 1945; Moore 1949–1958; Landolt–Börnstein 1951; Sutton 1965; Janz 1967; Herzberg 1970) and, where they are not, bond-contribution methods exist for their estimation (Howerton 1962). Tables of perfect-gas properties, based on a combination of theoretical and experimental work, are available in the literature (*Selected Values of Properties of Hydrocarbons and Related Compounds* 1977, 1978), but it is now much more convenient to make use of computer programs from which the properties may be evaluated routinely. Because of the difficulties with internal rotations and, to a lesser extent, vibration–rotation interaction, it is pragmatic to adopt empirical representations for some of the properties rather than to calculate everything directly from the partition function. One common approach (Assael et al. 1996) is to base perfect-gas property calculation on correlations of the perfect gas specific heat capacity.

We also note that real gases are composed of molecules between which the interactions fall off rapidly with increasing separation. When such a gas is very "dilute" (i.e., the density is low), the average molecular separation becomes large and the condition $U \approx 0$ is fulfilled if no external fields are present. Thus, all real gases exhibit ideal-gas behavior in the limit of zero density and can sometimes be modeled by the perfect-gas model with sufficient accuracy at nonzero but low densities.

One technique that makes use of this limiting behavior leads to the virial equation of state that we consider in the next section.

2.7 WHAT IS THE VIRIAL EQUATION AND WHY IS IT USEFUL?

As we have seen thermodynamic properties can be expressed as a function of the canonical partition function. This partition function is itself a product of two terms. The first term, called the *molecular part*, includes information about isolated molecules and therefore depends only on the molecular properties of the system such as mass, moment of inertia, and so on. The second term, known as the *configuration integral*, contains information about the interactions between the molecules of the system, and it is the only part that depends upon the density. Intermolecular forces in real systems enter calculations through the configuration integral. Unfortunately, even with the assumption that the total energy of interaction of a set of molecules can be described solely by the sum of interactions between pairs of molecules (pair additivity), evaluation of the configuration integral is very difficult for real systems. An alternative treatment that leads to results of great utility is to expand the configuration integral as a power series in the density about the zero-density limit. Then the mth coefficient of this series is rigorously related to molecular interactions in clusters of m molecules. Hence, provided that the series converges satisfactorily, the intractable N-body problem is transformed into a soluble series of 1-body, 2-body, 3-body, $\cdots$ problems.

The canonical partition function Q can be expressed as a product of a number of factors just as we did for the single particle partition function (Question 2.3) so that we can write for a system containing molecules with only vibrational and rotational energy

$$Q = Q^{\mathrm{T}}(N,T,V)\,Q^{\mathrm{V}}(N,T)\,Q^{\mathrm{R}}(N,T), \tag{2.98}$$

where Q^{T} is the translational partition function with obvious meanings for R and V as superscripts. Strictly, this product is an approximation because it assumes that the rotational motion of a molecule is unaffected by the density and not connected to the motion of the center of gravity of a molecule. While this may be true for small, nearly symmetric molecules, it is unlikely to be true for more complex asymmetric molecules since Q^{T} involves the energy connected with the motion of the molecules which, for a real fluid is of two kinds, the kinetic energy and the potential energy associated with the interactions between the molecules. This is conventionally expressed by splitting Q^{T} into two parts, one of which is the kinetic component identical to that of a perfect gas and the other known as the configurational integral Ω as

$$Q^{\mathrm{T}} = \frac{1}{N!}\left(\frac{2\pi mkT}{h^2}\right)^{3N/2} \Omega(N,V,T), \tag{2.99}$$

where

$$\Omega(N,V,T) = \int_V \cdots \int \exp\left(\frac{-U(\mathbf{r}_1 \cdots \mathbf{r}_N)}{kT}\right) \mathrm{d}r_1 \cdots \mathrm{d}r_N. \tag{2.100}$$

In Equation 2.99 the factor related to kinetic energy is easily seen to be the canonical analogue of the same component for the single particle partition function (Equation 2.24); the complete derivation of Equation 2.99 is beyond the scope of this text and the reader is referred to McQuarrie (2000).

The configurational integral is the only part that depends upon the volume V (or the density). It is an integral involving the potential energy $U(\mathbf{r}_1 \cdots \mathbf{r}_N)$ for the entire N molecules whose positions are described by vectorial positions $(\mathbf{r}_1 \cdots \mathbf{r}_N)$, thus Equation 2.93 for the pressure may be written as

$$p = kT\left(\frac{\partial \ln \Omega}{\partial V}\right)_{N,T}. \tag{2.101}$$

Given the earlier comments about the expansion of the configurational integral in density about the zero-density limit, we see from Equation 2.101 that the pressure will also be a power series of density

$$\frac{p}{\rho_n RT} = 1 + B\rho_n + C\rho_n^2 + D\rho_n^3 + \cdots, \tag{2.102}$$

Equation 2.102 is known as the virial equation of state and the coefficients of the virial series, $B, C, D, \cdots$, known as *virial coefficients*, are functions of temperature and composition but not of density. The importance of the virial equation of state lies in its rigorous theoretical foundation by which the virial coefficients appear not merely as empirical constants but with a precise relation to the intermolecular potential energy of groups of molecules. Specifically, the second virial coefficient B arises from the interaction between a pair of molecules, the third virial coefficient C depends upon interactions in a cluster of three molecules, D involves a cluster of four molecules, and so on. Consequently, experimental values of the virial coefficients can be used to obtain information about intermolecular forces or, conversely, virial coefficients may be calculated from known, or assumed, intermolecular potential-energy function. Moreover, exact relations can be derived for the virial coefficients of a gaseous mixture in terms of like- and unlike-molecular interactions.

The virial series converges only for sufficiently low densities. The radius of convergence is not well established theoretically except for hard spheres for which it encompasses all fluid densities. In real systems, the empirical evidence suggests that the series converges up to approximately the critical density. It certainly does not converge either for the liquid phase or in the neighborhood of the critical point. Furthermore, since not all of the coefficients of the virial series are known from theory or experiment, the series is usually limited in practice to densities much below the critical.

In the case when the potential energy of the system of N molecules is the sum of the interaction between all possible pairs we can express the configuration integral, Equation 2.100, in terms of Mayer function as (Assael et al. 1996)

$$\Omega = \int_V \cdots \int \left\{ \prod_{i=1}^{N-1} \prod_{j=i+1}^{N} (1 + f_{ij})\, d\mathbf{r}_1 \cdots d\mathbf{r}_N \right\}, \tag{2.103}$$

and expanding

$$\Omega = \int_V \cdots \int \left\{ 1 + \sum_{i=1}^{N-1} \sum_{j=i+1}^{N} f_{ij} + \cdots \right\} d\mathbf{r}_1 \cdots d\mathbf{r}_N, \tag{2.104}$$

where the Mayer function f_{ij} is

$$f_{ij} = \exp\left\{ -\frac{\phi_{ij}(\mathbf{r})}{kT} \right\} - 1, \tag{2.105}$$

in which $\phi_{ij}(\mathbf{r})$ represents the intermolecular pair potential between molecules i and j. One can show that the third term in the expansion involves summations over the product of two f_{ij}'s, the fourth term summations over the product of three f_{ij}'s, and so on. Since these higher terms involve the interaction of more than two molecules, the assumption of the pair additivity is an especially significant approximation. The full derivation of Equation 2.102 can be found elsewhere (Reed and Gubbins 1973; McQuarrie 2000).

We now integrate Equation 2.104 term by term. The first term is readily evaluated as V^N. The second term involves interactions between all distinct pairs of molecules in the system and there are $N(N-1)/2$ such terms. However, since all the molecules in a pure material interact with each other according to the same function ϕ, we can replace f_{ij} with, say, f_{12} and integrate over the coordinates $\mathbf{r}_3 \cdots \mathbf{r}_N$ one by one. Each such integration results in the factor V so that, approximating $N(N-1)/2$ by $N^2/2$ (for large N), and integrating over the coordinates $\mathbf{r}_1$, we finally obtain the configurational integral as

$$\Omega = V^N \left\{ 1 + 2\pi (N^2/V) \int_0^\infty f_{12} r_{12}^2 \, dr_{12} + \cdots \right\}. \tag{2.106}$$

The thermodynamic properties of the system all depend upon the logarithm of Ω and it is therefore useful to develop $\ln \Omega$ as a power series in $(1/V)$. This may be accomplished by noting that, at sufficiently low densities, the second and higher terms between brackets in Equation 2.106 are small so that

$$\ln \Omega = N \ln(V) + 2\pi \, (N^2/V) \int_0^\infty f_{12} r_{12}^2 \, dr_{12} + \cdots. \tag{2.107}$$

Expressions for the virial coefficients can be obtained by then inserting Equation 2.107 in Equation 2.101. Then, carrying out the differentiation with respect to volume, we obtain

$$p = \frac{NkT}{V} \left\{ 1 - 2\pi (N/V) \int_0^\infty f_{12} r_{12}^2 \, dr_{12} + \cdots \right\}. \tag{2.108}$$

Comparison of Equation 2.108 with Equation 2.102 shows that the second virial coefficient is given for a mole of substance by

$$B = 2\pi L \int_0^\infty \left\{1 - \exp\left(-\frac{\phi_{ij}}{k_B T}\right)\right\} r^2 \, dr. \tag{2.109}$$

In Equation 2.109, L denotes the Avogadro number and we note that B has the dimensions of molar volume.

One can show (Reed and Gubbins 1973) that, in the pair-additivity approximation, the third virial coefficient is given by

$$C = -\frac{8\pi^2}{3} L^2 \iiint f_{12} f_{13} f_{23} \, r_{12} r_{13} r_{23} \, dr_{12} \, dr_{13} \, dr_{23}. \tag{2.110}$$

Corrections to Equation 2.110 that allow for the fact that the energy of interaction of three molecules may not be the sum of that of all pairs have been evaluated (Reed and Gubbins 1973). Expressions can also be obtained for the higher virial coefficients, although they rapidly become complicated by the increasing number of coordinates over which integrations must be performed.

For a pure gas, values for these coefficients can be obtained in the following ways:

(a) Second virial coefficients may be represented rather well by one of the several model intermolecular potentials such as the square well or Lennard-Jones models introduced in Chapter 1. Tables of reduced second and third virial coefficients have been compiled for several model intermolecular potentials (Sherwood and Prausnitz 1964) and values of the scaling parameters σ and ε in the Lennard-Jones (12–6) potential are available for a large number of systems (Reid et al. 1988; Assael et al. 1996). Corrections to C for the effects of nonadditivity of the intermolecular forces have also been tabulated (Sherwood and Prausnitz 1964; Poling et al. 2001).

(b) It is possible to represent the first two coefficients empirically as a function of temperature by correlating values obtained from p-V-T measurements. This approach works well but it is obviously restricted to cases where measurements exist (Dymond and Smith 1980; Dymond et al. 2002; 2003).

(c) Although experimental data on second virial coefficients are abundant (Dymond and Smith 1980; Dymond et al. 2002; 2003), it is often necessary to estimate values of B for substances that have not been studied in sufficient detail. Several correlations have been developed for this purpose. One of the most common for nonpolar gases is the extended corresponding-states method of Pitzer and Curl (1958) and

Tsonopoulos and Prausnitz (1969) in which the virial coefficients are given as a function of the critical constants and the acentric factor, which may be evaluated easily from vapor-pressure data (Assael et al. 1996). Third virial coefficients of nonpolar gases have also been correlated using a similar model by Orbey and Vera (1983).

(d) In the special case of the hard-sphere potential, all of the virial coefficients are independent of temperature. The first eight virial coefficients have been evaluated (Maitland et al. 1981) for this system and the results are given in Table 2.1 wherein, σ represents the diameter of the rigid sphere.

2.7.1 What Happens to the Virial Series for Mixtures?

For gas mixtures Equation 2.102 remains formally the same but the interpretation of the terms is different. In order that the density is the molar density of the mixture, the second and third virial coefficients of a multicomponent gas mixture are given exactly by a quadratic and a cubic expression in the mole fractions, respectively, as

$$B_{\text{mix}}(T) = \sum_{i=1}^{\upsilon}\sum_{j=1}^{\upsilon} x_i x_j B_{ij}(T), \tag{2.111}$$

and

$$C_{\text{mix}}(T) = \sum_{i=1}^{\upsilon}\sum_{j=1}^{\upsilon}\sum_{k=1}^{\upsilon} x_i x_j x_k C_{ijk}(T). \tag{2.112}$$

In Equations 2.111 and 2.112, x_i is the mole fractions of species i in the mixture of υ components. In Equation 2.111, B_{ii} is the second virial coefficient of the pure species i, and B_{ij} is called the *interaction second virial coefficient.* B_{ij} is defined as the second virial coefficient corresponding to the potential-energy

TABLE 2.1 VIRIAL COEFFICIENTS FOR THE HARD-SPHERE POTENTIAL
$B = 2\pi N_A \sigma^3/3 = b_0$
$C = (5/8)\, b_0^2$
$D = 0.28695\, b_0^3$
$E = 0.11025\, b_0^4$
$F = 0.03888\, b_0^5$
$G = 0.01307\, b_0^6$
$H = 0.00432\, b_0^7$

function $\phi_{ij}(r)$ that describes the interaction of one molecule of species i with one of species j. B_{ij} is also referred to as the cross-virial coefficient, the cross-term virial coefficient, or the mixed virial coefficient.

Depending upon the availability of experimental (p, V, T) data, one of two general approaches may be adopted when dealing with multicomponent mixtures. If the (p, V, T) data for each pure component and for some compositions of each binary and ternary mixtures have been studied in great detail, one can fit the experimental data to the virial equation truncated after, say, the third virial coefficient and derive each of the possible pure component and interaction virial coefficients. The significant advantage provided by this approach is the use of the exact Equation 2.102 to generate the behavior of any mixture of the selected substances. An excellent example of this approach is offered by the GERG virial equation (Jaeschke et al. 1988; Jaeschke et al. 1991a; Jaeschke et al. 1991b) for natural gas type mixtures. For the 13 specified components, a total of 297 virial coefficients were required (B_{ii}, B_{ij}, C_{iii}, C_{iij}, C_{ijj}, and C_{ijk}). The resulting equation predicts the density of natural-gas mixtures of up to 13 components of arbitrary composition with an uncertainty of approximately 0.1 % at pressures up to 12 MPa and at temperatures between 265 and 335 K. The ubiquity and importance of natural gas in the world justifies the enormous effort represented by this program of measurement and analysis.

If, however, experimental measurements of second virial coefficients are not available, for example, for binary mixtures, then it is necessary to resort to predictive methods. To obtain interaction second virial coefficients, a wide range of empirical methods exist. Some apply combining rules to critical constants, while others use combining rules for the parameters of simple potential models, most of which are based on the Lorentz-Berthelot combining rules (Assael et al. 1996), as well as the formulae that relate the virial coefficients to intermolecular forces that are the subject of the next section. Similarly, several methods have been proposed for the estimation of the interaction third virial coefficients C_{ijk}, for example, Orbey and Vera (1983) who followed Chueh and Prausnitz (1967).

2.8 WHAT IS THE PRINCIPLE OF CORRESPONDING STATES?

So far we have discussed a perfect gas, a moderately dense gas, and we need to say something about the more general case and about the properties of a real fluid using Equation 2.97 and in particular address the residual component. There are a number of ways this can be done; we consider here only those that have a foundation in statistical mechanics we have covered. Of these methods we will thus only consider the principle of corresponding states. This is because it underpins in some form the very many thermodynamic models used in engineering. Here we will briefly describe its scientific basis for pure fluids. The extension to mixtures is discussed in Chapters 4 and 7 on the basis of very clear assumptions that provide a powerful predictive tool.

The principle of corresponding states establishes a connection between the configuration integrals of different substances and thereby allows each of the configurational thermodynamic properties of one fluid to be expressed in terms of those of another fluid. If one fluid can be selected as a reference fluid and the properties of all others related to it, then the basis for a powerful property prediction can be established. Since configurational and residual thermodynamic properties are related in a very simple way, the same results apply also to the latter.

The theoretical basis of the two-parameter corresponding-states principle is the assumption that the intermolecular potentials of two substances may be rendered identical by the suitable choice of two scaling parameters, the one applied to the separation and the other to the energy. Thus, the intermolecular potential of a substance that conforms to the principle is taken to be

$$\varphi(r) = \varepsilon F(r/\sigma), \tag{2.113}$$

where ε and σ are, respectively, scaling parameters for energy and distance, and F is a *universal* function among all relevant materials. Substances that obey Equation 2.113 are said to be *conformal*. One of the great strengths of the method is that the function F need not be known. Instead, a reference substance is introduced, identified by the subscript 0, for which the thermodynamic properties of interest are known and this is used to eliminate F from the problem. The configurational (and hence residual) properties of another conformal substance, identified by the subscript i, are thereby given in terms of those of the reference fluid. We shall also see that the parameters ε and σ may be eliminated in favor of measurable macroscopic quantities.

The consequences of conformality may be derived by means of the following "thought experiment." Consider two conformal substances, one of which is designated as the reference fluid, contained in separate vessels of the same shape but different volumes as illustrated in Figure 2.2. Let there be N molecules of type i contained in volume V at temperature T, while the N molecules of the reference fluid be contained in volume V/h_i at temperature T/f_i. Here, $h_i = (\sigma_i/\sigma_0)^3$ and $f_i = (\varepsilon_i/\varepsilon_0)$ are scaling ratios. We now suppose that the molecules are arranged in geometrical similar positions within their respective containers. Then, for each molecule in the system on the right with position vector $\mathbf{r}_i$ defined relative to the origin in that system, there is a corresponding molecule in the reference system with position vector $\mathbf{r}_0$ defined relative to the origin in that system and these position vectors are related by

$$\mathbf{r}_0 = \mathbf{r}_i / h_i^{1/3}. \tag{2.114}$$

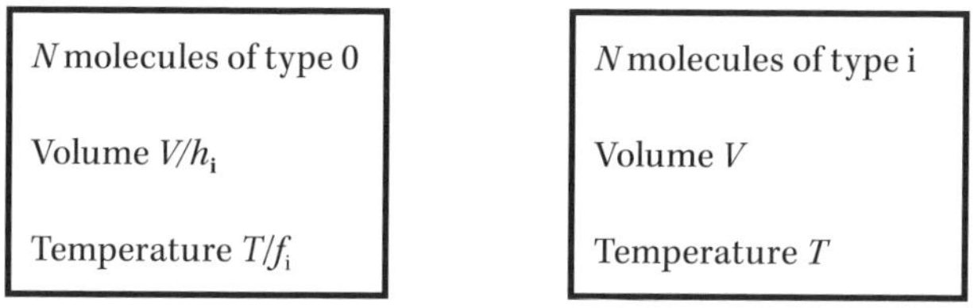

Figure 2.2 Corresponding-states principle.

Since it is assumed that the pair potentials are conformal and that either (i) the pair-additivity approximation is obeyed or (ii) that the N-body potentials are also conformal, the configurational energies of the two systems are related by

$$\Upsilon_0(\mathbf{r}_{0,1},\mathbf{r}_{0,2},\cdots,\mathbf{r}_{0,N}) = \frac{\Upsilon_\mathrm{i}(\mathbf{r}_{\mathrm{i},1},\mathbf{r}_{\mathrm{i},2},\cdots,\mathbf{r}_{\mathrm{i},N})}{f_\mathrm{i}}. \tag{2.115}$$

Equation 2.115 must apply to any configuration because for each configuration of the reference system a geometrically similar one exists for the second system.

The configuration integral Ω_0 for the reference system is given (Assael et al. 1996) by

$$\Omega_0\,(V/h_\mathrm{i},\,T/f_\mathrm{i}) = \int_{V/h_\mathrm{i}}^{\cdots}\int \exp\left(-\frac{f_\mathrm{i}\Upsilon_0}{kT}\right)\mathrm{d}\mathbf{r}_0^N, \tag{2.116}$$

while that for the other system is

$$\Omega_\mathrm{i}(V,T) = \int_V^{\cdots}\int \exp\left(-\frac{\Upsilon_\mathrm{i}}{kT}\right)\mathrm{d}\mathbf{r}_\mathrm{i}^N. \tag{2.117}$$

Upon changing the variables of integration from $\mathbf{r}_\mathrm{i}$ to $\mathbf{r}_0$, in accordance with Equation 2.115, and making use of Equation 2.115, Ω_i becomes

$$\Omega_\mathrm{i}(V,T) = \int_{V/h_\mathrm{i}}^{\cdots}\int \exp\left(-\frac{f_\mathrm{i}\,\Upsilon_0}{kT}\right)h_\mathrm{i}^N\,\mathrm{d}\mathbf{r}_0^N. \tag{2.118}$$

Then, comparing Equations 2.117 and 2.118, we see that the configuration integrals of the two systems are related by the simple equation

$$\Omega_\mathrm{i}(V,T) = h_\mathrm{i}^N\Omega_0\,(V/h_\mathrm{i},T/f_\mathrm{i}). \tag{2.119}$$

The compression factor is defined by Equation 2.95, it follows from Equation 2.101 that

$$Z = \left(\frac{V}{N}\right)\left(\frac{\partial \ln\Omega}{\partial V}\right)_T, \tag{2.120}$$

so that it is a purely configurational property. It then follows from Equation 2.119 that

$$Z_i(V,T) = Z_0(V/h_i, T/f_i). \tag{2.121}$$

Thus the compression factor of one conformal substance may be equated with that of another at a scaled volume and a scaled temperature. As this relation must hold also at the critical point, it follows that the scaling parameters are related to the critical constants by

$$f_i = T_{c,i}/T_{c,0} \tag{2.122}$$

and

$$h_i = V_{c,i}/V_{c,0}, \tag{2.123}$$

and that the reduced pressure $p_r = p/p^c$ is the same function of the reduced volume $V_r = V/V_c$ and the reduced temperature $T_r = T/T_c$ in all conformal systems. Consequently, the compression factor is a universal function of T_r and V_r or, alternatively, of T_r and p_r.

Generalized charts are available (Lee and Kesler 1975) giving the compression factor Z, as well as the residual enthalpy and entropy in terms of the residual molar enthalpy (H_m^{res}/RT_c) and molar entropy (S_m^{res}/R) as a function of reduced temperature and pressure. The principle of corresponding states does not provide the perfect-gas contribution to either the enthalpy or the entropy that must be evaluated by alternative means, which we have already discussed in Question 2.4.

The simple treatment outlined above will be applicable to substances that have conformal pair potentials. A group of substances for which it is nearly true is the monatomic gases Ar, Ke, and Xe, for which it works remarkably well. But for He, and to some extent Ne, deviation from the principle arise because at low temperatures quantum effects, which depend upon mass and not the potential, have to be considered. Several simple molecules, including N_2, CO, and CH_4, deviate only slightly from the principle but most other molecules depart considerably.

The reasons for the conformality of the monatomic gases and the relative failure for other species rest on the fact that the former group of systems are spherically symmetric (in agreement with the assumptions of the model) while the polyatomic molecules evidently do not have this symmetry. To apply the principle of corresponding states with any accuracy to molecular fluids, it is necessary to take into account the nonspherical nature of the molecules. The anisotropic nature of the intermolecular potential ϕ in these cases has already been briefly described in Question 1.4 and here we simply recall that ϕ is a function not only of the separation r but also of the relative orientation of the two molecules. Hence, in addition to the two scaling parameters described above, others are necessary in principle. In the first attempt to deal with this problem from an engineering perspective a third parameter was introduced, leading to a three-parameter corresponding-states principle. A third parameter was first proposed by Pitzer in 1955 (Pitzer et al. 1955; Pitzer 1955) who defined the acentric factor ω by

$$\omega = -1-\log_{10}\left\{\frac{p^{\mathrm{sat}}(T=0.7\cdot T_c)}{p_{\mathrm{c}}}\right\}, \tag{2.124}$$

where p_{c} is the critical pressure and p^{sat} is the vapor pressure.

Pitzer (1955) proposed a generalized thermodynamic property X can be written as a function of reduced temperature and pressure by

$$X(T_{\mathrm{r}},p_{\mathrm{r}}) = X_0(T_{\mathrm{r}},p_{\mathrm{r}}) + \omega X_1(T_{\mathrm{r}},p_{\mathrm{r}}). \tag{2.125}$$

In Equation 2.125, X_0 is known as the simple fluid term and X_1 is known as the correction term. Charts and equations representing the simple fluid and correction terms as functions of reduced temperature and pressure are available for the cases when $X=Z$, $(H_{\mathrm{m}}^{\mathrm{res}}/RT_{\mathrm{c}})$ and $(S_{\mathrm{m}}^{\mathrm{res}}/R)$.

Finally, to incorporate polar effects, four-parameter corresponding-states models are usually employed (Wu and Stiel 1985). In this case, the extra parameter is usually obtained experimentally.

2.8.1 How Can the Principle of Corresponding States Be Used to Estimate Properties?

To demonstrate the use of the principle of corresponding states we show a few simple examples that are chosen to provide readers with an exposure to the estimation methods commonly employed in chemical engineering practice or in software routines that inform chemical engineering practice. The methods are all exact in some hypothetical limit but are approximate in any real case so that the results of the application of these methods should always be used with circumspection about their uncertainty.

First, if we suppose that all substances conform to the same reduced pair potential $\phi(r)=\varepsilon F(r/\sigma)$ as set out in Equation 2.113, then it is easily shown from Equation 2.109 that the second virial coefficient for all such substances obeys the simple two-parameter law of corresponding states

$$B^{*} = \frac{B}{(2L\pi\sigma^3/3)} = \int_0^{\infty}\left[\frac{r}{\sigma}\right]^2\left\{1-\exp\left[-\frac{F(r/\sigma)}{T^{*}}\right]\right\}\mathrm{d}(r/\sigma) \tag{2.126}$$

where $T^{*} = kT/\varepsilon$. For an assumed functional form of the pair potential, tables of the reduced second virial coefficient can be calculated from which the real virial coefficient of any of the substances can be calculated from values for the two parameters σ and ε.

The power of the two-parameter principle of corresponding states can be demonstrated by estimating the density of argon from the density of krypton at some other temperature and pressure. In this example, the critical temperature and critical pressure are the scaling parameters. The density of krypton at $T = 348.15$ K and $p = 2$ MPa is 59.28 kg · m^{-3} (Evers et al. 2002). The critical temperature T_c, the critical pressure p_c, and the critical mass density ρ_c of krypton are 209.4 K, 5.5 MPa, and 918.8 kg · m^{-3}, respectively. For Kr the reduced temperature T_r, pressure p_r, and mass density ρ_r are as follows:

$$T_{r,Kr} = \frac{T}{T_{c,Kr}} = \frac{348.15\ \mathrm{K}}{209.4\ \mathrm{K}} = 1.663, \tag{2.127}$$

$$p_{r,Kr} = \frac{p}{p_{c,Kr}} = \frac{2\ \mathrm{MPa}}{5.5\ \mathrm{MPa}} = 0.364, \tag{2.128}$$

and

$$\rho_{r,Kr} = \frac{\rho}{\rho_{c,Kr}} = \frac{59.28\ \mathrm{kg \cdot m^{-3}}}{918.8\ \mathrm{kg \cdot m^{-3}}} = 0.0645. \tag{2.129}$$

For argon $T_{c,Ar} = 150.8$ K, $p_{c,Ar} = 4.87$ Mpa, and $\rho_{c,Ar} = 533.4$ kg · m^{-3} and when combined with Equations 2.127, 2.128, and 2.129, respectively, the temperature, pressure, and density of argon are given by the following:

$$T_{Ar} = T_{r,Kr} \cdot T_{c,Ar} = 1.663 \cdot 150.8\ \mathrm{K} = 250.8\ \mathrm{K}, \tag{2.130}$$

$$p_{Ar} = p_{r,Kr} \cdot p_{c,Ar} = 0.364 \cdot 4.87\ \mathrm{MPa} = 1.772\ \mathrm{MPa}, \quad \text{and} \tag{2.131}$$

$$\rho_{Ar} = \rho_{r,Kr} \cdot \rho_{c,Ar} = 0.0645 \cdot 533.4\ \mathrm{kg \cdot m^{-3}} = 34.40\ \mathrm{kg \cdot m^{-3}}. \tag{2.132}$$

The principle of corresponding states predicted the density for argon, from that of krypton, to be ρ_{Ar}(250.8 K, 1.772 MPa) = 34.40 kg · m^{-3}, which lies 0.9 % above the literature value of (Evers et al. 2002) 34.71 kg · m^{-3}. The comparison between the experiment and the corresponding states method is obviously quite good in this case, but this is not surprising because it is known that the pair potentials of argon and krypton are almost conformal (Maitland et al. 1981). A more significant test is provided by calculating the properties of methane at the same corresponding state.

We will now use Equations 2.127, 2.128, and 2.129 to estimate the density of methane for which $T_{c,CH4} = 190.55$ K, $p_{c,CH4} = 4.599$ MPa, and $\rho_{c,CH4} = 161.73$ kg · m^{-3} to give

$$T_{CH_4} = T_{r,Kr} \cdot T_{c,CH_4} = 1.663 \cdot 190.55\,\text{K} = 316.88\,\text{K}, \tag{2.133}$$

$$p_{CH_4} = p_{r,Kr} \cdot p_{c,CH_4} = 0.364 \cdot 4.599\,\text{MPa} = 1.674\,\text{MPa}, \quad \text{and} \tag{2.134}$$

$$\rho_{CH_4} = \rho_{r,Kr} \cdot \rho_{c,CH_4} = 0.0645 \cdot 161.73\,\text{kg} \cdot \text{m}^{-3} = 10.35\,\text{kg} \cdot \text{m}^{-3}. \tag{2.135}$$

The principle of corresponding states thus predicts the density for methane, from that of krypton, to be ρ(316.88 K, 1.674 MPa) = 10.35 kg · m^{-3} and it lies 4.3 % above the literature value of (Evers et al. 2002) 9.92 kg · m^{-3}. Methane is a nonspherical molecule, while krypton is spherical, and the greater difference between the estimated and actual density is thus not surprising.

These two examples demonstrate the two parameter principle of corresponding states generally written as

$$X(T_r, p_r) = X_0(T_r, p_r), \tag{2.136}$$

where a property X can easily be related to the same property of another fluid X_0 at the same reduced conditions.

As the complexity of the molecule's structure increases and consequently the intermolecular potential is no longer purely spherical the departure of the properties predicted with Equation 2.136 given by Equation 2.125 increase. Pitzer (1955) proposed a modification of the Equation 2.136 that included the acentric factor ω given by Equation 2.124.

Lee-Kesler (1975) with this approach produced a consistent scheme for the calculation of the density, enthalpy, entropy, and fugacity of hydrocarbons based on the properties of octane.

We will now use both Equations 2.125 and 2.136 to estimate (Assael et al. 1996) the density of dodecane at the following temperature and pressure: (a) $T = 298.15$ K and $p = 0.1$ MPa (b) $T = 358.15$ K and $p = 13.8$ MPa. The procedures required for Equation 2.136 follows those described for Equations 2.133, 2.134, and 2.135, while those for Equation 2.125 are provided elsewhere (Assael et al. 1996) and the results obtained are listed in Table 2.2, which also includes the accepted experimental values (Snyder and Winnick 1970) against which the predictions are compared. Clearly Equation 2.125 provides the best estimates when compared with the experiment. We also estimated from both Equations 2.125 and 2.136 the enthalpy change between condition (a) and (b). Equation 2.136 provided $\Delta H_m = 22.5$ kJ · mol^{-1}, while Equation 2.125 returned $\Delta H_m = 25.5$ kJ · mol^{-1}. The ΔH_m estimated with Equation 2.125 differs by less

TABLE 2.2 THE AMOUNT-OF-SUBSTANCE DENSITY ρ_N OF DODECANE ESTIMATED FROM BOTH EQUATION 2.136 AND EQUATION 2.125 AT THE FOLLOWING TEMPERATURE AND PRESSURE: (A) $T = 298.15$ K AND $p = 0.1$ MPA (B) $T = 358.15$ K AND $p = 13.8$ MPA

	$\rho_n/\text{mol} \cdot \text{m}^{-3}$		
	Equation 2.136	Equation 2.125	Ref
298.15 K, 0.10 MPa	3354	4446	4375
358.15 K, 13.8 MPa	3257	4277	4193
$\Delta T = 60$ K $\Delta p = 13.7$ MPa			

than 1.6 % from the measured value while that predicted with Equation 2.136 lies >10 % below the measured (Snyder and Winnick 1970) value of $\Delta H_m = 25.1$ kJ · mol^{-1}.

For further examples of the application of two and three-parameter corresponding-states the reader is referred to Assael et al. (1996).

2.9 WHAT IS ENTROPY *S*?

To conclude this chapter we deal with an issue that is somewhat different from estimating the physical properties of systems. We consider a question that is often asked by students of thermodynamics when the connection is first established for them between macroscopic properties and microscopic quantities and that is what is entropy? For some physical quantities, such as temperature and length, our feet provide us with a crude but not transferable measure of both, at least to answer the question "will the temperature of a bath cause my skin a burn?" There is also a tendency to ask for microscopic explanations for an observed macroscopic change: a molecular understanding of the change of the volume of a fluid mixture formed from two pure components can be provided by recourse to statistical mechanics that provides a quantum mechanical microscopic interpretation of thermodynamics functions albeit without simple pictures relating to everyday life. It is in this spirit that the concept of entropy raises difficult issues.

Clausius was the first to employ the word entropy taken from the Greek *εντροπία* whose translation is *"turning toward"*; Clausius preferred its interpretation as the energy of transformation (Clausius 1850). For the purpose of thermodynamics the term might be taken to mean the energy lost to dissipation (Clausius 1865a, 1865b) and as such it provides the definition for a reversible process

$$\Delta S = \int T^{-1} \mathrm{d}Q, \tag{2.137}$$

where Q is the heat (Clausius 1862) and arises from interaction with the surroundings.*

Clausius's now famous aphorism "Die Entropie der Welt strebt einem Maximum zu" that when translated yields "The entropy of the Universe tends toward a maximum" (Clausius 1865a, 1865b) might give rise to concepts of "mixed-upness" and indeed has been invoked to describe the fate of the universe. The latter has been argued because any process that takes place in the universe results in an increase of entropy and, is implied, albeit wrong, as an increase of "mixed-upness." Inevitably this pessimistic opinion envisages the universe in the end no longer consisting of stars and planets, of seas and land, but of structureless particles distributed uniformly throughout space: ultimately the universe will have a chaotic fate in the so-called "heat-death" (Landsberg 1961; Buchdahl 1966). Arguments of this type have assumed, for thermodynamics to apply, that the universe is bounded and isolated (from what?) and that the experimental science of thermodynamics applies to a system of the size of a Universe.

Both the 2nd law of thermodynamics and entropy have provoked great speculation on their own account and on their limitations but the law that includes entropy stands as correct until proven by experiment to be incorrect. An example of such an argument is provided by Maxwell (Maxwell 1872) who conceived a *being* (creature) that was capable of following the motion of every molecule in a vessel divided into two portions, A and B, by a partition. The partition has one hole that can be opened and closed by the creature without expenditure of work to permit the molecules of velocity greater than the mean of all the molecules to pass from chamber A to chamber B, and only the molecules with velocity lower than the mean of all the molecules in the box to pass from B to A. Acting in this manner the creature will raise the temperature of chamber B and lower the temperature of chamber A. This will contradict the 2nd law of thermodynamics. William Thomson gave this creature the name "Maxwell's demon" (Thomson 1874; Thomson 1879). The intention of the creature was to demonstrate that the 2nd law of thermodynamics has only a statistical certainty. Maxwell's statements about the demons were sufficiently brief to permit interpretations on which much has been written (Szilard 1929; Klein 1970; Bennett 1987; Collier 1990; Leff 1990; Skordos 1993; Corning 1998a, 1998b; Maddox 2002) and collations of the original scientific papers published (Leff 2003). This is not the place, therefore, to enter into a lengthy discussion of the problem posed by Maxell's demon or its resolution, but it is sufficient to note that to perform the task set the demon would need to measure the velocity of each particle and therefore interact with

* We will soon discover in Chapter 3 (Equation 3.2) one thermodynamic axiom that states the entropy of the system must increase if anything is happening in the system. This is referred to as a natural change and is the rate of internal entropy production on which more will be said in Chapter 5.

the system in a direct and nonstatistical fashion counter to the original proposition. Fortunately, Maxwell refrained from relating his *being* to entropy but he did predict Earth would become unfit for habitation by man (Thomson 1852).

Interestingly, an equation similar in form to Equation 2.137 is used in the subject of information theory (Brillouin 1961), useful for data transmission and cryptography, to describe how much information is produced by a discrete source and at what rate. This description led Shannon to utilize the term (information) entropy interchangeably with uncertainty (Shannon 1948a, 1948b).

2.9.1 How Can I Interpret Entropy Changes?

An increase of entropy is often stated to be equivalent to an increase of disorder or randomness or mixed-upness or probability when these are simply shorthand for the number of accessible eigenstates (energy) for an isolated system. The number of eigenstates is directly related to the concept of "mixed-upness" in two special cases which serve to illustrate the conceptual relationship. First we suppose we have two noninteracting gases each in an isolated container and separated from each other by an impermeable membrane. When the membrane is broken, an increased volume is available for each molecule and in addition the number of available combinations of translational energy eigenvalues increase, which we have seen make up its energy. Second, we consider crystals at temperatures close to zero where the geometrical orientations of the molecules on the lattice sites may be regular or irregular and may be ordered or disordered. In both cases, the number of accessible eigenstates is simply related to the purely geometrical or spatial "disorder" and also entropy.

However, for normal and realizable chemical processes there is no simple geometrical interpretation of the entropy change and it is not possible to extrapolate statistical-mechanical conclusions for systems of noninteracting particles or crystals at $T \rightarrow 0$ (discussed in Chapter 3 with Nernst's heat theorem) to beakers of liquids at $T = 293$ K.

Entropy is a state variable with the same status as temperature and pressure and is measurable (or at least differences are). Let us accept this simple and refreshing statement as a fact and, after introducing the second law in Chapter 3, review again the misconception of mixed-upness and better still ask another question: How would you measure the entropy change that accompanies the mixing of two gases?

2.10 REFERENCES

Assael M.J., Trusler J.P.M., and Tsolakis Th., 1996, *Thermophysical Properties of Fluids. An Introduction to their Prediction*, Imperial College Press, London.

Bennett C.H., 1987, "Demons, engines and the 2nd law," *Sci. Am.* **257**:108–116.

Brillouin L., 1961, "Thermodynamics, statistics and information," *Am. J. Phys.* **29**:318–328.

Buchdahl H.A., 1966, *The Concepts of Classical Thermodynamics*, Cambridge University Press, p. 17.

Chueh P.L., Prausnitz, J.M., 1967, "Vapor-liquid equilibria at high pressures. Vapor-phase fugacity coefficients in nonpolar and quantum-gas mixtures," *Ind. Eng. Chem. Fundam.* **6**:492–498.

Clausius R., 1850a, "Über die bewegende Kraft der Wärme, Part I," *Annalen der Physik* **79**:368–397 (also printed in 1851, "On the Moving Force of Heat, and the Laws regarding the Nature of Heat itself which are deducible therefrom. Part I," *Phil. Mag.* **2**:1–21).

Clausius R., 1850b, "Über die bewegende Kraft der Wärme, Part II," *Annalen der Physik* **79**:500–524 (also printed in 1851, "On the Moving Force of Heat, and the Laws regarding the Nature of Heat itself which are deducible therefrom. Part II," *Phil. Mag.* **2**:102–119).

Clausius R., 1862a, "The mechanical theory of heat," *Phil. Mag.* (series 4) **24**:201.

Clausius R., 1862b, "Sixth memoir on the application of the theorem of the equivalence of transformations," *Phil. Mag.* (series 4) **24**:81.

Clausius R., 1865a, "Über die Wärmeleitung gasförmiger Körper," *Annalen der Physik und Chemie* **125**:353–400.

Clausius R., 1865b, *"The Mechanical Theory of Heat—with Its Applications to the Steam Engine and to Physical Properties of Bodies,"* John van Voorst, London.

Collier J.D., 1990, "2 faces of Maxwells demon reveal the nature of irreversibility," *Stud. Hist. Philos.* Sci. **21**:257–268.

Corning P.A., and Stephen J.K., 1998a, "Thermodynamics, information and life revisited. Part I: To be or entropy," *Syst. Res. Behav. Sci.* **15**:273–295.

Corning P.A., and Stephen J.K., 1998b, "Thermodynamics, information and life revisited, Part II: 'Thermoeconomics' and 'control information'," *Syst. Res. Behav. Sci.* **15**:453–482.

Dymond D.H., Marsh K.N., and Wilhoit R.C., 2003, *Virial Coefficients of Pure Gases and Mixtures Group IV Physical Chemistry Vol. 21 Subvolume B Virial Coefficients of Mixtures.* Landolt-Börnstein *Numerical Data and Functional Relationships in Science and Technology*, eds. Martienssen W. (chief), Frenkel M., and Marsh K.N., Springer-Verlag, New York.

Dymond J.H., and Smith E.B., 1980, *The Virial Coefficients of Pure Gases and Mixtures. A Critical Compilation*, Clarendon Press, Oxford.

Dymond J.H., Marsh K.N., Wilhoit R.C., and Wong K.C., 2002, *Virial Coefficients of Pure Gases and Mixtures Group IV Physical Chemistry Vol. 21 Subvolume A Virial Coefficients of Pure Gases.* Landolt-Börnstein *Numerical Data and Functional Relationships in Science and Technology*, eds. Martienssen W. (chief), Frenkel M., and Marsh K.N., Springer-Verlag, New York.

Evers C., Losch H.W., and Wagner W., 2002, "An absolute viscometer-densimeter and measurements of the viscosity of nitrogen, methane, helium, neon, argon, and krypton over a wide range of density and temperature," *Int. J. Thermophys.* **23**:1411–1439.

Herzberg G., 1945, *Infrared and Raman Spectra of Polyatomic Molecules*, Van Nostrand, Princeton, NJ.

Herzberg G., 1970, *Molecular Spectra and Molecular Structure. Vol. 1. Spectra of Diatomic Molecules*, 2nd ed., Van Nostrand, Princeton, NJ.

Hill T.L., 1960, *An Introduction to Statistical Thermodynamics*, Addison Wesley, Reading, MA.

Howerton M.T., 1962, *Engineering Thermodynamics*, Van Nostrand, Princeton, NJ.

Hurly J.J., and Mehl J.B., 2007, "He-4 thermophysical properties: New ab initio calculations," *J. Res. Natl. Inst. Stand. Technol.* **112**:75–94.

Hurly J.J., and Moldover M.R., 2000, "Ab initio values of the thermophysical properties of helium as standards," *J. Res. Natl. Inst. Stand. Technol.* **105**:667–688.

Jaeschke M., Audibert S., van Caneghem P., Humphreys A. E., Janssen-van R., Pellei Q., Michels J.P.J., Schouten J.A., and ten Seldam C.A., 1988, *High Accuracy Compressibility Factor Calculation for Natural Gases and Similar Mixtures by Use of a Truncated Virial Equation, GERG*, Verlag des Vereins Deutscher Ingenieure, Dusseldorf.

Jaeschke M., Audibert S., van Caneghem P., Humphreys A.E., Janssen-van R., Pellei Q., Schouten J.A., and Michels J.P., 1991a, "Accurate prediction of compressibility factors by the GERG virial equation," *SPE Prod. Engng.* Aug., 343–349. SPE 17766-PA.

Jaeschke M., Audibert S., van Caneghem P., Humphreys A.E., Janssen-van R., Pellei Q., Schouten J.A., and Michels J.P., 1991b, "Simplified GERG virial equation for field use," *SPE Prod. Engng.* Aug., 350–355. SPE 17767-PA.

Janz G.J., 1967, *Thermodynamic Properties of Organic Compounds*, rev. ed., Academic Press, New York.

Klein M.J., 1970, "Maxwell, his Demon, and second law of thermodynamics," *Am. Sci.* **58**:84–94.

Landolt-Bornstein, 1951, *Band 1, Atom-und Molekularphysik. Teil 2. Molekulen, 1*, Springer-Verlag, Berlin, p. 328.

Landsberg P.T., 1961, *Thermodynamics*, Interscience, New York, p. 391.

Lee B.I., and Kesler M.G., 1975, "Generalized thermodynamic correlation based on 3-parameter corresponding states," *AIChE J.* **21**:510–527.

Leff H., and Rex A.F., 2003, *Editors of Maxwell's Demon 2: Entropy, Classical and Quantum Information, Computing for Inst. Phys. Pub.*, Philladelphia, PA.

Leff H.S., 1990, "Maxwell demon, power and time," *Am. J. Phys.* **58**:135–142.

Maddox J., 2002, "The Maxwell's demon: Slamming the door," *Nature* **417**:903.

Maitland G.C., Rigby M., Smith E.B., and Wakeham W.A., 1981, *Intermolecular Forces. Their Origin and Determination*, Clarendon Press, Oxford.

Maxwell J.C., 1872, *Theory of Heat*, 3rd ed., Longman and Green, London, pp. 307–309.

McQuarrie D.A., 2000, *Statistical Mechanics*, University Science Books, Sausalito, CA.

Mohr P.J., Taylor B.N., and Newel D., 2008, "CODATA recommended values of the fundamental physical constants: 2006," *J. Phys. Chem. Ref. Data* **3**:1187–1284.

Moore G.E., 1949–1958, *Atomic Energy States, Nat. Bur. Stand. Circ.* 467, vols.1–3.

Orbey M., and Vera J.M., 1983, "Correlation for the 3rd virial coefficient using T_c, P_c and omega as parameters," *A.I.Ch.E. J.* **29**:107–113.

Pitzer K.S., 1955, "The volumetric and thermodynamic properties of fluids. 1. Theoretical basis and virial coefficients," *J. Am. Chem. Soc.* **77**:3427–3433.

Pitzer K.S., and Curl R.F., 1958, "Volumetric and thermodynamic properties of fluids — Enthalpy, free energy and entropy," *Ind. Eng. Chem.* **50**:265–274.

Pitzer K.S., Lippman D.Z., Curl R.F., Huggins C.M., and Petersen D.E., 1955, "The volumetric and thermodynamic properties of fluids 2. Compressibility factor, vapor pressure and entropy of vaporization," *J. Am. Chem. Soc.* **77**:3433.

Poling B., Prausnitz J.M., and O'Connell J.P., 2001, *The Properties of Gases and Liquids*, 5th ed., McGraw-Hill, New York.

Prausnitz J.M., Lichtenthaler R.N., and Gomes de Azevedo E., 1986, *Molecular Thermodynamics of Fluid-Phase Equilibria*, 2nd ed., Prentice Hall.

Reed T.M., and Gubbins K.E., 1973, *Applied Statistical Mechanics,* McGraw-Hill, Kogakusha.

Reid R.C., Prausnitz J.M., and Poling B.E., 1988, *The Properties of Gases and Liquids*, 4th ed., McGraw-Hill, New York.

Selected Values of Properties of Hydrocarbons and Related Compounds, 1977, 1978, Thermodynamic Research Center, Texas A&M University.

Shannon C.E., 1948a, "A mathematical theory of communication," *Bell Sys. Tech. J.* **27**:379–423.

Shannon C.E., 1948b, "A mathematical theory of communication," *Bell Sys. Tech. J.* **27**:623–656.

Sherwood A.E., and Prausnitz J.M., 1964, "Virial coefficient for Kihara Exp-6 + Square well potentials," *J. Chem. Phys.* **41**:413–428.

Skordos P.A., 1993, "Compressible dynamics, time reversibility, Maxwell demon, and the 2nd law," *Phys. Rev. E* **48**:777–784.

Snyder P.S., and Winnick J., 1970, *Proc. of 5th Symp. Thermophys. Prop.*, ASME, Boston, p. 115.

Szilard L., 1929, "On the minimization of entropy in a thermodynamic system with interferences of intelligent beings," *Zeitschrift fuer Physik.* **53**:840–856.

Sutton L.E., 1965, *Tables of Interatomic Distances and Configuration in Molecules and Ions, Supplement*, The Chemical Society, London.

Thomson W., 1874, "Kinetic theory of the dissipation of energy," *Nature* **9**:441–444.

Thomson W., 1879, "The sorting demon of Maxwell," *Proc. R. Inst.* **9**:113.

Thomson W.P.R., 1852, "On a universal tendency in nature to the dissipation of mechanical energy," *Phil. Mag.* **4**:304–306.

Tsonopoulos C., and Prausnitz J., 1979, "A review for engineering applications," *Cryogenics* **9**:315–327.

van Ness H.C., and Abbott M.M., 1982, *Classical Thermodynamics of Nonelectrolyte Solutions*, McGraw-Hill, New York.

Wu G.Z.A., and Stiel L.I., 1985, "A generalized equation of state for the thermodynamic properties of polar fluids," *AIChE J.* **31**:1632–1644.

Chapter 3

2nd Law of Thermodynamics

3.1 INTRODUCTION

In Chapter 1 of this book we argue that thermodynamics is an experimental science consisting of a collection of axioms, derivable from statistical mechanics and in many circumstances from Boltzmann's distribution. So far we have introduced the 0th law and the 1st law of thermodynamics that interrelate physical quantities some of which are far more easily measured than others. We are also armed with two types of "thermodynamic-meter": (1) a thermometer to measure temperature and (2) a calorimeter used to measure differences in energy and enthalpy. Both of these will be put to good use in this chapter, which considers the 2nd law of thermodynamics. We will also introduce a third "meter": a chemical potentiometer used to measure differences in chemical potential.

Clausius provided the first broad statements of the 2nd law of thermodynamics (1850a, 1850b, and 1851) and these were refined by Thomson,* and those readers interested in the history of the formulation of the laws of thermodynamics should consider consulting the work of Atkins (2007) and Rowlinson (2003 and 2005).

3.2 WHAT ARE THE TWO 2ND LAWS?

The approach adopted here for the presentation of the second law follows Gibbs (1928), Guggenheim (1967), and McGlashan (1979) and uses axioms (or rules of the game) and states the second law in two equations, one an equality the other an inequality. These statements taken together are called the *second law* and will be discussed first for a homogeneous phase, throughout which all intensive properties are constant by definition, and then later extended to heterogeneous phases.

* Also known as Lord Kelvin, who described the absolute temperature scale, from 1892.

The first statement, which we will label 2a, is an equation concerning any infinitesimal change in the energy of a phase, while the second, which is then known as 2b, is an inequality. The later sections of this chapter will explore, with the use of auxiliary quantities introduced solely for convenience, some consequences of the second law as well as the techniques for the manipulation of the equations. We shall, therefore, be concerned with examples for practical applications in the hope that they will provide the reader with the set of tools necessary to apply thermodynamics appropriately to further practical examples.

An alternative to the axiomatic approach is to introduce the second law through Carnot's cycle (Denbigh 1971).

3.2.1 What Is Law 2a?

Let us start with the statement of part 2a for an infinitesimal change of state of a single phase consisting of a number of substances B

$$\mathrm{d}U = T\,\mathrm{d}S - p\,\mathrm{d}V + \sum_{\mathrm{B}} \mu_{\mathrm{B}}\,\mathrm{d}n_{\mathrm{B}}. \tag{3.1}$$

In Equation 3.1, the energy U is the characteristic function for the independent variables S, V, and n. Equation 3.1 assumes that the only work done arises from variations in pressure and volume ($\delta W = -p\,\mathrm{d}V$). Because thermodynamics is an experimental science Equation 3.1 can be regarded as an axiom to be tested with reference to practical experimentation. When tested in this way, Equation 3.1 has never been shown to be false. On the left hand side of Equation 3.1 we have $\mathrm{d}U$, the infinitesimal change in energy U, as we have seen in Chapter 1 differences in U can be measured with a calorimeter. On the far right hand side is a summation over all the substances B of the phase of the product of the chemical potential μ_{B}, which has been defined in Chapter 1 and for which only differences can be measured, and $\mathrm{d}n_{\mathrm{B}}$, any change in the amount of substance. The second term on the right hand side of Equation 3.1 contains pressure p and volume V. The first term on the right hand side contains the thermodynamic temperature T and an extensive quantity entropy S, which has been defined in Chapter 1 and described in Question 2.9.

3.2.2 What Is Law 2b?

The inequality of the second law states that if any measurable quantity changes perceptibly (if anything changes) in an isolated system, (which is one of constant energy U, volume V, and material content Σn_{B}, without regard to chemical state or any aggregation) the entropy of the system S must increase:

$$\left(\frac{\partial S}{\partial t}\right)_{U,V,\Sigma n_{\mathrm{B}}} \geq 0. \tag{3.2}$$

In Equation 3.2 t denotes time. It is a corollary that for an isolated system in which there are no changes in $T, p, V, U, \Sigma n_B$ there is nothing happening so that

$$\left(\frac{\partial S}{\partial t}\right)_{U,V,\Sigma n_B} = 0, \qquad (3.3)$$

and the system is in equilibrium. In reality, the term "nothing happening" means that anything that is happening does so either so slowly to be undetectable during the time of the observation, or so small as to be undetectable with the instruments used to measure the changes. This definition of equilibrium includes states that are otherwise known as *metastable* with respect to some change to another more stable state. For example, a mixture of hydrogen and oxygen held at room temperature in the absence of a catalyst matches these criteria, because the instruments used for typical observation times detect no changes. If the observation was made over a longer time or more sensitive instruments were used to monitor the system, changes might be observed that then reveal the system was not at equilibrium. The axiom of Equation 3.3 makes it clear that time is a relevant parameter in thermodynamics despite many comments to the contrary.

In view of Equation 3.2 we will digress and return to the discussion that we started in Chapter 2 regarding the question: what is entropy? As stated in Question 2.9.1, an increase of entropy is often stated to be equivalent to an increase of disorder or randomness or mixed-upness or probability; when these are simply shorthand for the number of accessible eigenstates for an isolated system. Two practical examples will be considered for systems that are realizable in a laboratory and demonstrate how the simple concepts break down.

First, let us consider what is happening when a supersaturated solution of aqueous Na_2SO_4 has another crystal of Na_2SO_4 added just before isolation in a Dewar flask. A process is underway in this isolated system and so something is happening and so the entropy must increase. In fact, the temperature of the system decreases as the anhydrous salt precipitates and the solute spatially separates from the solvent and there is a partial unmixing of the solution so the "mixed-upness" decreases in defiance of the simplistic interpretation of entropy.

Second, a mixture of (hydrogen + argon) is contained in an isolated vessel with a palladium membrane (which is permeable to hydrogen but not to argon), forming a barrier to an isolated evacuated volume. The hydrogen diffuses through the membrane, driven by the chemical potential gradient, and so something is happening $(\partial S/\partial t) > 0$. The circumstances and the process are sorting the hydrogen from the argon and not increasing the "mixed-upness" Indeed, for the mixing of two liquids at constant temperature and pressure the entropy change is not always positive. For example, the molar entropy of mixing of $\{0.5H_20 + 0.5(C_2H_5)_2NH\}$ at $T = 322.25$ K is -8.78 J $\cdot$ K^{-1} $\cdot$ mol^{-1} (Coop and Everett 1953).

We conclude by reiterating Question 2.9.1: entropy is a state variable with the same status as temperature and pressure and is measurable (or at least differences are). The second law states the entropy of a closed system never decreases (Margenau 1950). We accept this simply as a fact and ask another question: how would you measure the entropy change? This question will be addressed in Section 3.5.1.

For a system at uniform temperature and constant volume and amount of substance from Equation 3.1 it follows that

$$\left(\frac{\partial U}{\partial S}\right)_{V,n_\mathrm{B}} = T. \tag{3.4}$$

Use of the −1 rule (see Question 1.11.1) on Equations 3.2 and 3.3 for a system of constant entropy, volume, and content, it follows that

$$\left(\frac{\partial U}{\partial t}\right)_{S,V,\Sigma n_\mathrm{B}} < 0, \tag{3.5}$$

and the energy of the system is decreasing. Equation 3.5 will be used in Question 3.6.

3.3 WHAT DO I DO IF THERE ARE OTHER INDEPENDENT VARIABLES?

In Chapter 1 we saw in our discussion of calorimetry the repetitive and natural occurrence of $(U+pV)$, which is given the symbol H, and known as the enthalpy defined by:

$$H = U + pV. \tag{3.6}$$

When changes in H are measured rather than the changes in U then Equation 3.1 can be rewritten as

$$\mathrm{d}H = \mathrm{d}U + p\,\mathrm{d}V + V\,\mathrm{d}p, \tag{3.7}$$

and by replacing $\mathrm{d}U$ with Equation 3.1 we obtain

$$\mathrm{d}H = T\,\mathrm{d}S + V\,\mathrm{d}p + \sum_\mathrm{B} \mu_\mathrm{B}\,\mathrm{d}n_\mathrm{B}. \tag{3.8}$$

By analogy with Equation 3.5 it also follows that

$$\left(\frac{\partial H}{\partial t}\right)_{S,p,\Sigma n_\mathrm{B}} < 0. \tag{3.9}$$

According to Equation 3.9 the enthalpy decreases for a system of constant entropy, pressure, and content if anything is happening; Equation 3.9 will be used in Section 3.6.

As we saw in Chapter 1 there are several other thermodynamic quantities that arise when the variables of a particular problem are chosen in particular combinations. For the variables T, V, and n's the term $(U - TS)$ occurs naturally and so often that it is called the Helmholtz function A and is defined by the equation

$$A = U - TS. \tag{3.10}$$

The differential form of Equation 3.10 when combined with Equation 3.1 gives

$$\mathrm{d}A = -S\,\mathrm{d}T - p\,\mathrm{d}V + \sum_{\mathrm{B}} \mu_{\mathrm{B}}\,\mathrm{d}n_{\mathrm{B}}. \tag{3.11}$$

The inequality for a system of constant temperature, volume, and content is

$$\left(\frac{\partial A}{\partial t}\right)_{T,V,N} < 0. \tag{3.12}$$

That is, when anything happens in a system of constant temperature, volume, and content the Helmholtz function decreases.

Again according to Chapter 1, when the variables are T, p, and n's the combination $(U + pV - TS)$ arises naturally and it is called the Gibbs function G given by

$$G = U + pV - TS, \tag{3.13}$$

and by the use of a similar manipulation to that used for Equation 3.11 we obtain

$$\mathrm{d}G = -S\,\mathrm{d}T + V\,\mathrm{d}p + \sum_{\mathrm{B}} \mu_{\mathrm{B}}\,\mathrm{d}n_{\mathrm{B}}. \tag{3.14}$$

In addition, we find that for a system of constant temperature, pressure, and content

$$\left(\frac{\partial G}{\partial t}\right)_{T,p,\Sigma n_{\mathrm{B}}} < 0, \tag{3.15}$$

when something is happening; if there is some measurable change to a system of constant temperature, pressure, and amount of material then the Gibbs function must decrease. Equation 3.15 will be used in Section 3.6.

From Equation 3.14 the important relationship

$$\left(\frac{\partial G}{\partial n_{\mathrm{A}}}\right)_{T,p,n_{\mathrm{A}} \neq n_{\mathrm{B}}} = \mu_{\mathrm{B}} \tag{3.16}$$

can be obtained together with the definition of a partial molar quantity from Question 1.3.14 of Chapter 1 we find

$$G_{\mathrm{B}} = \left(\frac{\partial G}{\partial n_{\mathrm{A}}}\right)_{T,p,n_{\mathrm{A}}\neq n_{\mathrm{B}}} = \mu_{\mathrm{B}}. \tag{3.17}$$

3.3.1 Is Zero a Characteristic Thermodynamic Function?

Euler's theorem (provided in Chapter 1 Question 1.11.2) can be used to integrate Equations 3.1, 3.8, 3.11, and 3.14 because the T, p, and μ are intensive quantities, while S, V, n, U, H, A, and G are extensive quantities. The integrated form of Equation 3.1 is

$$U = TS - pV + \sum_{\mathrm{B}} n_{\mathrm{B}}\mu_{\mathrm{B}}, \tag{3.18}$$

while for Equation 3.8 it is

$$H = TS + \sum_{\mathrm{B}} n_{\mathrm{B}}\mu_{\mathrm{B}}. \tag{3.19}$$

Integrating Equation 3.11 yields

$$A = -pV + \sum_{\mathrm{B}} n_{\mathrm{B}}\mu_{\mathrm{B}}, \tag{3.20}$$

and for Equation 3.14 we obtain

$$G = \sum_{\mathrm{B}} n_{\mathrm{B}}\mu_{\mathrm{B}}. \tag{3.21}$$

Differentiation of Equation 3.21 gives

$$\mathrm{d}G = \sum_{\mathrm{B}} \mu_{\mathrm{B}}\,\mathrm{d}n_{\mathrm{B}} + \sum_{\mathrm{B}} n_{\mathrm{B}}\,\mathrm{d}\mu_{\mathrm{B}}. \tag{3.22}$$

Subtracting Equation 3.14 from Equation 3.22 yields

$$0 = S\,\mathrm{d}T - V\,\mathrm{d}p + \sum_{\mathrm{B}} n_{\mathrm{B}}\,\mathrm{d}\mu_{\mathrm{B}}. \tag{3.23}$$

Equation 3.23 is the Gibbs–Duhem equation and has 0 as the characteristic value. When Equation 3.23 is divided by the total amount of substance $\Sigma_{\mathrm{B}} n_{\mathrm{B}}$ we obtain

$$0 = S_{\mathrm{m}}\mathrm{d}T - V_{\mathrm{m}}\mathrm{d}p + \sum_{\mathrm{B}} x_{\mathrm{B}}\,\mathrm{d}\mu_{\mathrm{B}}. \tag{3.24}$$

For a phase at constant temperature and pressure Equation 3.23 can be written as

$$0 = \sum_{B} n_B \, d\mu_B. \tag{3.25}$$

The Gibbs–Duhem equation is particularly useful for treating phase equilibrium: it provides the first step in determining if some results are not thermodynamically consistent and a method for calculating the chemical potential differences or activity coefficients for a system from measurements.

The absolute activity λ_B of a substance B is another useful quantity that is defined in terms of the chemical potential by

$$\lambda_B = \exp\left(\frac{\mu_B}{RT}\right). \tag{3.26}$$

Equation 3.23 can be recast with Equations 3.6, 3.13, 3.21, and 3.26 as

$$0 = \left(\frac{H}{RT^2}\right) dT - \left(\frac{V}{RT}\right) dp + \sum_{B} n_B \, d\ln\lambda_B, \tag{3.27}$$

or

$$0 = \left(\frac{H_m}{RT^2}\right) dT - \left(\frac{V_m}{RT}\right) dp + \sum_{B} x_B \, d\ln\lambda_B. \tag{3.28}$$

3.4 WHAT HAPPENS WHEN THERE IS A CHEMICAL REACTION?

For a closed phase (see Question 1.3.4) the amount of substance of a species B n_B can only change if the extent of one or more chemical reactions changes. For a chemical reaction defined by

$$0 = \sum_{B} \nu_B B, \tag{3.29}$$

where ν is the stoichiometric number, the extent of reaction ξ for each B is defined by

$$dn_B = \nu_B \, d\xi. \tag{3.30}$$

It then follows that Equation 3.1 can be recast as

$$dU = T\,dS - p\,dV - A\,d\xi, \tag{3.31}$$

where we define the affinity $\boldsymbol{A}$ for the reaction of Equation 3.29 as

$$\boldsymbol{A} = -\sum_{\mathrm{B}} \nu_{\mathrm{B}} \mu_{\mathrm{B}}. \tag{3.32}$$

For an isolated phase for which $\mathrm{d}U = 0$ and $\mathrm{d}V = 0$ Equation 3.1 can be recast in view of Equation 3.31 as

$$T\,\mathrm{d}S = \boldsymbol{A}\,\mathrm{d}\xi. \tag{3.33}$$

According to Equation 3.2 if anything is happening in the system $\boldsymbol{A} > 0$ and $\mathrm{d}\xi/\mathrm{d}t > 0$ and the extent of reaction is increasing, while if $\boldsymbol{A} < 0$ and $\mathrm{d}\xi/\mathrm{d}t < 0$ the reaction of Equation 3.29 is reversing. If $\boldsymbol{A} = 0$ and $\mathrm{d}\xi/\mathrm{d}t = 0$ then $\sum_{\mathrm{B}} \nu_{\mathrm{B}} \mu_{\mathrm{B}} = 0$ and the system is in chemical equilibrium. We have here for the first time introduced one use of Equation 3.2 and will return to discuss further uses in Question 3.6.

The affinity $\boldsymbol{A}$ can be measured from differences in chemical potential for the species between the reaction conditions and an equilibrium state for which $\sum_{\mathrm{B}} \nu_{\mathrm{B}} \mu_{\mathrm{B}}^{\mathrm{eq}} = 0$ using

$$\boldsymbol{A} = -\sum_{\mathrm{B}} \nu_{\mathrm{B}} (\mu_{\mathrm{B}} - \mu_{\mathrm{B}}^{\mathrm{eq}}). \tag{3.34}$$

Equations 3.8, 3.11, and 3.14 can be written as follows:

$$\mathrm{d}H = T\,\mathrm{d}S + V\,\mathrm{d}p - \boldsymbol{A}\,\mathrm{d}\xi, \tag{3.35}$$

$$\mathrm{d}A = -S\,\mathrm{d}T - p\,\mathrm{d}V - \boldsymbol{A}\,\mathrm{d}\xi, \quad \text{and} \tag{3.36}$$

$$\mathrm{d}G = -S\,\mathrm{d}T + V\,\mathrm{d}p - \boldsymbol{A}\,\mathrm{d}\xi. \tag{3.37}$$

From Equations 3.31, 3.35, 3.36, and 3.37 the following series of equivalent expressions for $\boldsymbol{A}$ can be derived:

$$\begin{aligned} \boldsymbol{A} = -\sum_{\mathrm{B}} \nu_{\mathrm{B}} \mu_{\mathrm{B}} &= T\left(\frac{\partial S}{\partial \xi}\right)_{U,V} = T\left(\frac{\partial S}{\partial \xi}\right)_{H,p}, \\ &= -\left(\frac{\partial U}{\partial \xi}\right)_{S,V} = -\left(\frac{\partial H}{\partial \xi}\right)_{S,p}, \quad \text{and} \\ &= -\left(\frac{\partial A}{\partial \xi}\right)_{T,V} = -\left(\frac{\partial G}{\partial \xi}\right)_{T,p}. \end{aligned} \tag{3.38}$$

Since most chemical reactions are studied at constant temperature (provided they are substantially neither exo- nor endothermic) and constant pressure (as they mostly are) then the most important form of Equation 3.38 is

$$\boldsymbol{A} = -\left(\frac{\partial G}{\partial \xi}\right)_{T,p}. \tag{3.39}$$

The change of entropy with respect to composition arising from mixing $\Delta_{\text{mix}}S$ or a chemical reaction $\Delta_{\text{r}}S$ are given by

$$\Delta_{\text{mix}}S = \frac{\Delta_{\text{mix}}H - \Delta_{\text{mix}}G}{T} = -\left(\frac{\partial \Delta_{\text{mix}}G}{\partial T}\right)_p, \tag{3.40}$$

and

$$\Delta_{\text{r}}S = \frac{\Delta_{\text{r}}H - \Delta_{\text{r}}G}{T} = -\left(\frac{\partial \Delta_{\text{r}}G}{\partial T}\right)_p, \tag{3.41}$$

respectively. Equations 3.40 and 3.41 will be used in Chapters 4 and 5, respectively.

3.5 WHAT AM I ABLE TO DO KNOWING LAW 2a?

As a result of the formulations that flow from Equation 3.1 we are able to do a number of things that prove useful in an engineering and experimental context. In this section we illustrate some of these applications in the field of thermodynamics and the measurement of properties for, both, fluid systems and solids. We use the speed of sound as an example of a property of a material that can be measured with great precision by modern means and relate it to the thermodynamic properties of fluids and solids.

3.5.1 How Do I Calculate Entropy, Gibbs Function, and Enthalpy Changes?

The second law provides relationships that permit the determination of the dependence on pressure of the Gibbs function, entropy, and enthalpy for a non-reacting material of constant composition. For example, the two partial derivatives of Equation 3.14 can be written for a constant composition as

$$\left(\frac{\partial G}{\partial T}\right)_p = -S, \tag{3.42}$$

and

$$\left(\frac{\partial G}{\partial p}\right)_T = V. \tag{3.43}$$

Equation 3.43 provides a means of determining differences in Gibbs function arising from a pressure change through

$$G(T_1, p_2) - G(T_1, p_1) = \int_{p_1}^{p_2} V\,\mathrm{d}p. \tag{3.44}$$

The definition

$$G = H - TS, \tag{3.45}$$

can be recast with Equation 3.42 to be

$$H = G - T\left(\frac{\partial G}{\partial T}\right)_p, \tag{3.46}$$

and is called the *Gibbs–Helmholtz equation*; $U = A - T(\partial A/\partial T)_V$ is also unfortunately referred to as the Gibbs–Helmholtz equation.

Differentiation of Equation 3.42 with respect to p at constant T gives

$$\left\{\frac{\partial(\partial G/\partial T)_p}{\partial p}\right\}_T = -\left(\frac{\partial S}{\partial p}\right)_T, \tag{3.47}$$

and differentiation of Equation 3.43 with respect to T at constant p gives

$$\left\{\frac{\partial(\partial G/\partial p)_T}{\partial T}\right\}_p = \left(\frac{\partial V}{\partial T}\right)_p. \tag{3.48}$$

Using the rule of cross-differentiation from Question 1.11.1 on Equations 3.47 and 3.48 gives

$$\left(\frac{\partial S}{\partial p}\right)_T = -\left(\frac{\partial V}{\partial T}\right)_p, \tag{3.49}$$

which is called a *Maxwell equation*. Integration of Equation 3.49 gives

$$S(T_1, p_2) - S(T_1, p_1) = -\int_{p_1}^{p_2} \left(\frac{\partial V}{\partial T}\right)_p \mathrm{d}p. \tag{3.50}$$

Thus, measurements of V as a function of temperatures over a range around T_1 at pressures from p_1 to p_2 yield values of $(\partial V/\partial T)_p$ over the pressure range p_1 to p_2. The entropy difference of Equation 3.50 is then determined from the area beneath a plot of $(\partial V/\partial T)_p$ on the ordinate as a function of p on the abscissa. Another form of the Maxwell equation can be obtained by the same procedure starting with Equation 3.11 at constant composition to give

$$\left(\frac{\partial S}{\partial V}\right)_T = \left(\frac{\partial p}{\partial T}\right)_V. \tag{3.51}$$

Equation 3.51 can be applied to determine entropy changes with respect to volume at constant temperature.

For variations of temperature at constant pressure the Gibbs function is given by Equation 3.42 as

$$G(T_2, p_1) - G(T_1, p_1) = -\int_{T_1}^{T_2} S\,\mathrm{d}T, \tag{3.52}$$

but because only differences in entropy can be measured and not absolute values Equation 3.52 is of no practical value. The enthalpy difference associated with a temperature change can be measured with a flow calorimeter but there is another way to obtain the same information.

Differentiating

$$H = G + TS, \tag{3.53}$$

with respect to pressure at constant temperature gives

$$\left(\frac{\partial H}{\partial p}\right)_T = \left(\frac{\partial G}{\partial p}\right)_T + T\left(\frac{\partial S}{\partial p}\right)_T, \tag{3.54}$$

Substitution of Equations 3.43 and 3.49 into Equation 3.54 gives

$$\left(\frac{\partial H}{\partial p}\right)_T = V - T\left(\frac{\partial V}{\partial T}\right)_p, \tag{3.55}$$

and integration gives

$$H(T_1, p_2) - H(T_1, p_1) = \int_{p_1}^{p_2} \left\{ V - T\left(\frac{\partial V}{\partial T}\right)_p \right\} \mathrm{d}p. \tag{3.56}$$

We have seen, for example, in Chapter 1 how the left hand side of Equation 3.56 can be determined with a flow calorimeter in Question 1.8.7, while the right hand can be estimated from direct measurements of p, V, and T. Thus, one is able to either perform measurements that confirm the thermodynamic consistency expressed in Equation 3.56 or determine one unknown quantity given the knowledge of others.

Differentiation of Equation 3.46, the Gibbs–Helmholtz equation, with respect to temperature at constant pressure leads to the result

$$\left(\frac{\partial H}{\partial T}\right)_p = -T\left(\frac{\partial^2 G}{\partial T^2}\right)_p, \tag{3.57}$$

and with Equation 3.42 becomes

$$\left(\frac{\partial S}{\partial T}\right)_p = \frac{(\partial H/\partial T)_p}{T} = \frac{C_p}{T}. \tag{3.58}$$

Equation 3.58 provides the basis for a calorimetric method for the measurement of entropy difference because

$$S(T_2, p) - S(T_1, p) = \int_{T_1}^{T_2} \left(\frac{C_p}{T}\right) \mathrm{d}T. \tag{3.59}$$

The variation of entropy with pressure for a phase of fixed composition is given by Equation 3.50 combined with Equation 3.59 to obtain the dependence of entropy on both T and p from

$$S(T_2, p_2) - S(T_1, p_1) = \int_{T_1}^{T_2} \left(\frac{C_p}{T}\right) \mathrm{d}T + \int_{p_1}^{p_2} \left(\frac{\partial V}{\partial T}\right)_p \mathrm{d}p. \tag{3.60}$$

Expressions for the change in chemical potential with respect to pressure $\{\mu_B(T_1, p_2) - \mu_B(T_1, p_1)\}$ and the change in chemical potential with respect to temperature $\{\mu_B(T_2, p_1) - \mu_B(T_1, p_1)\}$ illustrate how these differences will now be measured. From Equations 3.17, 3.43, 1.7, and the rule given by Equation 1.143 we obtain

$$\left(\frac{\partial \mu_B}{\partial p}\right)_T = \left\{\frac{\partial(\partial G/\partial n_B)_{T,p.n_A \neq n_B}}{\partial p}\right\}_T = \left\{\frac{\partial(\partial G/\partial p)_T}{\partial n_B}\right\}_{T,p.n_A \neq n_B}$$

$$= \left(\frac{\partial V}{\partial n_B}\right)_{T,p.n_A \neq n_B} = V_B, \tag{3.61}$$

where V_B is the partial molar volume of B. Integration of Equation 3.61 then gives

$$\mu_B(T_1, p_2) - \mu_B(T_1, p_1) = \int_{p_1}^{p_2} V_B \, \mathrm{d}p. \tag{3.62}$$

From Equation 3.26

$$\left(\frac{\partial \ln \lambda_B}{\partial p}\right)_T = \frac{V_B}{RT}, \tag{3.63}$$

thus the ratio of absolute activities is

$$\ln\left\{\frac{\lambda_B(T_1, p_2)}{\lambda_B(T_1, p_1)}\right\} = \int_{p_1}^{p_2} \left(\frac{V_B}{RT_1}\right) \mathrm{d}p. \tag{3.64}$$

The difference $\{\mu_B(T_2, p_1) - \mu_B(T_1, p_1)\}$ and ratio ln $\{\lambda_B(T_2, p_1)/\lambda_B(T_1, p_1)\}$ are given by

$$\left(\frac{\partial \mu_B}{\partial T}\right)_p = -S_B, \tag{3.65}$$

or

$$\mu_B(T_2, p_1) - \mu_B(T_1, p_1) = -\int_{T_1}^{T_2} S_B \, dT, \tag{3.66}$$

and

$$\left(\frac{\partial \ln \lambda_B}{\partial T}\right)_p = -\frac{H_B}{RT^2}, \tag{3.67}$$

or

$$\ln\left\{\frac{\lambda_B(T_2, p_1)}{\lambda_B(T_1, p_1)}\right\} = -\int_{T_1}^{T_2} \left(\frac{H_B}{RT^2}\right) dT, \tag{3.68}$$

respectively.

The routes to obtain the difference $\{\mu_B(T_2, p_1) - \mu_B(T_1, p_1)\}$ and ratio ln $\{\lambda_B(T_2, p_1)/\lambda_B(T_1, p_1)\}$ provided by Equations 3.66 and 3.68 are of no immediate use because neither S_B nor H_B can be measured, however, the equations themselves are useful as we shall see in Chapter 4.

3.5.2 How Do I Calculate Expansivity and Compressibility?

The isobaric (constant pressure) expansivity or coefficient of thermal expansion α is defined by

$$\alpha = \frac{1}{V}\left(\frac{\partial V}{\partial T}\right)_p = \left(\frac{\partial \ln V}{\partial T}\right)_p. \tag{3.69}$$

α is usually positive but for water at temperatures between 273.15 K and 277.13 K it is negative. The isothermal compressibility κ_T is defined by

$$\kappa_T = -\frac{1}{V}\left(\frac{\partial V}{\partial p}\right)_T = -\left(\frac{\partial \ln V}{\partial p}\right)_T, \tag{3.70}$$

and by the −1 rule (provided by Equation 1.142) is related to α by

$$\frac{\alpha}{\kappa_T} = \left(\frac{\partial p}{\partial T}\right)_V. \tag{3.71}$$

The isentropic (constant entropy) compressibility κ_S is defined by

$$\kappa_S = -\frac{1}{V}\left(\frac{\partial V}{\partial p}\right)_S . \tag{3.72}$$

The difference between $(\partial V/\partial p)_T$ and $(\partial V/\partial p)_S$ can be written using the rule for changing a variable held constant (given by Equation 1.141) as

$$\left(\frac{\partial V}{\partial p}\right)_T - \left(\frac{\partial V}{\partial p}\right)_S = -\left(\frac{\partial V}{\partial T}\right)_p \left(\frac{\partial T}{\partial p}\right)_S . \tag{3.73}$$

Using −1 rule (Equation 1.142) on $(\partial T/\partial p)_S$, Equation 3.73 becomes

$$\left(\frac{\partial V}{\partial p}\right)_T - \left(\frac{\partial V}{\partial p}\right)_S = \frac{(\partial V/\partial T)_p(\partial S/\partial p)_T}{(\partial S/\partial T)_p} . \tag{3.74}$$

Substituting Equations 3.49 and 3.58 into Equation 3.74 gives

$$\left(\frac{\partial V}{\partial p}\right)_T - \left(\frac{\partial V}{\partial p}\right)_S = -\frac{T\{(\partial V/\partial T)_p\}^2}{C_p}, \tag{3.75}$$

that with the definitions of Equations 3.69, 3.70, and 3.72 gives

$$\kappa_T - \kappa_S = \frac{T\alpha^2 V}{C_p} . \tag{3.76}$$

If we had independent means of measuring κ_S, κ_T, α, T, V, and C_p of a phase then Equation 3.76 could be used to test the measurements for thermodynamic consistency. If one parameter of Equation 3.76 cannot or has not been measured then it can be calculated from measurements of the others from the same equation. Because α^2, T, V, and C_p are all positive from Equation 3.76, κ_T is always greater then κ_S.

For a closed phase of fixed composition, for which Equation 3.1 becomes

$$dU = T\,dS - p\,dV, \tag{3.77}$$

and with the first law given by

$$dU = \delta Q + \delta W, \tag{3.78}$$

then Equation 3.1 can be recast as

$$T\,dS = \delta Q + (\delta W + p\,dV), \tag{3.79}$$

and it follows from Equation 3.79 that if the process is adiabatic, so that $\delta Q = 0$ and reversible so that $\delta W = -p\,dV$ it must also be isentropic $dS = 0$. This process can be realized for a closed phase of fixed composition. An expansion (or compression) of a known volume of fluid in a thermally insulated vessel can be achieved adiabatically by quickly changing the pressure and remeasuring the pressure and volume. The expansion can also be performed reversibly by changing the pressure slowly, and provided the thermal insulation is good then the process is both adiabatic and reversible and so isentropic. This is easily achieved particularly with liquids to provide direct measurements of κ_S. For fluids no one quantity in Equation 3.76 is more difficult to measure than the other but for solids κ_T is hard to determine and is therefore obtained from Equation 3.76 from measurements of κ_S, α, T, V, and C_p. Measurements of the speed of sound are used to determine κ_S, which will be discussed in Question 3.5.3, and α is determined from the temperature dependence of the lattice constant by X-ray diffraction. We will return to expansion in Question 3.5.5 and 3.5.6. Expansion and compression were considered in Question 1.7.4 through 1.7.6.

3.5.3 What Can I Gain from Measuring the Speed of Sound in Fluids?

While the speed of sound in a phase is important in its own right in a number of applications, most of the interest in this quantity arises from its relation with the thermodynamic properties of isotropic, Newtonian fluids and isotropic elastic solids. For fluid phases, as these usually support only a single longitudinal sound mode, the sound propagation speed u is given by (Herzfeld and Litovitz 1959)

$$u^2 = \left(\frac{\partial p}{\partial \rho}\right)_s = \frac{1}{\rho \kappa_S} = \frac{\gamma}{\rho \kappa_T}. \qquad (3.80)$$

In Equation 3.80 all the symbols have been previously defined, including $\gamma = C_p/C_V$, which is now given in form of C_p and C_V, the molar isobaric and isochoric heat capacities, respectively (see Question 1.7.6). Equation 3.80 is strictly valid only in the limits of vanishing amplitude and vanishing frequency (Herzfeld and Litovitz 1959; Morse and Ingard 1968; Goodwin and Trusler 2003) of the sound wave. The situation corresponding to the first of these limits is extremely easy to approach in practice, while that corresponding to the second is usually, but not always, realized. Equation 3.80 shows that the isentropic compressibility may be obtained from measurements of the speed of sound and the density, and that the isothermal compressibility may also be obtained if γ is known. Equation 3.80 forms the basis of almost all experimental determinations of the isentropic compressibility and is a convenient route to γ.

For independent variables of either (T, p) or (T, ρ_n), where ρ_n is the amount-of-substance density (which we here distinguish from the mass density ρ), Equation 3.80 can be recast for (T, p) as

$$u^2 = \frac{1}{M}\left[\left(\frac{\partial \rho_n}{\partial p}\right)_T - \frac{T}{\rho_n^2 C_p}\left(\frac{\partial \rho_n}{\partial T}\right)_p^2\right]^{-1}, \qquad (3.81)$$

and for (T, ρ_n) as

$$u^2 = \frac{1}{M}\left[\left(\frac{\partial p}{\partial \rho_n}\right)_T + \frac{T}{\rho_n^2 C_V}\left(\frac{\partial p}{\partial T}\right)_{\rho_n}^2\right]. \qquad (3.82)$$

In Equations 3.81 and 3.82, M is the molar mass, C_p is the isobaric molar heat capacity, and C_V is the isochoric molar heat capacity; to adhere strictly to the *International Union of Pure and Applied Chemistry* (IUPAC) (Quack et al. 2007) a subscript m should be included to indicate a molar quantity to give $C_{p,\mathrm{m}}$ and $C_{V,\mathrm{m}}$. We have not included them here to preserve simplicity and to avoid confusion.

In principle, these equations allow one to compute the speed of sound u from an equation of state in the form $\rho_n = \rho_n(T, p) = V_\mathrm{m}^{-1}$ or $p = p(T, \rho_n)$, although one requires a knowledge of the heat capacity in some reference state. An equation for C_p can be obtained by differentiation of Equation 3.55 with respect to T at constant p that gives

$$\left\{\frac{\partial(\partial H/\partial p)_T}{\partial T}\right\}_p = -T\left(\frac{\partial^2 V}{\partial T^2}\right)_p, \qquad (3.83)$$

and use of the rule for cross-differentiation from Chapter 1 and in view of Equation 1.142 we obtain

$$\left(\frac{\partial C_p}{\partial p}\right)_T = -T\left(\frac{\partial^2 V}{\partial T^2}\right)_p, \qquad (3.84)$$

which when integrated becomes

$$C_p = C_p^\circ(T) - \int_{p^\circ}^{p} T\left(\frac{\partial^2 \rho_n^{-1}}{\partial T^2}\right) \mathrm{d}p, \qquad (3.85)$$

where C_p° is the molar isobaric specific heat capacity on the reference isobar $p = p^\circ$ and $\rho_n = 1/V_\mathrm{m}$. The analogous expression for C_V is

$$C_V = C_V^\circ(T) - \int_{\rho_n^\circ}^{\rho_n}\left(\frac{T}{\rho_n^2}\right)\left(\frac{\partial^2 p}{\partial T^2}\right) \mathrm{d}\rho_n, \qquad (3.86)$$

where ρ_n^{o} is the amount-of-substance density on a reference isochore (line of constant amount of substance density).

3.5.4 What Can I Gain from Measuring the Speed of Sound in Solids?

The elastic properties of an isotropic solid may be specified by a pair of quantities such as the bulk modulus K and the shear modulus G. Other commonly used parameters are Young's modulus E, Poisson's ratio σ, and the Lamé constants λ and μ. The shear modulus G is identical with the second Lamé constant μ and the other parameters are interrelated as follows (Landau and Lifshitz 1987):

$$\left.\begin{aligned} E &= \frac{9GK}{3K+G} = 3(1-2\sigma)K \\ \sigma &= \frac{3K-2G}{6K+2G} = \frac{E}{2G} - 1 \\ \lambda &= K - \frac{2G}{3} \end{aligned}\right\}. \tag{3.87}$$

The elastic constants relate various types of stress and strain under isothermal conditions. In the case of pure shear stress, the resulting strain takes place without change of volume and so an isothermal and reversible shear process is also isentropic. Consequently, the shear modulus is the same for both static and dynamic processes in an elastic body. However, compressive stress gives rise to a change in volume so that an isothermal compression is not generally isentropic. The isothermal bulk modulus $K = 1/\kappa_T$ therefore differs from the isentropic bulk modulus, which is $K_S = 1/\kappa_S$, and the two are related by Equation 3.76

$$\frac{1}{K_S} = \frac{1}{K} - \frac{T\alpha^2}{\rho c_p}, \tag{3.88}$$

where α is the coefficient of thermal expansivity, ρ the mass density, and c_p is the specific heat capacity. The isentropic analogue E_S of Young's modulus is given by

$$E_S = \frac{E}{\{1 - ET\alpha^2/9\rho c_p\}}. \tag{3.89}$$

Solids generally have both longitudinal or compressive sound modes, in which the direction of stress and strain is parallel to the direction of propagation, and two orthogonal shear or transverse wave modes in each of which the direction of shear stress is perpendicular to the direction of propagation. In an isotropic solid, the two shear modes are degenerate, each propagating with speed u_S given by

$$u_S^2 = \frac{G}{\rho}. \tag{3.90}$$

The speed u_L of longitudinal sound waves in a bulk specimen is given by

$$u_L^2 = \frac{(K_S + 4G/3)}{\rho}. \tag{3.91}$$

The actual phase speed u of a compression wave propagating along the axis of a solid bar generally depends upon the lateral dimension of the bar: u approaches u_L when the lateral extent (the dimension normal to the direction of propagation) of the bar is much greater than the wavelength of the sound. For bars of smaller cross-section, the phase speed u is generally smaller than u_L and, when the lateral dimensions are much smaller than the wavelength, it reaches a limit u_E given by

$$u_E^2 = \frac{E_S}{\rho}, \tag{3.92}$$

in an isotropic elastic solid $u_L > u_E > u_S$.

Acoustic, especially ultrasonic, methods are the most common means of determining the elastic constants of solids. High frequency (that is 10 MHz) ultrasonic measurements typically provide directly values of u_S and u_L (Papadakis 1998). When combined with a measurement of the density, G and K_S may then be determined. The difference between κ_S and κ_T in a solid material is typically very small (<1 %; see Ledbetter 1982) and so the isothermal bulk modulus K is easily obtained from K_S by means of a calculated correction according to Equation 3.88. Very approximate values of α and c_p will suffice for that purpose. Once G and K have been obtained, σ follows from Equation 3.87.

Low frequency (i.e., 100 kHz) ultrasonic resonance experiments typically provide directly values of u_S and u_E from which G and E_S are obtained (Weston 1975), which then permits determination of the isothermal Young's modulus E.

3.5.5 Can I Evaluate the Isobaric Heat Capacity from the Isochoric Heat Capacity?

The difference $(C_p - C_V)$ is given by Equation 1.148 as

$$C_p - C_V = -\frac{T\{(\partial V/\partial T)_p\}^2}{(\partial V/\partial p)_T} = \frac{T\alpha^2 V}{\kappa_T}, \tag{3.93}$$

where Equations 3.69 and 3.70 were used to obtain the second part of the equality. Equation 3.93 is important because it interrelates independently measurable quantities, which can be used to both experimentally verify the laws and permit the use of results for the most easily measured quantity to evaluate the least easily measured. For a perfect gas $pV_m = RT$ and substitution into Equation 3.93 gives $C_p^{pg} - C_V^{pg} = R$.

3.5.6 Why Use an Isentropic Expansion to Liquefy a Gas?

In Question 1.7.6 we discussed with specific quantities the work required for the adiabatic compression of a gas. Here we consider a fast, but not explosively fast (for which the pressure is neither uniform nor defined and the expansion irreversible) expansion or compression in a thermally insulated vessel. This process is isentropic and the temperature varies with pressure according to

$$\left(\frac{\partial T}{\partial p}\right)_S = -\frac{(\partial S/\partial p)_T}{(\partial S/\partial T)_p} = \frac{T(\partial V/\partial T)_p}{C_p} = \frac{T\alpha V}{C_p}. \tag{3.94}$$

For a perfect gas, integration of Equation 3.94 gives

$$T_2 = T_1\left(\frac{p_2}{p_1}\right)^{R/C_p}. \tag{3.95}$$

Question 1.7.6 provides an alternative derivation of Equation 3.95. For a perfect monatomic gas, for example, argon, for which the molar heat capacity is $C_p = 5R/2$ then starting at $T_1 = 300$ K, the expansion from $p_1 = 3.2$ MPa to $p_2 = 0.1$ MPa yields a final temperature $T_2 = 75$ K; this is a vivid explanation of why isentropic expansion is used to liquefy so-called permanent gases. The converse of this expansion, namely the isentropic compression of air leads to a temperature increase familiar to those who inflate their bicycle tires with a hand pump.

3.5.7 Does Expansion of a Gas at Constant Energy Change Its Temperature?

The dependence of the temperature of a gas on its volume at constant energy (Joule 1845) $(\partial T/\partial V)_U$ can be obtained from the −1 rule (Equation 1.142) as

$$\left(\frac{\partial T}{\partial V}\right)_U = -\frac{(\partial U/\partial V)_T}{(\partial U/\partial T)_V}. \tag{3.96}$$

The numerator of Equation 3.96 can be obtained from Equation 3.1 as

$$\left(\frac{\partial U}{\partial V}\right)_T = T\left(\frac{\partial S}{\partial V}\right)_T - p. \tag{3.97}$$

Substitution of Equation 3.51 into Equation 3.97, and using Equation 1.89 $\{(\partial U/\partial T)_V = C_V\}$, Equation 3.96 becomes

$$\left(\frac{\partial T}{\partial V}\right)_U = \frac{p - T(\partial p/\partial T)_V}{C_V}. \quad (3.98)$$

Substitution of Equations 3.71 and 3.93 into Equation 3.98 gives

$$\left(\frac{\partial T}{\partial V}\right)_U = \frac{p - T\alpha/\kappa_T}{C_p - T\alpha^2 V/\kappa_T}. \quad (3.99)$$

An expression for the rate of change of temperature with respect to pressure at constant energy $(\partial T/\partial p)_U$ can also be obtained by first using the −1 rule (Equation 1.142) to give

$$\left(\frac{\partial T}{\partial p}\right)_U = -\frac{(\partial U/\partial p)_T}{(\partial U/\partial T)_p}. \quad (3.100)$$

To find an expression for $(\partial U/\partial T)_p$ requires the use of the rule of change of variable held constant (Equation 1.141) to give

$$\left(\frac{\partial U}{\partial T}\right)_p = \left(\frac{\partial U}{\partial T}\right)_V + \left(\frac{\partial U}{\partial V}\right)_T \left(\frac{\partial V}{\partial T}\right)_p. \quad (3.101)$$

The first derivative on the right hand side of Equation 3.101 is defined by Equation 1.89, the second term requires Equation 3.97 for $(\partial U/\partial V)_T$ and Equation 3.69 to give

$$\left(\frac{\partial U}{\partial T}\right)_p = C_V + \left(\frac{T\alpha}{\kappa_T} - p\right)\alpha V = C_p - p\alpha V, \quad (3.102)$$

where Equation 3.93 has been used. The next requirement to solve Equation 3.100 is to find an expression for $(\partial U/\partial p)_T$. Combining Equations 3.6, 3.10, and 3.13 gives

$$U = H - G + A, \quad (3.103)$$

which when differentiated with respect to p at constant T gives

$$\left(\frac{\partial U}{\partial p}\right)_T = \left(\frac{\partial H}{\partial p}\right)_T - \left(\frac{\partial G}{\partial p}\right)_T + \left(\frac{\partial A}{\partial p}\right)_T. \quad (3.104)$$

The first derivative on the right hand side of Equation 3.104 is given by Equation 3.55 and the second derivative is given by Equation 3.43. However, an expression is required for $(\partial A/\partial p)_T$ that can be obtained by differentiation of

$$A = G - pV, \quad (3.105)$$

with respect to pressure at constant temperature to give

$$\left(\frac{\partial A}{\partial p}\right)_T = \left(\frac{\partial G}{\partial p}\right)_T - V - p\left(\frac{\partial V}{\partial p}\right)_T = pV\kappa_T. \tag{3.106}$$

Equations 3.43 and 3.70 were also substituted to obtain the second equality of Equation 3.106. Substitution of Equations 3.43, 3.55, and 3.106 into Equation 3.104, using Equations 3.69 and 3.70 gives

$$\left(\frac{\partial U}{\partial p}\right)_T = V - T\left(\frac{\partial V}{\partial T}\right)_p - V + pV\kappa_T = V(p\kappa_T - \alpha T). \tag{3.107}$$

Finally, substitution of Equations 3.102 and 3.107 into Equation 3.100 gives

$$\left(\frac{\partial T}{\partial p}\right)_U = \frac{T\alpha V - p\kappa_T V}{C_p - p\alpha V}. \tag{3.108}$$

For a perfect gas it is readily shown that $(\partial T/\partial V)_U = 0$ and also $(\partial U/\partial V)_T = 0$, $(\partial U/\partial p)_T = 0$, $(\partial H/\partial p)_T = 0$, and $(\partial H/\partial V)_T = 0$, so that changes of volume produce no temperature changes. Joule carried out measurements of the effect of an expansion at constant energy on the real gas air; he employed a thermometer of uncertainty ±0.01 K and observed no temperature change. However, a modern thermometer with uncertainty of $< \pm 0.001$ K would reveal a temperature decrease of 0.003 K under the conditions he employed because for N_2(g) at a temperature of 293 K and pressure of 0.1 MPa $(\partial T/\partial p)_U \approx 0.003\ \mathrm{K \cdot kPa^{-1}}$. The effects are evidently very small relative to the large heat capacity of any practical container and the method is not used to study the properties of gases. It should be remarked in the context of the discussions in Chapter 1 that the origin of the small temperature change in a real gas on expansion arises from the work done against the intermolecular potential at the expense of the kinetic energy.

3.5.8 What Is a Joule-Thomson Expansion?

In a Joule-Thomson isenthalpic expansion the temperature difference across a throttle or porous plug is measured as gas flows through the device from a higher pressure to a lower pressure. In the limit of very low pressures this experiment serves to determine the Joule-Thomson coefficient μ_{JT} defined by

$$\mu_{\mathrm{JT}} = \left(\frac{\partial T}{\partial p}\right)_H = -\frac{V - T(\partial V/\partial T)_p}{C_p}, \tag{3.109}$$

which is zero for a perfect gas. As was indicated in Question 1.8.7 an alternative experiment is possible in which the gas exiting the porous plug is heated and restored to the inlet temperature by the application of electrical power. In that case the experiment is isothermal and the change of enthalpy under those conditions in the limit of low pressure is defined as the isothermal Joule-Thomson coefficient ϕ_{JT} defined by

$$\phi_{\mathrm{JT}} = \left(\frac{\partial H}{\partial p}\right)_T = V - T\left(\frac{\partial V}{\partial T}\right)_p . \tag{3.110}$$

For a perfect gas Equation 3.110 is zero.

3.6 WHAT AM I ABLE TO DO KNOWING LAW 2b?

3.6.1 How Are Thermal Equilibrium and Stability Ensured?

For an isolated system consisting of two phases α and β, of temperature T^{α} and T^{β}, separated by a rigid impermeable diathermic wall we have

$$\mathrm{d}U^{\alpha} + \mathrm{d}U^{\beta} = 0, \tag{3.111}$$

and $\mathrm{d}V^{\alpha} = 0$, $\mathrm{d}V^{\beta} = 0$, $\mathrm{d}n^{\alpha} = 0$, and $\mathrm{d}n^{\beta} = 0$ for all substances. From Equation 3.1

$$\mathrm{d}S^{\alpha} = \frac{\mathrm{d}U^{\alpha}}{T^{\alpha}} \quad \text{and} \quad \mathrm{d}S^{\beta} = \frac{\mathrm{d}U^{\beta}}{T^{\beta}}, \tag{3.112}$$

and the sum of the entropy for both phases is

$$\mathrm{d}S = \left(\frac{\mathrm{d}U^{\alpha}}{T^{\alpha}}\right) + \left(\frac{\mathrm{d}U^{\beta}}{T^{\beta}}\right) = \mathrm{d}U^{\alpha}\left(\frac{T^{\beta} - T^{\alpha}}{T^{\beta}T^{\alpha}}\right). \tag{3.113}$$

If $T^{\beta} > T^{\alpha}$ and a process is underway (something is happening) then from inequality given by Equation 3.2, which states $(\partial S/\partial t)_{U,V,\Sigma n} > 0$, it follows that $\mathrm{d}U^{\alpha}/\mathrm{d}t > 0$ from Equation 3.113 and $\mathrm{d}U^{\beta}/\mathrm{d}t < 0$ from Equation 3.111. Thus the energy flows from the higher temperature phase to the lower temperature one. If $T^{\alpha} = T^{\beta}$ then nothing is happening and the system is in thermal equilibrium and $(\partial S/\partial t)_{U,V,\Sigma n} = 0$. This is an important statement because it is the reassurance that the temperature is the thermodynamic temperature.

The condition that ensures the thermal stability of an isolated phase can be determined by considering a phase that is initially of energy $2U$ and volume $2V$. This phase is then divided in two with one part of energy $(U + \delta U)$ and volume V and the other part of energy $(U - \delta U)$ and also volume V. The entropy change is given by

$$\delta S = S(U + \delta U, V) + S(U - \delta U, V) - S(2U, 2V). \tag{3.114}$$

Use of Taylor's theorem Equation 1.161 on Equation 3.114 gives

$$\delta S = \left(\frac{\partial^2 S}{\partial U^2}\right)_V (\delta U)^2 + \mathrm{O}(\delta U)^4, \tag{3.115}$$

where $\mathrm{O}(\delta U)^4$ means terms $(\delta U)^4$ and higher powers of δU. From Equation 3.3 we find $(\partial S/\partial U)_V = 1/T$ and by further substitution of Equation 1.89 we find that Equation 3.115 becomes

$$\left(\frac{\partial^2 S}{\partial U^2}\right)_V = \left(\frac{\partial T^{-1}}{\partial U}\right)_V = -\frac{1}{TC_V}. \tag{3.116}$$

From Equation 3.116 and 3.115 δS can only be positive and permit the proposed change if C_V is negative. The condition of thermal stability of a phase is

$$C_V > 0. \tag{3.117}$$

As a result, for example, the temperature of an isolated metal bar which is initially in thermal equilibrium will not spontaneously increase in temperature at one end, while the temperature at the other decreases. If energy is added to a phase of constant volume and fixed composition the temperature always increases.

3.6.2 How Are Mechanical Equilibrium and Stability Ensured?

For an isolated system at uniform temperature T, consisting of two phases α and β separated by a moveable impermeable diathermic wall, where the pressures of the phases are p^α and p^β, we have

$$\mathrm{d}U^\alpha + \mathrm{d}U^\beta = 0 \tag{3.118}$$

and

$$\mathrm{d}V^\alpha + \mathrm{d}V^\beta = 0, \tag{3.119}$$

$\mathrm{d}n^\alpha = 0$ and $\mathrm{d}n^\beta = 0$ for all substances. From Equation 3.1

$$T\,\mathrm{d}S^\alpha = \mathrm{d}U^\alpha + p^\alpha \mathrm{d}V^\alpha \quad \text{and} \quad T\,\mathrm{d}S^\beta = \mathrm{d}U^\beta + p^\beta \mathrm{d}V^\beta, \tag{3.120}$$

and the sum of the entropy for both phases is

$$T\,\mathrm{d}S = (p^\alpha - p^\beta)\,\mathrm{d}V^\alpha. \tag{3.121}$$

If $p^\beta > p^\alpha$ and a process is underway (something is happening) then because of the inequality given by Equation 3.2 $\mathrm{d}V^\alpha/\mathrm{d}t < 0$ and $\mathrm{d}V^\beta/\mathrm{d}t > 0$ and the volume of the phase with the lower pressure decreases and that of the higher pressure increases and the partition moves. If $p^{\alpha} = p^{\beta}$ then nothing is happening and the system is in hydrostatic equilibrium. This is a purely mechanical result derived from the second law.

For an isolated system consisting of a phase separated by a moveable impermeable diathermic wall, with each part initially at the same T and V, the condition that prevents a difference in the pressure of each part can be derived. To do so we assume that the partition moves so that one part has a volume of $(V + \delta V)$ and the other has a volume of $(V - \delta V)$ both at the same temperature T. For variables of T and V the Helmholtz function A should be used, and similarly to Equation 3.114 the change δA is given by

$$\delta A = A(V+\delta V,T) + A(V-\delta V,T) - A(T,2V). \qquad (3.122)$$

Use of Taylor's theorem given by Equation 1.161 on Equation 3.122 gives

$$\delta A = \left(\frac{\partial^2 A}{\partial V^2}\right)_T (\delta V)^2 + O(\delta V)^4. \qquad (3.123)$$

From Equation 3.11 the derivative $(\partial A/\partial V)_T = -p$ so that

$$\left(\frac{\partial^2 A}{\partial V^2}\right)_T = -\left(\frac{\partial p}{\partial V}\right)_T. \qquad (3.124)$$

If the partition were to move, Equations 3.123 and 3.124 show that δA is negative and would require $(\partial p/\partial V)_T > 0$. It follows that the condition that ensures hydrostatic stability of an isolated phase in thermal equilibrium is

$$\left(\frac{\partial p}{\partial V}\right)_T < 0 \qquad (3.125)$$

or

$$-\frac{1}{V}\left(\frac{\partial V}{\partial p}\right)_T = \kappa_T > 0. \qquad (3.126)$$

so that when the pressure of a phase of fixed composition is increased at constant temperature the volume always decreases.

3.6.3 How Are Diffusive Equilibrium and Stability Ensured?

When we consider an isolated system of uniform temperature T with two phases α and β separated by a rigid diathermic wall permeable to substance A of chemical potential μ_A^{α} and μ_A^{β}, the relation

$$dU^{\alpha} + dU^{\beta} = 0, \qquad (3.127)$$

holds, and $dV^{\alpha} = 0$, $dV^{\beta} = 0$, $dn_A^{\alpha} + dn_A^{\beta} = 0$, and $dn_{B \neq A}^{\alpha} = 0$ and $dn_{A \neq B}^{\beta} = 0$. From Equation 3.1

$$T\,dS^{\alpha} = dU^{\alpha} + p^{\alpha}\,dV^{\alpha} - \mu_A^{\alpha} dn_A^{\alpha} \quad \text{and} \quad T\,dS^{\beta} = dU^{\beta} + p^{\beta}\,dV^{\beta} - \mu_A^{\beta}\,dn_A^{\beta}, \tag{3.128}$$

so the entropy for the system is given by

$$T\,dS = (\mu_A^{\beta} - \mu_A^{\alpha})\,dn_A^{\alpha}. \tag{3.129}$$

If $\mu_A^{\beta} > \mu_A^{\alpha}$ and a process is underway (something is happening) then because of Equation 3.2 $dn_A^{\alpha}/dt > 0$ and $dn_A^{\beta}/dt < 0$ so that the substance A is flowing from the phase with higher chemical potential μ_A^{β} to the one with lower chemical potential μ_A^{α}. If $\mu_A^{\alpha} = \mu_A^{\beta}$ then $(\partial S/\partial t)_{U,V,\Sigma n} = 0$ and nothing is happening and the system is said to be in diffusive equilibrium. When $\mu_A^{\alpha} = \mu_A^{\beta}$ for some but not all substances and $p^{\alpha} \neq p^{\beta}$, which is permitted since nothing was stated about pressure, the system is said to be in osmotic equilibrium.

What is the condition that prevents a substance being concentrated in one part of the system and depleted in another? To address this question we propose a system that is initially at temperature T, pressure p, and contains amount of substance $2n_A$ of A and $2n_B$ of B; the number of components can be infinite but for simplicity we have chosen two. As we have before we will assume the phase can be split into two parts both with the same temperature T and pressure p with the following: (1) amount of substance of A of $(n_A + \delta n_A)$ and n_B of B and (2) amount of substance of A of $(n_A - \delta n_A)$ and n_B of B. For variables n, T, and p the Gibbs function G should be used, and the change arising from the proposed movement of substance A is given by

$$\delta G = G(T,p,n_A + \delta n_A, n_B) + G(T,p,n_A - \delta n_A, n_B) - G(T,p,2n_A,2n_B). \tag{3.130}$$

Use of Taylor's theorem given by Equation 1.161 on Equation 3.130 gives

$$\delta G = \left(\frac{\partial^2 G}{\partial n_A^2}\right)_{T,p,n_B} (\delta n_A)^2 + O(\delta n_A)^4. \tag{3.131}$$

From Equation 3.14 we have

$$\left(\frac{\partial^2 G}{\partial n_A^2}\right)_{T,p,n_B} = \left(\frac{\partial \mu_A}{\partial n_A}\right)_{T,p,n_B}. \tag{3.132}$$

The amount of substance A can be greater in one part of a phase if δG is negative and

$$\left(\frac{\partial \mu_A}{\partial n_A}\right)_{T,p,n_B} < 0. \tag{3.133}$$

Thus, for an isolated phase at constant temperature and pressure, diffusional stability requires

$$\left(\frac{\partial \mu_A}{\partial n_A}\right)_{T,p,n_B} > 0, \tag{3.134}$$

and when substance A is added to a phase at constant temperature, pressure, and fixed composition the chemical potential of A must increase.

3.7 IS THERE A 3RD LAW?

Some have argued that there is a 3rd law of thermodynamics that states that the entropy of a system tends to zero as the temperature tends to zero. It is the view of the current authors that there is no formal basis for such a law as the following argument indicates.

Nernst's heat theorem states that as the temperature tends to absolute zero the entropy change for a chemical reaction vanishes:

$$\sum_B \nu_B S_B(T \to 0) = 0. \tag{3.135}$$

This rule is not obeyed for every substance because it relies upon the proposition that as the temperature tends to zero a perfectly ordered solid (p.o.s.) is formed. Not all substances fit this constraint; for example CO, N_2O, NO, and H_2O. However, in the p.o.s case it is true that

$$\sum_B \nu_B S_{B,p.o.s.}(T \to 0) = 0. \tag{3.136}$$

Both Equations 3.135 and 3.136 are independent of pressure and we have simply no way of knowing if $S = 0$ at $T = 0$. For some of the exceptions, but not for supercooled liquids, it may be that Equation 3.135 might become identical to Equation 3.136 if the temperature was made low enough.

Perhaps the greatest use of Nernst's heat theorem can be found by considering the entropy change of a reaction as the temperature tends toward zero. This change can be measured from the sum of the standard molar entropy difference

$$S^{\ominus}_{B,state}(T) - S_{B,s}(T \to 0), \tag{3.137}$$

for each reacting substance and the standard molar entropy change

$$\sum_B \nu_B S^{\ominus}_{B,state}(T), \tag{3.138}$$

for the reaction at a temperature T that is convenient (and often 298.15 K) to give

$$\sum_B \nu_B S_{B,s}(T \to 0) = -\sum_B \nu_B \left\{S^{\ominus}_{B,state}(T) - S_{B,s}(T \to 0)\right\} + \sum_B \nu_B S^{\ominus}_{B,state}(T). \tag{3.139}$$

We have considered the term $\{S^{\ominus}_{\mathrm{B,state}}(T) - S_{\mathrm{B,s}}(T \to 0)\}$ for a pure substance in Equation 3.59 and the corrections to standard pressure and thus standard molar entropy are addressed in Question 1.10 and by Equation 3.50. Molar heat capacity measurements can be made to the lowest realizable temperature and then extrapolated to $T = 0$ by the Debye rule provided in Equation 2.41, that is, $C_{p,\mathrm{m}}$ is proportional to T^3, so that

$$\{S^{\ominus}_{\mathrm{B,s}}(T) - S_{\mathrm{B,s}}(T \to 0)\} = \frac{1}{3} C_{p,\mathrm{m}}(T), \tag{3.140}$$

where T is close to zero. Practically, this means at about 10 K for liquid H_2 and at about 1 K for liquid He.

A method of determining $\sum_{\mathrm{B}} \nu_{\mathrm{B}} S^{\ominus}_{\mathrm{B,state}}(T)$ of Equations 3.138 and 3.139 must be found, and to this we now turn from Equation 1.116 for the standard equilibrium constant and Equation 1.112 for the standard molar entropy and the standard molar enthalpy of Equation 1.113 given by

$$H^{\ominus}_{\mathrm{i}} = \mu^{\ominus}_{\mathrm{i}} - T\frac{\mathrm{d}\mu^{\ominus}_{\mathrm{i}}}{\mathrm{d}T} = -RT^2 \frac{\mathrm{d}\ln \lambda^{\ominus}_{\mathrm{i}}}{\mathrm{d}T}, \tag{3.141}$$

we can write

$$\sum_{\mathrm{i}} \nu_{\mathrm{i}} S^{\ominus}_{\mathrm{i}} = R\ln K^{\ominus} + RT\frac{\mathrm{d}\ln K^{\ominus}}{\mathrm{d}T}, \tag{3.142}$$

and

$$\sum_{\mathrm{i}} \nu_{\mathrm{i}} H^{\ominus}_{\mathrm{i}} = RT^2\frac{\mathrm{d}\ln K^{\ominus}}{\mathrm{d}T}. \tag{3.143}$$

Eliminating $\mathrm{d}\ln K^{\ominus}/\mathrm{d}T$ from Equations 3.142 and 3.143 gives

$$-RT\ln\{K^{\ominus}(T)\} = \sum_{\mathrm{i}} \nu_{\mathrm{i}} H^{\ominus}_{\mathrm{i}}(T) - T\sum_{\mathrm{i}} \nu_{\mathrm{i}} S^{\ominus}_{\mathrm{i}}(T). \tag{3.144}$$

Rearranging Equation 3.144 gives

$$-RT\ln\{K^{\ominus}(T)\} = \Delta H^{\ominus}_{\mathrm{m}}(T) - T\sum_{\mathrm{i}} \nu_{\mathrm{i}}\{S^{\ominus}_{\mathrm{i}}(T) - S_{\mathrm{i}}(s, T \to 0)\}$$

$$- T\sum_{\mathrm{i}} \nu_{\mathrm{i}} S_{\mathrm{i}}(s, T \to 0). \tag{3.145}$$

For a reaction that obeys Nernst's theorem Equation 3.145 becomes

$$-RT\ln\{K^{\ominus}(T)\} = \Delta H^{\ominus}_{\mathrm{m}}(T) - T\sum_{\mathrm{i}} \nu_{\mathrm{i}}\{S^{\ominus}_{\mathrm{i}}(T) - S_{\mathrm{i}}(s, T \to 0)\}, \tag{3.146}$$

which shows that calorimetric measurements may be used to obtain the standard equilibrium constant for a chemical reaction.

As mentioned previously there are examples where Nernst's heat theorem is not obeyed but they are confined to relatively few molecules. Even when the theorem is not obeyed the difference is only about $R\ln 3$ and this means the $K^{\ominus}(T)$ from Equation 3.146 is about a factor of 4 in error. Although this may seem to be a large error it is not as serious in practice as it might appear.

One exception is CO for which, from Equation 2.70, the difference $S^{\ominus}_{\mathrm{m,CO,g}}(81.61\ \mathrm{K})/R - \ln\{g(\varepsilon_{0,\mathrm{CO}})\} = 19.22$. The available calorimetric measurements give $\{S^{\ominus}_{\mathrm{m,CO,g}}(81.61\ \mathrm{K}) - S_{\mathrm{m,CO,s}}\ (T \to 0)\}/R = 18.6$. Thus, the difference $S_{\mathrm{m,CO,s}}(81.61\ \mathrm{K})/R - \ln\{g(\varepsilon_{0,\mathrm{CO}})\} = -0.6 \approx -\ln 2$ and not zero as the Nernst theorem would require; we could replace the subscript CO, s in these expressions with CO(s). Explanations for this difference have been given that include the failure of Nernst's theorem. However, the correct explanation begins with the realization that when the symmetry number of 2 is used rather than 1, as is strictly correct for CO, Nernst's theorem Equation 3.135 is obeyed because there is now an additional term ln 2; the difference disappears. This rather indicates that CO behaves as if it were the symmetric molecule O_2. In CO the implied degeneracy might arise from random arrangements of CO and OC in the solid at low temperature but the precise source of this observation remains obscure.

Equation 3.146 is also useful when written in the form

$$\sum_{\mathrm{i}} \nu_{\mathrm{i}} S_{\mathrm{i,state}}(T) = R\ln\{K^{\ominus}(T)\} + \frac{\Delta H^{\ominus}_{\mathrm{m}}(T)}{T}, \qquad (3.147)$$

to test Nernst's theorem from calorimetric measurement of $\Delta H^{\ominus}_{\mathrm{m}}(T)$ and direct determination of the standard equilibrium constant $K^{\ominus}(T)$.

Nernst's heat theorem is still not well understood on a statistical basis. Only such an understanding could form the foundations of a third law: formulae derived from statistical mechanics that cannot otherwise be obtained from the zeroth, 1st, or 2nd laws of classical thermodynamics. In particular, it is possible to obtain expressions for entropy changes in disperse systems, such as gases, at $T \to 0$ and for the mixing of similar compounds such as isotopes (Guggenheim 1967; Münster 1974). Nernst's heat theorem cannot be a law because it is not universally obeyed and so there is no third law. Equation 3.147 is an important test of Nernst's heat theorem. It is with these comments that we state there is no third law of thermodynamics independently of statistical mechanical arguments that also includes Nernst's heat theorem.

3.8 HOW IS THE 2ND LAW CONNECTED TO THE EFFICIENCY OF A HEAT ENGINE?

We know from everyday life that not all forms of energy are of equal "utility". For example, electrical or mechanical work (from a resistance heater or a stirrer) can—without restriction—be used to increase the internal energy of a water reservoir. It is not possible, however, to transform all the energy stored within a warm water reservoir into useful work. This statement holds for all processes. In a Clausius-Rankine process as an example of a heat engine discussed in Chapter 6, the heat provided by the combustion of coal can only partially be converted into work. The basic principle of a heat engine is depicted in Figure 3.1: though part of the heat Q from a heat source at temperature T may be used in the form of (shaft) work W_s, the rest of the heat Q_0 is rejected to the surroundings at temperature T_0.

As we will detail in Question 6.3 the thermal efficiency η_{th} of a heat engine quantifies just to what extent the conversion of heat to work can be performed, it is defined as the ratio between the net work output and the heat input and may be rewritten in the form

$$\eta_{th} = 1 - \frac{|Q_0|}{Q}. \tag{3.148}$$

The thermal efficiency can never take a value of one, as always some heat has to be rejected and it is this that is a consequence of the second law.

The underlying reason is that the system has to "get rid" of the entropy that is inevitably provided with the heat added. Assuming that the heat is added at constant system temperature T for simplicity, the entropy provided may be quantified as $\Delta S = Q/T$. Because in any cyclic process each state variable has to take up the same value after the cycle as in the beginning, that is no entropy can be accumulated, the only way to do so is to discharge this additional entropy in the process by rejecting heat at a lower temperature T_0, which may be expressed

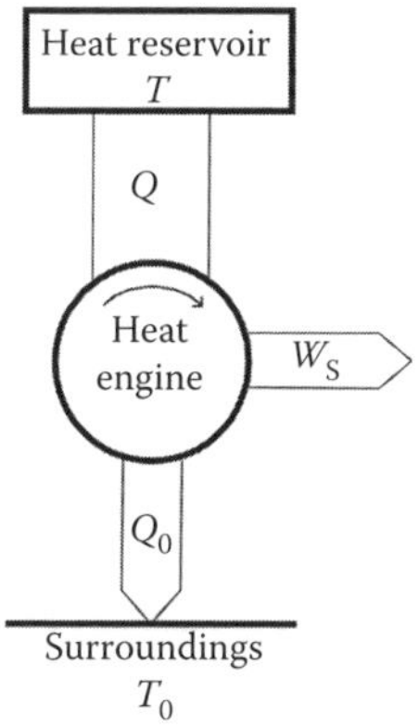

Figure 3.1 Basic scheme of a heat engine: heat Q provided is partially converted to work W_s, another part Q_0 is discarded to the surroundings.

as $-\Delta S = \Delta S_0 = Q_0/T_0$ (the heat rejected is negative). The consequence is that even for an ideal process there is an upper limit for the thermal efficiency of a heat engine. The exemplary process to illustrate this optimum efficiency of an ideal, reversible process is the so-called *Carnot* process, which will also be described in Question 6.3, and accordingly this maximum thermal efficiency for a heat engine is often termed the *Carnot efficiency,* which is

$$\eta_{\mathrm{th,rev}} = 1 - \frac{T_0}{T}. \tag{3.149}$$

To put the argument in a more formal framework, we consider that the change of energy for a working fluid of fixed chemical composition from Equation 3.1 may be written as follows:

$$\mathrm{d}U = T\,\mathrm{d}S - p\,\mathrm{d}V, \tag{3.150}$$

which when combined with the first law

$$\mathrm{d}U = \delta Q + \delta W, \tag{3.151}$$

gives

$$\mathrm{d}S = \frac{\delta Q}{T} + \frac{\delta W}{T} + \frac{p\,\mathrm{d}V}{T}. \tag{3.152}$$

If the process is reversible, then

$$\delta W = -p\,\mathrm{d}V \tag{3.153}$$

and

$$\mathrm{d}S = \frac{\delta Q}{T}. \tag{3.154}$$

For a closed cycle the final and initial values of S must be equal or

$$\oint \mathrm{d}S = 0. \tag{3.155}$$

Because $T > 0$ Equation 3.155 can only be fulfilled for a heat engine that takes heat from a heat source, if there are also steps that reject heat to the surroundings, that is, where $\delta Q < 0$. It is thus an inevitable consequence of the second law that not all of the heat provided may be turned into useful work.

A real process is not reversible so that the form of Equation 3.154 that is used is as follows:

$$\mathrm{d}S > \frac{\delta Q}{T}. \tag{3.156}$$

In consequence, the additional entropy "generated" by the irreversibility must be disposed of by rejecting heat to the surroundings with the same result that only a fraction of the heat provided is used to produce work.

3.9 WHAT IS EXERGY GOOD FOR?

While the *Carnot efficiency* poses a fundamental upper limit for the thermal efficiency of a heat engine, practical machines of course exhibit lower efficiencies, and it is the art of the engineer to come as close to the ideal limit as possible. The resulting actual efficiency is a product of two factors, namely the efficiency determined by the second law and, of necessity, a factor that represents the departures of the real machine from perfection. To better balance the actual technical achievement against what is possible in principle and to identify sources within a process where useful energy is destroyed or wasted, it has turned out to be beneficial to introduce a specific term for the useful energy. According to a suggestion of Rant (1956) this property is called *exergy E*. The term *exergy* does not introduce anything physically new that is different from the second law, yet it has proven to be a useful concept for technical purposes.

Thus *exergy* is defined as that part of energy that—relative to a given reference state—can, without any restriction, be transformed into any other form of energy. Because the work that can be extracted from a process is the form of energy that is of primary interest, an alternative formulation of the definition is as follows: *exergy* is that part of energy that—relative to a given reference state—can, without any restriction, be transformed into useful work.

It is important to note that the definition relies on the specification of a reference state, and this is normally taken to be the environment. For example, a cold reservoir with temperature below ambient contains exergy or useful work. However, when a system is brought to the pressure and temperature of the environment, no potential exists for the extraction of useful work. Thus the environmental state is a *dead state*. An instructive and more detailed discussion of the terms immediate surroundings, surroundings, and environment is given by Çengel and Boles (2006).

To be complete, the other part of the energy, namely the one that in principle cannot be transformed into useful work, should also have a name: it is termed *anergy B*. Thus: *energy* = *exergy E* + *anergy B.*

The central question now is how to determine the exergy (and anergy) inherent in the different forms of energy. From the definition it is obvious that *mechanical* (*or electrical*) *energy* is pure exergy. Note that the limited conversion efficiency of an electrical motor that certainly is below unity does not contradict this statement, because ultimately this is an indication of the departure of the machine from perfection.

In the case of *heat* the question is simple to answer, too; the maximum fraction of heat that can be transferred into other forms of energy and especially work is given by the Carnot efficiency:

$$E_Q = \eta_{\text{th,rev}} \cdot Q = \left(1 - \frac{T_0}{T}\right) \cdot Q. \tag{3.157}$$

Like heat itself, exergy may be expressed as a specific property $e = E/m$, in this case

$$e_q = \eta_{\text{th,rev}} \cdot q = \left(1 - \frac{T_0}{T}\right) \cdot q. \tag{3.158}$$

The anergy is then

$$B_Q = \left(\frac{T_0}{T}\right) \cdot Q \quad \text{or} \quad b_q = \left(\frac{T_0}{T}\right) \cdot q. \tag{3.159}$$

Adopting the concept of exergy one may regard an ideal (i.e., reversible) heat engine as a machine that separates the heat provided to it into two parts: the useful part, exergy, is transferred into work, the remainder, anergy, is rejected to the surroundings as shown in Figure 3.2. In a real process, entropy is generated, which results in a loss of exergy (useful energy). We shall consider the connection of exergy loss and entropy generation later.

As another important example, we consider the exergy E_{FS} connected with a flowing fluid. E_{FS} is the maximum amount of work that can be extracted when bringing the fluid stream from state 1 to the dead state, that is, to equilibrium with the environmental state 0. Again, the maximum work can only be realized in an ideal, reversible process. We start with the first law using specific quantities

$$q_{10} + w_{s10} = h_0 - h_1 + \frac{1}{2}(c_0^2 - c_1^2) + g(z_0 - z_1), \tag{3.160}$$

which states that heat q and (shaft) work w_s crossing the system boundaries result in a change of the total energy of the system, namely of the sum of enthalpy, kinetic (velocity c) and potential energy (acceleration of free fall g and height z).

The exergy is, in magnitude, identical to the maximum work that can be extracted from the process:

$$e_{\text{FS}} = -w_{s10}. \tag{3.161}$$

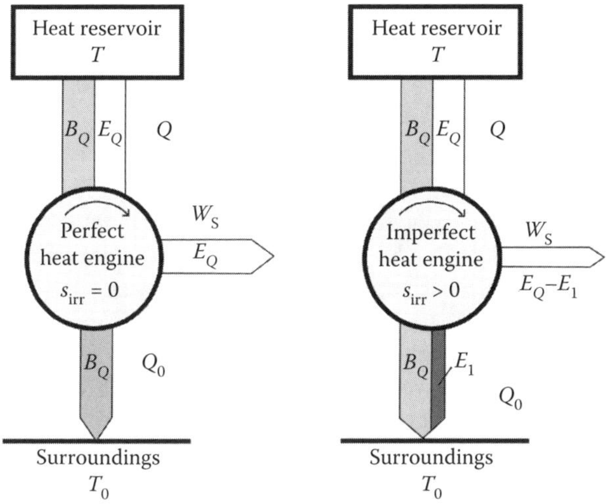

Figure 3.2 A heat engine may be regarded as a "separator" for exergy and anergy: In a perfect process (left), all the exergy of the heat provided is turned into useful work, anergy as the remainder is rejected to the surroundings; in a real heat engine with irreversibilities part of the exergy is turned into anergy, it must be rejected as additional waste heat.

Taking into account that both kinetic and potential energy at the dead state are zero, we obtain

$$e_{FS} = h_1 - h_0 + q_{10} + \frac{1}{2}c_1^2 + gz_1. \tag{3.162}$$

Because the process is to be performed in a reversible manner, the heat must be transferred at a vanishing temperature difference, that is, at the temperature T_0 of the surroundings. From the second law:

$$q_{10} = T_0(s_0 - s_1). \tag{3.163}$$

Finally, after dropping the index 1 to distinguish the specific initial state:

$$e_{FS} = h - h_0 - T_0(s - s_0) + \frac{1}{2}c^2 + gz. \tag{3.164}$$

Neglecting the contributions of kinetic and potential energy, the exergy e_h associated with the enthalpy h may be expressed as

$$e_h = h - h_0 - T_0(s - s_0). \tag{3.165}$$

Consequently, the corresponding anergy b_h is

$$b_h = h_0 + T_0(s - s_0). \tag{3.166}$$

In a similar manner, the exergy e_u connected with the internal energy u may be obtained as

$$e_u = u - u_0 - T_0(s - s_0) + p_0(v - v_0). \tag{3.167}$$

One of the motivations for the introduction of exergy was to identify whether energy is properly used within a process. This can be achieved by comparing the exergy provided to a process with the exergy available after the process has been performed. Ideally, the amount of exergy withdrawn from a process should equal the useful work extracted from this process.

As an example, let us consider the provision of a stream of hot water heated by an electrical resistance heater. Raising the temperature of the water stream from an ambient temperature of 298 K (or 25 °C) to 333 K (or 60 °C) increases the exergy connected with enthalpy from zero to an amount we denote by e_h, as given by Equation 3.162. With

$$(h - h_0) = c_p(T - T_0), \quad (s - s_0) = c_p \ln\left(\frac{T}{T_0}\right), \tag{3.168}$$

and $c_p \approx 4.2\ \mathrm{kJ \cdot kg^{-1} \cdot K^{-1}}$, we obtain

$$e_h = c_p(T - T_0) - T_0 c_p \ln\left(\frac{T}{T_0}\right) = 147\,\mathrm{kJ \cdot kg^{-1}} - 139\,\mathrm{kJ \cdot kg^{-1}} = 8\ \mathrm{kJ \cdot kg^{-1}} \tag{3.169}$$

The exergy of $8\ \mathrm{kJ \cdot kg^{-1}}$ available after this process e_{out} must be compared with the exergy input e_{in}. In the case of an electric heater, this input is simply given by the work required to heat up the water (remember that electrical energy is made up of exergy only):

$$e_{\mathrm{in}} = w_s = (h - h_0) = c_p(T - T_0) = 147\ \mathrm{kJ \cdot kg^{-1}}. \tag{3.170}$$

How efficiently exergy is used may be judged by the exergetic (or second law) efficiency η_{ex}, defined as

$$\eta_{\mathrm{ex}} = \frac{e_{\mathrm{out}}}{e_{\mathrm{in}}}. \tag{3.171}$$

In the present case, $\eta_{\mathrm{ex}} = 8/147 = 5\ \%$. This poor result reflects the fact that energy of "high quality" (pure exergy) is (mis-) used to provide energy at a low temperature. For that purpose waste heat from an engine or district heat from a power plant would suffice. The poor exergetic efficiency also implicitly takes into account the fact that electricity itself can be produced only with a certain thermal efficiency (ultimately limited by the Carnot factor) at an electrical power plant.

Because the concept of exergy is closely connected with the second law, it is obvious that it must be linked to another central term in connection with the second law, namely entropy. This fact can be seen, for example, from Equation 3.165, we can, however, obtain a more general relation between the two properties. In an irreversible process, entropy is generated, and at the same time, useful energy, that is, exergy is destroyed. For the derivation of such a relation we consider a steady-flow process, Figure 3.3 (in a similar way an identical relation may also be obtained for other systems).

The specific exergy loss e_l may be simply obtained by setting up a control volume and balancing the exergy that enters into the system e_{in} against that leaving the system e_{out},

$$e_l = e_{in} - e_{out}. \tag{3.172}$$

In our example a steady flow enters into and leaves the control volume, associated with specific enthalpies h_{in} and h_{out}, respectively (kinetic and potential energies are neglected for simplicity because taking them into account would not alter the result). Heat entering and leaving the system is summarized into one specific quantity q, the same holds for all forms of work, resulting in a term w_s. Thus

$$e_l = e_{in} - e_{out} = e_{h,in} - e_{h,out} + e_w + e_q. \tag{3.173}$$

From Equation 3.165,

$$e_{h,in} - e_{h,out} = h_{in} - h_{out} - T_0\left(s_{in} - s_{out}\right), \tag{3.174}$$

and using the first law, again ignoring kinetic and potential energy,

$$q + w_s = h_{out} - h_{in}, \tag{3.175}$$

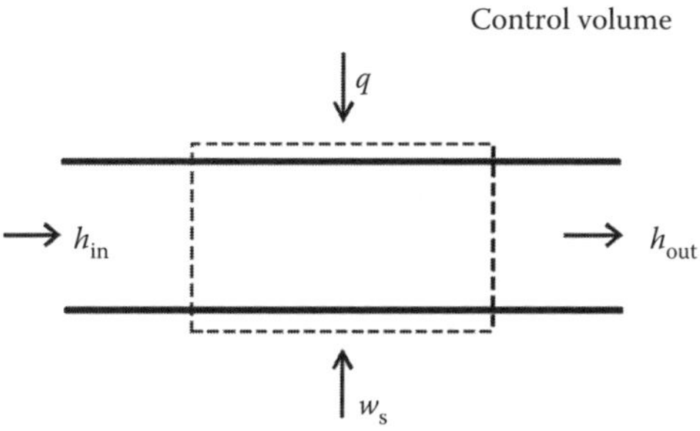

Figure 3.3 Schematic for determining the exergy loss for an open system: a steady flow associated with specific enthalpies h_{in} and h_{out} crosses the control volume, heat, and shaft work transferred are summarized in the resulting quantities q and w_s, respectively.

we obtain

$$e_l = -q - w_s - T_0(s_{in} - s_{out}) + e_w + e_q. \quad (3.176)$$

Because the shaft work w_s is pure exergy, $w_s = e_w$, and from Equation 3.158 for e_q, we obtain

$$e_l = -q - T_0(s_{in} - s_{out}) + \left(1 - \frac{T_0}{T}\right) \cdot q. \quad (3.177)$$

Performing an entropy balance,

$$s_{out} - s_{in} = \frac{q}{T} + s_{irr}, \quad (3.178)$$

where q/T is the entropy connected with the net heat q and s_{irr} summarizes all sources of irreversibility (including that of heat transfer), we finally obtain

$$e_l = -q - T_0\left(-\frac{q}{T} - s_{irr}\right) + \left(1 - \frac{T_0}{T}\right) \cdot q = T_0 \cdot s_{irr}. \quad (3.179)$$

This result, which is of general applicability, demonstrates that exergy loss is directly proportional to the entropy generated within a process.

Returning to our example of the electric resistance heater above, this exergy loss shows up in the second term of Equation 3.169 with a magnitude of 139 kJ · kg^{-1}. Using the exergy loss we may also write the exergetic efficiency of Equation 3.172 in a different form

$$\eta_{ex} = \frac{e_{out}}{e_{in}} = 1 - \frac{e_l}{e_{in}}. \quad (3.180)$$

3.10 REFERENCES

Atkins P., 2007, *Four Laws That Drive the Universe*, Oxford University Press, Oxford.

Çengel Y.A., and Boles M.A., 2006, *Thermodynamics—an Engineering Approach*, McGraw-Hill, Boston.

Clausius R., 1850a, "Über die bewegende Kraft der Wärme, Part I," *Annalen der Physik* **79**:368–397 (also printed in 1851, "On the Moving Force of Heat, and the Laws regarding the Nature of Heat itself which are deducible therefrom. Part I," *Phil. Mag.* **2**:1–21).

Clausius R., 1850b, "Über die bewegende Kraft der Wärme, Part II," *Annalen der Physik* **79**:500–524 (also printed in 1851, "On the Moving Force of Heat, and the Laws regarding the Nature of Heat itself which are deducible therefrom. Part II," *Phil. Mag.* **2**:102–119).

Copp J.I., and Everett D.H., 1953, "Thermodynamics of binary mixtures containing amines," *Discuss. Faraday Soc.* **15**:174–188.

Denbigh K.G., 1971, *The Principles of Chemical Equilibrium*, 3rd ed., Cambridge University Press, Cambridge.

Gibbs J.W., 1928, *The Collected Works. Volume I. Thermodynamics*, Longman Green, New York.

Goodwin A.R.H., and Trusler J.P.M., 2003, *Sound Speed*, Chapter 6, in *Experimental Thermodynamics, Volume VI, Measurement of the Thermodynamic Properties of Single Phases*, eds. Goodwin A.R.H., Marsh K.N., and Wakeham W.A., for IUPAC, Elsevier, Amsterdam.

Guggenheim E.A., 1967, *Thermodynamics*, 5th ed., North-Holland, Amsterdam.

Herzfeld K.F., and Litovitz T.A., 1959, *Pure and Applied Physics, Volume 7, Absorption and Dispersion of Ultrasonic Waves*, ed. Massey H.S.W., Academic Press, London.

Joule J.P., 1845, "LIV. On the changes of temperature produced by the rarefaction and condensation of air," *Phil. Mag.* (series 3) **26**:369–383.

Landau L.D., and Lifshitz E.M., 1987, *Theory of Elasticity*, 2nd ed., Pergamon, Oxford.

Ledbetter H.M., 1982, "The temperature behavior of Young moduli of 40 engineering alloys," *Cryogenics* **22**:653–656.

Margenau H., 1950, *The Nature of Physical Reality*, McGraw-Hill, New York, p. 215.

McGlashan M.L., 1979, *Chemical Thermodynamics*, Academic Press, London.

Morse P.M., and Ingard K.U., 1968, *Theoretical Acoustics*, McGraw-Hill, New York, p. 233.

Münster A., 1974, *Statistical Thermodynamics. Volume II*, Springer-Verlag, Berlin and Academic Press, New York, p. 79.

Papadakis E.P., 1998, "Ultrasonic wave measurements of elastic moduli E, G, and MU for product development and design calculations," *J. Test. Eval.* **26**:240–246.

Quack M., Stohner J., Strauss H.L., Takami M., Thor A.J., Cohen E.R., Cvitas T., Frey J.G., Holström B., Kuchitsu K., Marquardt R., Mills I., and Pavese F., 2007, *Quantities, Units and Symbols in Physical Chemistry*, 3rd ed., RSC Publishing, Cambridge.

Rant Z., 1956, "Exergie, ein neues Wort für technische Arbeitsfähigkeit," *Forsch. Ingenieurwes.* **22**:36–37.

Rowlinson J.S., 2003, "The work of Thomas Andrews and James Thomson on the liquefaction of gases," *Notes Rec. R. Soc.* **57**:143–159.

Rowlinson J.S., 2005, "Which Kelvin?, Book Review for Degrees Kelvin: A tale of genius, invention, and tragedy," *Notes Rec. R. Soc.* **59**:339–341.

Weston W.F., 1975, "Low-temperature elastic-constants of a superconducting coil composite," *J. Appl. Phys.* **46**:4458–4465.

Chapter 4

Phase Equilibria

4.1 INTRODUCTION

This chapter introduces the thermodynamic concepts required for the treatment of the equilibrium of any system with independent variables of temperature, pressure, and amount of substance of the components within it; this includes a pure substance and multicomponent mixtures. When these are combined with the rules of thermodynamics provided in Chapter 3 we then have methods to measure changes of entropy, energy, and enthalpy with temperature, pressure, and composition and also methods to determine changes in Gibbs function and chemical potential (i.e., also absolute activity) with respect to pressure and composition (Guggenheim 1959; McGlashan 1979). In principle, these are sufficient to determine the equilibrium between phases of a pure substance or mixtures; however, other methods will need to be introduced to expedite such calculations and that is the purpose of this introductory section.

In the previous chapters we have been concerned with the thermodynamic relationships for a homogeneous phase. This chapter extends our questions, examples, and discussion to a heterogeneous system that is one containing more than one phase. Thus we are to discuss phase equilibrium and include the variation of thermodynamic functions with composition. Equations introduced in Chapter 3 will be used and extended to a heterogeneous system of phases. In particular, for the case when temperature and pressure are the independent variables, as is most often the case both for experiments performed to determine thermodynamic properties and in a chemical process plant, the Gibbs function and Equation 3.14 is the appropriate function and, for multiple phases, can be written as

$$\sum_{\alpha} \mathrm{d}G^{\alpha} = -\sum_{\alpha} S^{\alpha}\,\mathrm{d}T + \sum_{\alpha} V^{\alpha}\,\mathrm{d}p + \sum_{\alpha}\sum_{\mathrm{B}} \mu_{\mathrm{B}}\,\mathrm{d}n_{\mathrm{B}}^{\alpha}, \tag{4.1}$$

where the Σ means the sum over all phases included in the system. In Equation 4.1, we have purposely omitted the superscript α on the uniform intensive properties of T, p, μ_B, and n_B because, for now, we will only consider systems that are in thermal, hydrostatic, and diffusive equilibrium.

Removing the summation over all phases in Equation 4.1 and for simplicity replacing it with a superscript Σ for the system Equation 4.1 can be cast as follows:

$$dG^{\Sigma} = -S^{\Sigma}\,dT + V^{\Sigma}\,dp + \sum_{B} \mu_B\,dn_B^{\Sigma}. \tag{4.2}$$

Equation 4.2 does not include one situation that arises when two phases are separated by a partition permeable to some substances but not others in the system. In this case $p^{\alpha} \neq p^{\beta}$ and this special case is called *osmotic equilibrium* for which the absolute difference $|p^{\alpha} - p^{\beta}|$ is the osmotic pressure that will be discussed further in Question 4.3.3.

When the system is of fixed chemical composition $dn_B^{\Sigma} = 0$ and in that case Equation 4.2 becomes

$$dG^{\Sigma} = -S^{\Sigma}\,dT + V^{\Sigma}\,dp. \tag{4.3}$$

Equation 4.3 can be modified for a closed system with a chemical reaction by addition of the term $A\,d\xi = -(\Sigma_B\, \nu_B\, \mu_B)\,d\xi$, where, as discussed in Chapter 3, ξ is the extent of reaction and A is the affinity for a chemical reaction.

4.1.1 What Is the Phase Rule?

The Gibbs–Duhem equation for a phase in thermal, hydrostatic, and diffusive equilibrium is, according to Equation 3.24, given by

$$0 = S_m^{\alpha}\,dT - V_m^{\alpha}\,dp + \sum_{B} x_B^{\alpha}\,d\mu^{\alpha}. \tag{4.4}$$

The number of independent intensive variables in Equation 4.4 is $(C+1)$, where C is the number of components in the phase and that equals the number of terms in the summation. If there are P phases present in the system then there are P Equations 4.4 that provide $(P-1)$ restrictions. Thus, the number of independent intensive variables or degrees of freedom of the system are

$$F = (C+1) - (P-1) = C + 2 - P. \tag{4.5}$$

If there are R chemical reactions occurring in the system and if these are all at equilibrium then there is one additional equation $\Sigma_B\, \nu_B\, \mu_B = 0$ for each, and so the F must be reduced by R so that Equation 4.5 becomes

$$F = C + 2 - P - R. \tag{4.6}$$

Equation 4.6 is the Phase Rule. We are now armed with sufficient information to start the discussion of the phase equilibrium of a pure substance. Before doing so we draw some conclusions from Equation 4.4 that at constant temperature and pressure becomes

$$\sum_{\mathrm{B}} x_{\mathrm{B}}^{\alpha}\,\mathrm{d}\mu^{\alpha} = 0, \tag{4.7}$$

and for a binary mixture $(1 - x)\mathrm{A} + x\mathrm{B}$ Equation 4.7 is

$$(1 - x)\,\mathrm{d}\mu_{\mathrm{A}} + x\,\mathrm{d}\mu_{\mathrm{B}} = 0, \tag{4.8}$$

or

$$\mu_{\mathrm{A}}(T, p, x^{\beta}) - \mu_{\mathrm{A}}(T, p, x^{\alpha}) = \int_{x^{\alpha}}^{x^{\beta}} \frac{x}{1 - x}\,\mathrm{d}\mu_{\mathrm{B}}. \tag{4.9}$$

Equation 4.9 provides a route to $\mu_{\mathrm{A}}(T, p, x^{\beta}) - \mu_{\mathrm{A}}(T, p, x^{\alpha})$ from measurements of the difference $\mu_{\mathrm{B}}(T, p, x) - \mu_{\mathrm{B}}^{*}(T, p)$, at mole fractions x that include x^{α} and x^{β}. Here, the superscript asterisk denotes a pure substance, and we should note that the measurements are to be performed with x^{α} constant. Because only $(C - 1)$ of the differences in chemical potential are independent, where C is the number of components, the necessary work is slightly reduced.

4.2 WHAT IS PHASE EQUILIBRIUM OF A PURE SUBSTANCE?

For two phases α and β of a pure, nonreacting substance in equilibrium $C = 1$, $P = 3$, and $R = 0$ so that according to Equation 4.6 $F = 1$; the equilibrium of three phases of a pure substance results in $F = 0$, and the system has no independent intensive variables. The equilibrium temperature is called the *triple point temperature*, while the equilibrium pressure is the triple point pressure and both are fixed. A $p(T)$ projection for the phase equilibrium of a pure substance is shown schematically in Figure 4.1 (Goodwin and Ambrose 2005).

The curves AB, BD, and BC meet at the triple point B; for a solid with more than one solid phase there is more than one triple point. The curve AB depicting the s = g equilibrium tends to zero pressure at low temperatures, while the curve BD (representing the s = l equilibrium) continues upward indefinitely

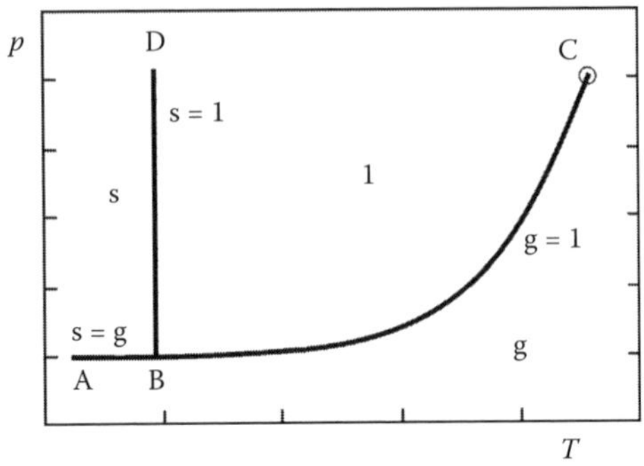

Figure 4.1 Pressure p of a pure substance as function of temperature T. The figure shows the solid (s), liquid (l), and gaseous (g) phases. The lines are defined as follows: A to B sublimation line, where solid is in equilibrium with vapor (s = g); B to C liquid in equilibrium with vapor (l = g); B to D the melting line, where solid is in equilibrium with liquid (s = l); and, O: the critical point. B is the triple point where solid, liquid, and vapor coexist (s = l = g).

and has a large and positive slope for most substances; water is an exception and for this the slope is large and negative. If vapor pressure is plotted as a function of temperature, as it is represented schematically in Figure 4.1, the curves for the solid (AB) and liquid (BC) intersect at the triple point (point B in Figure 4.1), with a discontinuity of slope and terminates at higher temperature at the critical point C where the properties of vapor and liquid become identical, and at this temperature the vapor pressure is known as the critical pressure; at $T = T_c$ $(\partial p/\partial V_m)_T = 0$ and $(\partial^2 p/\partial V_m^2)_T = 0$. The critical temperature is the highest temperature at which two fluid phases of liquid and gas for a pure substance can coexist. Supercooled liquid, which is metastable, has a higher vapor pressure than that of the stable solid.

For a pure substance Equation 3.45 is

$$G_m = H_m - TS_m = \mu, \tag{4.10}$$

where the definition from Chapter 3 $G_m = \mu$ of Equation 3.17 has been used. For the equilibrium of two phases α and β $\mu^\alpha = \mu^\beta$ so that $G_m^\alpha = G_m^\beta$ and thus

$$\Delta_\alpha^\beta H_m = T\Delta_\alpha^\beta S_m. \tag{4.11}$$

The $\Delta_\alpha^\beta H_m$ of Equation 4.11 can be obtained experimentally from Equations 1.93 and 3.56, while $\Delta_\alpha^\beta S_m$ can be obtained from Equation 3.60.

For two phases (solid, liquid, or gas) α and β of a pure substance in equilibrium Equation 4.6 gives $F = 1$ so there is a relationship between, for example, the temperature T and pressure p. If the temperature T is chosen as the independent variable then the pressure is dependent, and in this text it will be denoted by p^{sat} for the case of liquid and gas equilibrium (written as $\text{l} = \text{g}$). For two phases α and β, which could be solid and liquid ($\text{s} = \text{l}$), solid and gas ($\text{s} = \text{g}$), or liquid and gas ($\text{l} = \text{g}$), of a pure substance in equilibrium the Gibbs–Duhem equation (Equation 3.24) for each of the phases are

$$0 = S_{\text{m}}^{\alpha}\,\mathrm{d}T - V_{\text{m}}^{\alpha}\,\mathrm{d}p + \mathrm{d}\mu, \tag{4.12}$$

and

$$0 = S_{\text{m}}^{\beta}\,\mathrm{d}T - V_{\text{m}}^{\beta}\,\mathrm{d}p + \mathrm{d}\mu. \tag{4.13}$$

The chemical potential μ of the substance B, and therefore the partial molar Gibbs function must be equal in both phases so that Equations 4.12 and 4.13 can be written as

$$\frac{\mathrm{d}p^{\text{sat}}}{\mathrm{d}T} = \frac{S_{\text{m}}^{\beta} - S_{\text{m}}^{\alpha}}{V_{\text{m}}^{\beta} - V_{\text{m}}^{\alpha}} = \frac{\Delta_{\alpha}^{\beta} S_{\text{m}}}{\Delta_{\alpha}^{\beta} V_{\text{m}}}, \tag{4.14}$$

or in view of Equation 4.11 can be written as

$$\frac{\mathrm{d}p^{\text{sat}}}{\mathrm{d}T} = \frac{\Delta_{\alpha}^{\beta} H_{\text{m}}}{T\Delta_{\alpha}^{\beta} V_{\text{m}}}. \tag{4.15}$$

Equation 4.15 is called *Clapeyron's equation* and describes the slope of any one of the three saturation lines shown in Figure 4.1: $\text{s} = \text{l}$, $\text{s} = \text{g}$, and $\text{l} = \text{g}$. If, for example, phase α represents the solid phase ($\alpha = \text{s}$) and phase β the liquid ($\beta = \text{l}$) $\Delta_{\text{s}}^{\text{l}} H_{\text{m}}$ is the molar enthalpy of fusion, which for chemists should, according to International Union of Pure and Applied Chemistry (IUPAC) nomenclature (Quack et al. 2007), be written as $\Delta_{\text{fus}} H_{\text{m}}$. Similarly, $\Delta_{\text{s}}^{\text{l}} V_{\text{m}}$ should be written as $\Delta_{\text{fus}} V_{\text{m}}$. If one of the phases is a dilute gas, so that it can be considered perfect with $pV_{\text{m}} = RT$, and the molar volume of the gas phase $V(\text{g}) \gg V(\text{l})$ or $V(\text{g}) \gg V(\text{s})$ so that $V(\text{l})$ or $V(\text{s})$ can be neglected then Equation 4.15 becomes

$$\frac{\mathrm{d}p^{\text{sat}}}{\mathrm{d}T} \approx \frac{p^{\text{sat}}\Delta_{\alpha}^{\beta} H_{\text{m}}}{RT^2}. \tag{4.16}$$

Further simplification can be obtained by assuming $\Delta_{\alpha}^{\beta} H_{\text{m}}$ is independent of temperature over a range of temperatures $T_1 - T_2$ then Equation 4.16 can be written as

$$\ln\left\{\frac{p^{\text{sat}}(T_2)}{p^{\text{sat}}(T_1)}\right\} \approx \frac{\Delta_{\alpha}^{\beta} H_{\text{m}}(T_2 - T_1)}{RT_1T_2}. \tag{4.17}$$

When T_1 is fixed by selecting p_1, for example, $p_1 = p^\ominus = 0.1$ MPa,* then Equation 4.17 reduces to

$$\ln\left\{\frac{p^{\mathrm{sat}}(T_2)}{p^\ominus}\right\} \approx \frac{\Delta_\alpha^\beta H_\mathrm{m}}{RT_1} - \frac{\Delta_\alpha^\beta H_\mathrm{m}}{RT_2} = a - \frac{b}{T_2}. \quad (4.18)$$

Over a range of temperature close to the normal boiling temperature the observed vapor pressure may be fitted to an equation of the form suggested by Equation 4.18, that is,

$$\ln\left(\frac{p^{\mathrm{sat}}}{p^\ominus}\right) = a + \frac{b}{T}, \quad (4.19)$$

where a and b are substance-dependent parameters for each phase. The Antoine equation is given by

$$\ln\left(\frac{p^{\mathrm{sat}}}{p^\ominus}\right) = e + \frac{f}{g+T}, \quad (4.20)$$

where e, f, and g, are also substance-dependent parameters for each phase, and provides a better representation over a slightly wider temperature range about the normal boiling temperature.

To represent measurements of the vapor pressure within experimental error from the triple point temperature to the critical temperature requires a complex equation. One representation that has been extremely successful is the so-called *Wagner equation*

$$\ln\left(\frac{p}{p_\mathrm{c}}\right) = (n_1\tau + n_2\tau^{1.5} + n_3\tau^\mathrm{c} + n_4\tau^\mathrm{d})\left(\frac{T_\mathrm{c}}{T}\right), \quad (4.21)$$

where T_c and p_c are the critical temperature and pressure, respectively, $\tau = (1 - T/T_\mathrm{c})$, the n_i, with $i = 1, 2, 3$, and 4, are parameters for each substance that are adjusted to the available measurements, and, typically, $c = 2.5$ and $d = 5$. Equation 4.21 reduces to Equation 4.19 when truncated after the first term $n_1\tau$. Vapor pressure is affected by the curvature of the surface from which evaporation takes place, and the vapor pressure of microscopic droplets is higher than the normal value; this affects the formation of clouds and rain.

* The value for $p^\ominus$ is 10^5 Pa and has been the IUPAC recommendation since 1982 and should be used to tabulate thermodynamic data. Before 1982 the standard pressure was usually taken to be $p^\ominus = 101\,325$ Pa (=1.01325 bar or 1 atm), called the standard atmosphere. In any case, the value for $p^\ominus$ should be specified.

Engineers prefer the use of specific quantities and Equation 4.15 can be written as

$$\frac{\mathrm{d}p^{\mathrm{sat}}}{\mathrm{d}T}=\frac{\Delta_{\alpha}^{\beta}h}{T\Delta_{\alpha}^{\beta}v}. \tag{4.22}$$

It is also common practice in engineering problems to use the specific gas constant $R_s = R/M$ so that the same approximations used for Equation 4.16 results in

$$\frac{\mathrm{d}p^{\mathrm{sat}}}{\mathrm{d}T}=\frac{p^{\mathrm{sat}}\Delta h_0^{\mathrm{v}}}{R_sT^2}. \tag{4.23}$$

The vapor pressures of different substances vary widely. At T = 298.15 K, for example, the vapor pressures of many involatile substances are too low ($<10^{-5}$ Pa) to be measurable, whereas that of a volatile substance such as carbon dioxide is about 6 MPa. The temperature at which the vapor pressure of a substance is 0.101325 MPa is defined as its normal boiling temperature T^{b}; normal boiling temperatures range from 4.2 K for helium up to, for example, 6,000 K for tantalum.

For l = g the vapor pressure at a temperature T close to the normal boiling temperature T^{b} Equation 4.18 can be obtained from a modified form of Equation 4.18

$$\frac{p^{\mathrm{sat}}(T)}{\mathrm{MPa}}\approx 0.1\exp\left\{\frac{10(T-T^{\mathrm{b}})}{T}\right\}, \tag{4.24}$$

where it is assumed $\Delta_{\mathrm{l}}^{\mathrm{g}}H_{\mathrm{m}}/(RT^{\mathrm{b}})=10$. Equation 4.24 is called *Trouton's rule* and with the assumption that the critical pressure of all substances is the same is a result of the principle of corresponding states discussed in Chapter 2.

From a plot of T^{sat} as a function of amount of substance density $\rho_n\,(=1/V_{\mathrm{m}})$ for a l = g phase boundary it is found that $\{\rho_{n,\mathrm{l}}+\rho_{n,\mathrm{g}}\}/2$ lies on a straight line and this is referred to as the law of the rectilinear diameter that is obeyed for all pure substances. For mixtures the reduced orthobaric densities $\rho_{\mathrm{g}}/\rho_{\mathrm{c}} = V_{\mathrm{m,c}}/V_{\mathrm{m,g}}$ and $\rho_{\mathrm{l}}/\rho_{\mathrm{c}} = V_{\mathrm{m,c}}/V_{\mathrm{m,l}}$ plotted against reduced temperature T/T_{c} follow the Cailleter and Mathias's law of the rectilinear diameter of

$$\frac{1}{2}\frac{\{\rho_{\mathrm{l}}+\rho_{\mathrm{g}}\}}{\rho_{\mathrm{c}}}=1+0.797\left(1-\frac{T}{T_{\mathrm{c}}}\right), \tag{4.25}$$

a further consequence of the principle of corresponding states for a two-phase fluid mixture.

The critical pressures of the majority of substances do not exceed 5 MPa, although a few are much higher than this, for example, water with $p_{\mathrm{c}} = 22.05$ MPa.

Critical properties of only a few hundred elements and compounds have been measured because for many, particularly the involatile elements, the temperatures are too high to be experimentally accessible, and most compounds decompose before the critical temperature is reached. So far consideration has been restricted to substances that vaporize without decomposition. If the substance vaporizing decomposes irreversibly, as do many inorganic compounds at high temperatures, there are different chemical species in the liquid and vapor phases.

The classical equations of state, for example, the van der Waals equation described later, can at least qualitatively provide estimates of the $p(\rho)$ isotherms at temperatures close to critical. The van der Waals equation is an analytic equation, that is, one for which the expansion as a Taylor series about the critical point converges about that point. Analytic equations are unable to predict the behavior observed at the critical point and indeed they cannot even fit observations near the critical point. By this we mean that the analytic equation must yield, for differences in the coexisting densities, at temperatures $T \to T^{\mathrm{c}}$ $\{\rho_{\mathrm{l}} - \rho_{\mathrm{g}}\} \propto (T^{\mathrm{c}} - T)^{\beta}$, where $\beta = 0.5$ and is called the *critical exponent* and is in this case the so-called classical value. All analytic equations of state predict a finite value of C_V at the critical temperature, and that is inconsistent with direct measurement that shows divergence at T^{c}. The observations require critical exponents that differ from the classical values and this topic is outside the scope of this text; and the interested reader is referred to, for example, Behnejard et al. (2010).

Close to the critical point there is a characteristic opalescence in the fluid; this arises because the correlation length or mean distance over which the molecules' order increases by several orders of magnitudes from about 1 nm through the wavelengths of visible light (between about 380 to 780 nm) and these correlated fluctuations scatter the light.

4.2.1 What Does Clapeyron's Equation Have to Do with Ice-Skating?

Now, how is Clapeyron's equation (Equation 4.22) related or at least, allegedly, related to ice-skating? It seems obvious that if a film of liquid water forms at the solid surface between a skate and ice then it can explain the lubrication process that makes it so simple to slip on ice. However, the origin of this liquid film is not that obvious, and historically there have been debates on the underlying mechanism (Rosenberg 2005; Dash et al. 2006). One potential explanation for the observation is the decrease of the melting temperature of water with pressure; indeed this has become a myth; "pressure melting became the standard textbook explanation, and it has been propagated through generations of students" (Dash et al. 2006). We can quickly demonstrate, however, that the resulting effect is minimal. An estimate of the decrease of melting temperature with pressure may be directly obtained from Clapeyron's equation.

To that end, we apply Equation 4.22 to the equilibrium between liquid and solid water and obtain

$$\frac{\mathrm{d}p^{\mathrm{sat}}}{\mathrm{d}T} = \frac{\Delta_{\mathrm{s}}^{\mathrm{l}}h}{T\Delta_{\mathrm{s}}^{\mathrm{l}}v}, \tag{4.26}$$

where $\Delta_{\mathrm{s}}^{\mathrm{l}}h$ is the specific enthalpy of the transition from solid to liquid (melting) and $\Delta_{\mathrm{s}}^{\mathrm{l}}v$ is the difference in the specific volumes v^{l} and v^{s} of the liquid and solid phases, respectively. Inserting into Equation 4.26 the respective values for these properties from Feistel and Wagner (2006) at a temperature 273.15 K and ambient pressure of 0.101325 MPa we obtain

$$\frac{\mathrm{d}p^{\mathrm{sat}}}{\mathrm{d}T} = \frac{333\cdot 10^{3}\ \mathrm{J\cdot kg^{-1}}}{-90.7\cdot 10^{-6}\ \mathrm{m^{3}\cdot kg^{-1}}273\ \mathrm{K}} \approx -1.34\cdot 10^{7}\ \mathrm{Pa\cdot K^{-1}} \approx -13.4\ \mathrm{MPa\cdot K^{-1}}. \tag{4.27}$$

Because water exhibits the peculiarity that the specific volume of the solid phase is larger than that of the liquid phase the melting curve has a negative slope, meaning that the melting point is shifted toward lower temperatures with increasing pressure. For the sake of completeness it should be mentioned that the estimate provided in Equation 4.27 refers only to a specific temperature and one form of ice (ice exhibits a variety of phases); of importance here is the common hexagonal structure. From Equation 4.27 we deduce that a considerable pressure of 13.4 MPa is needed to lower the melting temperature by only 1 K. It is certainly difficult to give a precise estimate for the pressure exerted by an ice-skater as the contact area might be a topic of debate. However, if we consider a skater of mass 120 kg (that would be considered rather heavy) skating on two blades each with dimensions of about 40 cm × 1.5 mm, we obtain an area of $6\cdot 10^{-4}$ m² and a force of about 600 N (on each blade) resulting in a pressure of 1 MPa. Even when skating just on one leg or when using special blades it becomes obvious that the resulting effect on the melting point will be well below 1 K. If pressure melting was the decisive effect for making ice skating possible no one could skate at Celsius temperatures a few degrees below 0. Yet we know—perhaps from our own painful experience—that one may also slip on ice with normal shoes and that also skiing is possible on "solid water" with skis that have a much larger area than ice-skates.

In consequence, we may answer the question from the headline by stating that Clapeyron's equation has little to do with ice-skating. There are other mechanisms mainly responsible for a liquid film on the ice surface, namely frictional melting, see, for example, Colbeck (1995), and most importantly "premelting," indicating that a liquid-like layer is formed on the ice surface well below the normal bulk melting temperature. The thickness and structure of this layer have been topics of intense research during the last few years, including the effects of impurities and confinement, for recent reviews the reader may consult Dash et al. (2006) and Wettlaufer and Grae Worster (2006).

4.2.2 How Do I Calculate the Chemical Potential?

Now let us return to the issue of determining chemical potential and thus estimating phase equilibria. To address this question we assume that the van der Waals equation of state given by

$$p = \frac{RT}{V_{\mathrm{m}} - b} - \frac{a}{V_{\mathrm{m}}^2}, \tag{4.28}$$

represents the properties of the fluid. In Equation 4.28 the parameters b and a are given by

$$b = \frac{V_{\mathrm{m}}^{\mathrm{c}}}{3}, \tag{4.29}$$

and

$$a = \frac{9RT^{\mathrm{c}}V_{\mathrm{m}}^{\mathrm{c}}}{8}. \tag{4.30}$$

Equations 4.28, 4.29, and 4.30 can be used to determine the chemical potential for a pure gas from

$$\mu_{\mathrm{B,g}}(T, V_{\mathrm{m}}) = \mu_{\mathrm{B,g}}^{\ominus}(T) + pV_{\mathrm{m}} - RT - RT\ln\left(\frac{p^{\ominus}V_{\mathrm{m}}}{RT}\right) + \int_{\infty}^{V_{\mathrm{m}}}\left(p - \frac{RT}{V_{\mathrm{m}}}\right)\mathrm{d}V_{\mathrm{m}}, \tag{4.31}$$

with the result

$$\mu_{\mathrm{B,g}}(T, V_{\mathrm{m}}) = \mu_{\mathrm{B,g}}^{\ominus}(T) + \frac{RTV_{\mathrm{m}}}{V_{\mathrm{m}} - b} - \frac{2a}{V_{\mathrm{m}}} - RT - RT\ln\left(\frac{p^{\ominus}V_{\mathrm{m}}}{RT}\right) + RT\ln\left(\frac{V_{\mathrm{m}}}{V_{\mathrm{m}} - b}\right), \tag{4.32}$$

and the equilibria of the gas with the liquid are obtained by solving the simultaneous equations

$$\mu_{\mathrm{B,g}}(T, V_{\mathrm{m,g}}) = \mu_{\mathrm{B,l}}(T, V_{\mathrm{m,l}}), \tag{4.33}$$

and

$$p(T, V_{\mathrm{m,g}}) = p(T, V_{\mathrm{m,l}}), \tag{4.34}$$

with

$$\left(\frac{\partial p}{\partial V_{\mathrm{m,g}}}\right)_T < 0, \tag{4.35}$$

and

$$\left(\frac{\partial p}{\partial V_{\mathrm{m,l}}}\right)_T < 0. \tag{4.36}$$

In this chapter and in Equations 4.31 through 4.36 we have used the nomenclature introduced in Question 1.10. The van der Waals Equation 4.28 when substituted into Equation 4.33 gives

$$\frac{(V_{\mathrm{m,g}} - V_{\mathrm{m,l}})b}{(V_{\mathrm{m,g}} - b)(V_{\mathrm{m,l}} - b)} + \ln\left(\frac{V_{\mathrm{m,g}} - b}{V_{\mathrm{m,l}} - b}\right) = \frac{2a(V_{\mathrm{m,g}} - V_{\mathrm{m,l}})}{RTV_{\mathrm{m,g}}V_{\mathrm{m,l}}}, \tag{4.37}$$

and Equation 4.34 provides

$$\frac{V_{\mathrm{m,g}}V_{\mathrm{m,l}}}{(V_{\mathrm{m,g}} - b)(V_{\mathrm{m,l}} - b)} = \frac{2a(V_{\mathrm{m,g}} + V_{\mathrm{m,l}})}{RTV_{\mathrm{m,g}}V_{\mathrm{m,l}}}. \tag{4.38}$$

Use of Equation 4.28 in Equations 4.35 and 4.36 gives

$$\frac{RT}{(V_{\mathrm{m,g}} - b)^2} > \frac{2a}{(V_{\mathrm{m,g}})^3}, \tag{4.39}$$

and

$$\frac{RT}{(V_{\mathrm{m,l}} - b)^2} > \frac{2a}{(V_{\mathrm{m,l}})^3}. \tag{4.40}$$

For mixtures, to use the van der Waals equation, parameters are required for each substance in a phase.

4.3 WHAT IS THE CONDITION OF EQUILIBRIUM BETWEEN TWO PHASES OF A MIXTURE OF SUBSTANCES?

The mole fractions x^{α} and x^{β} of two coexisting phases α and β of a binary mixture for which the independent variables are temperature T and pressure p are determined by solution of the simultaneous equations

$$\mu_{A,\alpha}(T,p,x^{\alpha}) = \mu_{A,\beta}(T,p,x^{\beta}) \tag{4.41}$$

and

$$\mu_{B,\alpha}(T,p,x^{\alpha}) = \mu_{B,\beta}(T,p,x^{\beta}). \tag{4.42}$$

The diffusional stability conditions are

$$\left(\frac{\partial \mu_A}{\partial x}\right)_{T,p}^{\alpha} < 0, \quad \left(\frac{\partial \mu_B}{\partial x}\right)_{T,p}^{\alpha} > 0, \quad \left(\frac{\partial \mu_A}{\partial x}\right)_{T,p}^{\beta} < 0, \quad \text{and} \quad \left(\frac{\partial \mu_B}{\partial x}\right)_{T,p}^{\beta} > 0. \tag{4.43}$$

The mole fractions x^{α} and x^{β} of two coexisting phases α and β of a binary mixture with independent variables of temperature T and molar volume V_m are determined by solution of the simultaneous equations

$$\mu_{A,\alpha}(T,V_m^{\alpha},x^{\alpha}) = \mu_{A,\beta}(T,V_m^{\beta},x^{\beta}), \tag{4.44}$$

$$\mu_{B,\alpha}(T,V_m^{\alpha},x^{\alpha}) = \mu_{B,\beta}(T,V_m^{\beta},x^{\beta}), \tag{4.45}$$

and

$$p(T,V_m^{\alpha},x^{\alpha}) = p(T,V_m^{\beta},x^{\beta}). \tag{4.46}$$

The diffusional stability is given by, for example,

$$\left(\frac{\partial \mu_A}{\partial x}\right)_{T,p}^{\alpha} = \left(\frac{\partial \mu_A}{\partial x}\right)_{T,V_m}^{\alpha} - \left(\frac{\partial \mu_A}{\partial V_m}\right)_{T,x}^{\alpha} \frac{(\partial p/\partial x)_{T,V_m}^{\alpha}}{(\partial p/\partial V_m)_{T,x}^{\alpha}} < 0 \tag{4.47}$$

and others analogous to Equation 4.43. However, as either x or p tend to zero the chemical potential tends to infinity. This fact makes the treatment of phase equilibrium in fluids using chemical potential rather inconvenient and leads to the use of a quantity known as fugacity that replaces the chemical potential in the absence of this undesirable characteristic.

4.3.1 What Is the Relationship between Several Chemical Potentials in a Mixture?

For a binary mixture $(1-x)\mathrm{A}+x\mathrm{B}$ at constant temperature and at equilibrium $\mu_{\mathrm{A,g}}=\mu_{\mathrm{A,l}}=\mu_{\mathrm{A}}$, $\mu_{\mathrm{B,g}}=\mu_{\mathrm{B,l}}=\mu_{\mathrm{B}}$, and $p=p^{\mathrm{sat}}$ the Gibbs–Duhem Equation 4.4 becomes

$$(1-x)(\mathrm{d}\mu_{\mathrm{A}}-V_{\mathrm{A,l}}\,\mathrm{d}p^{\mathrm{sat}})+x(\mathrm{d}\mu_{\mathrm{B}}-V_{\mathrm{B,l}}\,\mathrm{d}p^{\mathrm{sat}})=0, \tag{4.48}$$

which may be solved for x to give

$$x=\frac{\mathrm{d}\mu_{\mathrm{A}}-V_{\mathrm{A,l}}\,\mathrm{d}p^{\mathrm{sat}}}{\mathrm{d}\mu_{\mathrm{A}}-V_{\mathrm{A,l}}\,\mathrm{d}p^{\mathrm{sat}}-\mathrm{d}\mu_{\mathrm{B}}+V_{\mathrm{B,l}}\,\mathrm{d}p^{\mathrm{sat}}}. \tag{4.49}$$

4.3.2 What Can Be Done with the Differences in Chemical Potential?

The chemical potential difference $\mu_{\mathrm{B}}-\mu_{\mathrm{B}}^{\ominus}$ or the corresponding ratio of absolute activities $\lambda_{\mathrm{B}}/\lambda_{\mathrm{B}}^{\ominus}$ occurs frequently and it is called the *relative activity a* defined by

$$RT\ln a=\mu_{\mathrm{B}}-\mu_{\mathrm{B}}^{\ominus}, \tag{4.50}$$

or

$$a=\frac{\lambda_{\mathrm{B}}}{\lambda_{\mathrm{B}}^{\ominus}}. \tag{4.51}$$

In Equation 4.50 $\mu^{\ominus}$ is the standard chemical potential, defined in Question 1.10, while in Equation 4.51 $\lambda_{\mathrm{B}}^{\ominus}$ is the standard absolute activity and evidently depends on the choice of the standard state.

4.3.3 How Do I Measure Chemical Potential Differences (What Is Osmotic Pressure)?

Consider a perfect gas mixture (i.e., a gas mixture at low pressure) denoted as phase α that is separated by a membrane permeable only to one substance B from another perfect gas mixture containing the same C components and denoted as phase β. At equilibrium the chemical potentials of substance B in α and β are equal and are given by

$$\mu_{\mathrm{B}}^{\mathrm{pg}}(T,p^{\beta},x_{\mathrm{C}}^{\beta})=\mu_{\mathrm{B}}^{\mathrm{pg}}(T,p^{\alpha},x_{\mathrm{C}}^{\alpha}) \tag{4.52}$$

and are related by the expression

$$x_{\mathrm{B}}^{\beta} p^{\beta} = x_{\mathrm{B}}^{\alpha} p^{\alpha}. \tag{4.53}$$

In Equation 4.52, x_{C}^{α} represents the $(C - 1)$ independent mole fractions in the phase α and similarly x_{C}^{β} refers to the $(C - 1)$ independent mole fractions in phase β. In Equation 4.53 x_{B}^{α} and x_{B}^{α} are the mole fractions of B in phases α and β. Equation 4.53 is a result of the statistical mechanical treatment of perfect gas mixtures on the assumption that perfect gas mixtures behave as if it were an assembly of noninteracting particles (see Question 2.4 and Equation 2.62).

The pressures p^{β} and p^{α} are according to Equation 4.53 related by the mole fraction of B in phases α and β.

The difference

$$\mu_{\mathrm{B}}^{\mathrm{pg}}(T, p, x_{\mathrm{C}}^{\beta}) - \mu_{\mathrm{B}}^{\mathrm{pg}}(T, p, x_{\mathrm{C}}^{\alpha}) \tag{4.54}$$

can be obtained by addition and subtraction of measurable differences and use of Equation 4.53 to give

$$\begin{aligned} \mu_{\mathrm{B}}^{\mathrm{pg}}(T, p, x_{\mathrm{C}}^{\beta}) - \mu_{\mathrm{B}}^{\mathrm{pg}}(T, p, x_{\mathrm{C}}^{\alpha}) &= RT \ln\left(\frac{p^{\alpha}}{p^{\beta}}\right) + \mu_{\mathrm{B}}^{\mathrm{pg}}(T, p^{\beta}, x_{\mathrm{C}}^{\beta} \\ &= \mu_{\mathrm{B}}^{\mathrm{pg}}(T, p^{\alpha}, x_{\mathrm{C}}^{\alpha}), \end{aligned} \tag{4.55}$$

and from Equations 4.52 and 4.53 we have

$$\mu_{\mathrm{B}}^{\mathrm{pg}}(T, p, x_{\mathrm{C}}^{\beta}) - \mu_{\mathrm{B}}^{\mathrm{pg}}(T, p, x_{\mathrm{C}}^{\alpha}) = RT \ln\left(\frac{x_{\mathrm{B}}^{\beta}}{x_{\mathrm{B}}^{\alpha}}\right). \tag{4.56}$$

Thus, we see that for two perfect gas mixtures the difference in chemical potential depends only on the mole fractions of the substance B in each of the phases and not on the other components.

For a real gas mixture the chemical potential difference $\mu_{\mathrm{B,g}}(T, p, x_{\mathrm{C}}^{\beta}) - \mu_{\mathrm{B,g}}(T, p, x_{\mathrm{C}}^{\alpha})$ is given by

$$\begin{aligned} \mu_{\mathrm{B,g}}(T, p, x_{\mathrm{C}}^{\beta}) - \mu_{\mathrm{B,g}}(T, p, x_{\mathrm{C}}^{\alpha}) &= \int_0^p \{V_{\mathrm{B,g}}(T, p, x_{\mathrm{C}}^{\beta}) - V_{\mathrm{B,g}}(T, p, x_{\mathrm{C}}^{\alpha})\}\, \mathrm{d}p \\ &\quad + RT \ln\left(\frac{x_{\mathrm{B}}^{\beta}}{x_{\mathrm{B}}^{\alpha}}\right), \end{aligned} \tag{4.57}$$

where $V_{\text{B,g}}$ is the partial molar volume of gaseous substance B. Measurements of partial molar volume can be used to determine $\mu_{\text{B,g}}(T,p,x_{\text{C}}^{\beta}) - \mu_{\text{B,g}}(T,p,x_{\text{C}}^{\alpha})$ that is equivalent to

$$\left(\frac{\partial \ln \lambda_{\text{B}}}{\partial p}\right)_T = \frac{V_{\text{B}}}{RT}. \tag{4.58}$$

The use of the relationship

$$\left(\frac{\partial \ln \lambda_{\text{B}}}{\partial T}\right)_p = -\frac{H_{\text{B}}}{RT^2}, \tag{4.59}$$

is discussed in Question 4.6.1.

Consider two-liquid mixtures denoted as phases α and β containing the same components separated by a membrane permeable solely to substance B. At equilibrium the chemical potentials of substance B in phases α and β are equal and are given by

$$\mu_{\text{B,l}}(T,p^{\beta},x_{\text{C}}^{\beta}) = \mu_{\text{B,l}}(T,p^{\alpha},x_{\text{C}}^{\alpha}), \tag{4.60}$$

the difference $\mu_{\text{B,l}}(T,p,x_{\text{C}}^{\beta}) - \mu_{\text{B,l}}(T,p,x_{\text{C}}^{\alpha})$ is given by

$$\mu_{\text{B,l}}(T,p,x_{\text{C}}^{\beta}) - \mu_{\text{B,l}}(T,p,x_{\text{C}}^{\alpha}) = \int_{p^{\beta}}^{p} V_{\text{B,l}}(T,p,x_{\text{C}}^{\beta})\,\mathrm{d}p + \int_{p^{\alpha}}^{p} V_{\text{B,l}}(T,p,x_{\text{C}}^{\alpha})\,\mathrm{d}p. \tag{4.61}$$

If phase α is pure B, denoted by superscript *, Equation 4.60 can be written as

$$\mu_{\text{B}}(T,p+\Pi,x_{\text{C}}^{\beta}) = \mu_{\text{B}}^{*}(T,p), \tag{4.62}$$

where Π is the pressure that must be applied to ensure diffusive equilibrium and is called the *osmotic pressure* of the mixture for substance B. In this case, the chemical potential difference is given by

$$\mu_{\text{B}}(T,p+\Pi,x_{\text{C}}^{\beta}) - \mu_{\text{B}}^{*}(T,p) = -\int_{p}^{p+\Pi} V_{\text{B}}(T,p,x_{\text{C}}^{\beta})\,\mathrm{d}p. \tag{4.63}$$

The problem of obtaining the chemical potential difference requires selection of a membrane permeable solely to substance B.

4.4 DO I HAVE TO USE CHEMICAL POTENTIALS? WHAT IS FUGACITY?

The chemical potential for a substance B in a gas mixture is given by

$$\mu_{B,g}(T,p,y_C) = \mu^{\ominus}_{B,g}(T) + RT\ln\left(\frac{y_B}{p^{\ominus}}\right) + \int_0^p V_{B,g}(T,p,y_C)\,dp, \tag{4.64}$$

which uses the expression

$$\mu_{B,g}(T,p\to 0,y_C^{\beta}) - \mu_{B,g}(T,p\to 0,y_C^{\alpha}) = RT\ln\left(\frac{y_C^{\beta}}{y_C^{\alpha}}\right) \tag{4.65}$$

for a mixture of perfect gases. The integral in Equation 4.64 diverges as $p\to 0$ and is eliminated by rearranging Equation 4.64 to give

$$\mu_{B,g}(T,p,y_C) = \mu^{*}_{B,g}(T,p\to 0) + \int_0^{p^{\ominus}}\left(\frac{RT}{p}\right)dp + \int_{p^{\ominus}}^{p}\left(\frac{RT}{p}\right)dp + RT\ln y_B$$

$$+\int_0^p\left\{V_{B,g}(T,p,y_C) - \frac{RT}{p}\right\}dp$$

$$= \mu^{\ominus}_{B,g}(T) + RT\ln\left(\frac{y_B p}{p^{\ominus}}\right) + \int_0^p\left\{V_{B,g}(T,p,y_C) - \frac{RT}{p}\right\}dp. \tag{4.66}$$

In terms of absolute activity λ, Equation 4.66 is

$$\frac{p^{\ominus}\lambda_{B,g}(T,p,y_C)}{\lambda^{\ominus}_{B,g}(T)} = (y_B p)\exp\left[\int_0^p\left\{\frac{V_{B,g}(T,p,y_C)}{RT} - \frac{1}{p}\right\}dp\right]. \tag{4.67}$$

In Equations 4.64, 4.66, and 4.67 y_C denotes the $(C-1)$ independent mole fractions $y_B, y_C, \cdots$, of substances B, C, $\cdots$. Equation 4.67 provides, for a substance B in a gas mixture, a measure of the departure of the gas from perfection and as such is defined as the *fugacity* $\tilde{p}_B$ given by

$$\tilde{p}_{B,g}(T,p,y_C) \overset{\text{def}}{=} (y_B p)\exp\left\{\int_0^p\left(\frac{V_{B,g}(T,p,y_C)}{RT} - \frac{1}{p}\right)dp\right\}, \tag{4.68}$$

and $\tilde{p}_{\mathrm{B}}$ has the dimension of pressure. The fugacity coefficient, $\phi_{\mathrm{B,g}}(T, p, y_{\mathrm{C}})$, is defined by

$$\phi_{\mathrm{B,g}}(T,p,y_{\mathrm{C}}) \overset{\mathrm{def}}{=} \frac{\tilde{p}_{\mathrm{B,g}}(T,p,y_{\mathrm{C}})}{y_{\mathrm{B}}p}. \tag{4.69}$$

The fugacity coefficient is preferred to the fugacity because the fugacity coefficient varies less than the fugacity with respect to changes in temperature, pressure, and composition. Equation 4.68 can be cast as

$$\ln\phi_{\mathrm{B}} = \ln\left(\frac{\tilde{p}_{\mathrm{B,g}}}{y_{\mathrm{B}}p}\right) = \int_0^p \left\{\frac{V_{\mathrm{B,g}}(T,p,y_{\mathrm{C}})}{RT} - \frac{1}{p}\right\} \mathrm{d}p. \tag{4.70}$$

For a pure substance $y_{\mathrm{B}} = 1$, Equation 4.70 becomes

$$\ln\left(\frac{\tilde{p}^{*}_{\mathrm{B,g}}}{p}\right) = \int_0^p \left\{\frac{V_{\mathrm{m,g}}(T,p)}{RT} - \frac{1}{p}\right\} \mathrm{d}p. \tag{4.71}$$

For a perfect gas, for which $pV_{\mathrm{m}} = RT$, Equation 4.71 reduces to

$$\tilde{p}^{\mathrm{pg}}_{\mathrm{B}} = p. \tag{4.72}$$

For a gas mixture at sufficiently low pressure the properties can be represented by the virial equation (Chapter 2) truncated after the second virial coefficient in which the second virial coefficient of the mixture is given by

$$B(T,x) = (1-x)^2 B_{\mathrm{A}} + 2(1-x)xB_{\mathrm{AB}} + x^2 B_{\mathrm{B}}. \tag{4.73}$$

In the absence of sufficient measurements to determine B_{AB} this quantity may be obtained from molar volumes of mixing through

$$\Delta_{\mathrm{mix}}V_{\mathrm{m}} = 2(1-x)x\delta_{\mathrm{AB}}, \tag{4.74}$$

where

$$\delta_{\mathrm{AB}} = B_{\mathrm{AB}} - \frac{1}{2}(B_{\mathrm{A}} + B_{\mathrm{B}}). \tag{4.75}$$

It is often a good approximation to assume

$$B_{\mathrm{AB}} = \frac{1}{2}(B_{\mathrm{A}} + B_{\mathrm{B}}), \tag{4.76}$$

and then $\delta_{AB} = 0$. Equation 4.68 can then be written as

$$\tilde{p}_{B,g}(T, p, y_C) = (y_B p) \exp\left(\frac{B_B p}{RT}\right), \tag{4.77}$$

which is often used to correct equilibrium constants to different thermodynamic conditions.

4.4.1 Can Fugacity Be Used to Calculate (Liquid + Vapor) Phase Equilibrium?

For the equilibrium of the liquid and gaseous phases of a mixture the fugacity is given by Equation 4.67 or 4.68 with the pressure of p^{sat}. At equilibrium, $\mu_{B,g}(T, p, y_C) = \mu_{B,l}(T, p, x_C)$ or $\ln\{\lambda_{B,g}(T, p, y_C)\} = \ln\{\lambda_{B,l}(T, p, x_C)\}$ so that for all B

$$\tilde{p}_{B,g}(T, p^{sat}, y_C) = \tilde{p}_{B,l}(T, p^{sat}, x_C). \tag{4.78}$$

The chemical potential of the liquid is given by

$$\begin{aligned} \mu_{B,l}(T, p, x_C) &= \mu^{*}_{B,g}(T, p \to 0) + RT \ln\left\{\frac{x_C p^{sat}}{p^{\ominus}}\right\} + \int_{p^{\ominus}}^{p} \left(\frac{RT}{p}\right) dp \\ &\quad + \int_{0}^{p^{sat}} \left\{V_{B,g}(T, p, x_C) - \frac{RT}{p}\right\} dp + \int_{p^{sat}}^{p} V_{B,l}(T, p, x_C)\, dp \\ &= \mu^{\ominus}_{B,g}(T) + RT \ln\left(\frac{y_B p^{sat}}{p^{\ominus}}\right) + \int_{0}^{p} \left\{V_{B,g}(T, p, y_C) - \frac{RT}{p}\right\} dp \\ &\quad + \int_{p^{sat}}^{p} V_{B,l}(T, p, x_C)\, dp, \end{aligned} \tag{4.79}$$

or is recast using the ratio of absolute activities (that are much more convenient to use than the chemical potential in this case) as

$$\frac{p^{\ominus} \lambda_{B,l}(T, p, x_C)}{\lambda^{\ominus}_{B,g}(T)} = (x_B p^{sat}) \exp\left[\begin{aligned} &\int_{0}^{p} \left\{\frac{V_{B,g}(T, p, y_C)}{RT} - \frac{1}{p}\right\} dp \\ &+ \frac{1}{RT} \int_{p^{sat}}^{p} V_{B,l}(T, p, x_C)\, dp \end{aligned}\right]. \tag{4.80}$$

Equations precisely analogous to Equations 4.79 and 4.80 apply to equilibrium with a solid. Substitution of Equation 4.68 into Equation 4.80 gives

$$\frac{\tilde{p}_{\mathrm{B,l}}(T,p^{\mathrm{sat}},x_{\mathrm{C}})}{x_{\mathrm{B}}p^{\mathrm{sat}}}=\frac{\tilde{p}_{\mathrm{B,g}}(T,p^{\mathrm{sat}},y_{\mathrm{C}})}{y_{\mathrm{B}}p^{\mathrm{sat}}}\exp\left\{\frac{1}{RT}\int_{p^{\mathrm{sat}}}^{p}V_{\mathrm{B,l}}(T,p,x)\,\mathrm{d}p\right\},\qquad(4.81)$$

and with use of the definition of a fugacity coefficient of substance B in a mixture given by Equation 4.69

$$\tilde{p}_{\mathrm{B,l}}(T,p^{\mathrm{sat}},x_{\mathrm{C}})=\phi_{\mathrm{g}}(T,p^{\mathrm{sat}},y_{\mathrm{C}})x_{\mathrm{B}}p^{\mathrm{sat}}\exp\left[\frac{1}{RT}\int_{p^{\mathrm{sat}}}^{p}V_{\mathrm{B,l}}(T,p,x_{\mathrm{C}})\,\mathrm{d}p\right].\qquad(4.82)$$

The fugacity coefficient of the liquid is given by

$$\phi_{\mathrm{B,l}}(T,p,x_{\mathrm{C}})\overset{\mathrm{def}}{=}\frac{\tilde{p}_{\mathrm{B,l}}(T,p,x_{\mathrm{C}})}{x_{\mathrm{B}}p^{\mathrm{sat}}}.\qquad(4.83)$$

The problem of calculating phase equilibria has been changed from estimating chemical potentials to the determination of saturation pressure, liquid volumes, and the fugacity coefficient that itself can be determined from the compression factor. The fugacity coefficients are usually more slowly varying functions of temperature, pressure, and composition than the fugacity. This indicates the power of equations of state in this context because they provide what is at least a thermodynamically consistent means of evaluating these quantities.

The fugacity coefficient of Equation 4.81 requires an equation of state to evaluate the partial molar volume. However, equations of state usually have temperature and volume as the independent variables, and the fugacity coefficient of a substance in a mixture at constant temperature and composition can be obtained from

$$RT\ln\phi_{\mathrm{B}}(x)=\int_{V}^{\infty}\left[\left(\frac{\partial p}{\partial n_{\mathrm{B}}}\right)_{T,V,n_{\mathrm{C}\neq\mathrm{B}}}-\frac{RT}{V}\right]\mathrm{d}V-RT\ln Z\qquad(4.84)$$

in which $Z=pV/nRT$ is the compression factor of the mixture; the partial derivative in Equation 4.84 can be obtained from the equation of state. For a pure substance Equation 4.84 reduces to

$$RT\ln\phi_{\mathrm{B}}^{*}=RT\ln\frac{f}{p}=\int_{V}^{\infty}\left[\frac{p}{n_{\mathrm{B}}}-\frac{RT}{V}\right]\mathrm{d}V-RT\ln Z+RT(Z-1).\qquad(4.85)$$

For an extensive treatment of the fugacity concept, the reader should refer to Modell and Reid (1983), van Ness and Abbott (1982), Smith et al. (2004), and Prausnitz et al. (1986).

The exponential term in Equation 4.82 is called the *Poynting factor* and is widely used by engineers; it is denoted by F_B and is given by

$$F_B(T,x) = \exp\left[\int_{p^{sat}}^{p}\left\{\frac{V_{B,1}(T,p,x)}{RT}\right\}dp\right]. \tag{4.86}$$

The liquid molar volume $V_{B,1}(T,p,x)$ in Equation 4.86 is usually a weak function of pressure at temperature below critical and thus

$$\int_{p^{sat}}^{p} V_{A,l}(T,p,x)\,dp \approx V_{A,l}(T,x)(p-p^{sat}) \tag{4.87}$$

and therefore

$$F_B = \exp\left\{\frac{V_{A,l}(T,x)(p-p^{sat})}{RT}\right\}. \tag{4.88}$$

At constant temperature and pressure, from Equation 4.4 the Gibbs–Duhem equation for fugacity coefficients can be obtained as

$$\sum_B x_B d\left[\ln\{\phi_B(T,p,x)\}\right] = 0. \tag{4.89}$$

In summary, an equation of state provides a thermodynamically consistent route to the evaluation of the fugacity of components in both vapor and liquid phases. It thus offers a very convenient basis for phase-equilibrium calculations. Indeed, the most well-known application of such methods in chemical engineering lies in the field of high-pressure vapor + liquid equilibria (VLE) where the equation-of-state approach is the method of choice for the vast majority of systems.

4.5 WHAT ARE IDEAL LIQUID MIXTURES?

In Chapter 1, the molar function for mixing of extensive quantities was introduced for $\{(1-x)A + xB\}$. Here we define for the same system $\{(1-x)A + xB\}$ an ideal mixture by

$$\lambda_{A,id}(T,p,x) = (1-x)\lambda_A^*(T,p), \tag{4.90}$$

and

$$\lambda_{\rm B,id}(T,p,x) = x\lambda_{\rm B}^{*}(T,p), \tag{4.91}$$

where λ is the absolute activity and we remind the reader that the superscript asterisk denotes a pure substance. Here we only consider binary mixtures because the generalization is straightforward and the understanding can be obscured by the complexity. The molar functions of mixing for an ideal mixture are then given by

$$\Delta_{\rm mix}G_{\rm m,id} = RT\{(1-x)\ln(1-x) + x\ln x\}, \tag{4.92}$$

$$\Delta_{\rm mix}S_{\rm m,id} = -R\{(1-x)\ln(1-x) + x\ln x\}, \tag{4.93}$$

$$\Delta_{\rm mix}H_{\rm m,id} = 0, \tag{4.94}$$

and

$$\Delta_{\rm mix}V_{\rm m,id} = 0. \tag{4.95}$$

No real mixture is ideal but those formed from similar substances do behave in large measure as ideal. The excess molar functions $X_{\rm m}^{\rm E}$ are defined by

$$X_{\rm m}^{\rm E} = \Delta_{\rm mix}X_{\rm m} - \Delta_{\rm mix}X_{\rm m,id}, \tag{4.96}$$

where the excess is over the ideal mixture denoted by id and is defined by Equations 4.92 through 4.95 thus the excess molar functions for a binary mixture are

$$G_{\rm m}^{\rm E} = \Delta_{\rm mix}G_{\rm m} - RT\{(1-x)\ln(1-x) + x\ln x\} = (1-x)\mu_{\rm A}^{\rm E} + x\mu_{\rm B}^{\rm E}, \tag{4.97}$$

$$S_{\rm m}^{\rm E} = \Delta_{\rm mix}S_{\rm m} + R\{(1-x)\ln(1-x) + x\ln x\}, \tag{4.98}$$

$$H_{\rm m}^{\rm E} = \Delta_{\rm mix}H_{\rm m}, \tag{4.99}$$

and

$$V_{\rm m}^{\rm E} = \Delta_{\rm mix}V_{\rm m}. \tag{4.100}$$

4.6 WHAT ARE ACTIVITY COEFFICIENTS?

Many mixtures of interest in the chemical industry exhibit strong nonideality, for example, acetone + water, and these have traditionally been described by activity coefficients (or as we will see equivalently the excess molar Gibbs function) for the liquid phase, and an equation of state for the vapor phase. It is to the description of the activity coefficient that we now turn. However, the activity-coefficient description has numerous drawbacks that include (1) the inability to define standard states for supercritical components; (2) the critical phenomena cannot be predicted because a different model is used for the vapor and liquid phases; (3) the model parameters are highly temperature dependent; and (4) it cannot predict values for the density, enthalpy, and entropy from the same model.

The chemical potential of substance A in a binary liquid mixture is obtained from

$$\mu_{\mathrm{A,l}}(T,p,x) = \mu_{\mathrm{A,l}}(T,p^{\mathrm{sat}},x) + \int_{p^{\mathrm{sat}}}^{p} V_{\mathrm{A,l}}(T,p,x)\,\mathrm{d}p. \tag{4.101}$$

At equilibrium, the liquid and gas phase chemical potentials of substance A are equal and are given by

$$\mu_{\mathrm{A,l}}(T,p^{\mathrm{sat}},x) = \mu_{\mathrm{A,g}}(T,p^{\mathrm{sat}},y), \tag{4.102}$$

where we have used y as the mole fraction of substance A in the gas phase and x denotes the mole fractions of the substance A in the liquid phase. In view of Equation 1.125, Equation 4.102 can be written as

$$\mu_{\mathrm{A,g}}(T,p^{\mathrm{sat}},y) = \mu_{\mathrm{A,g}}^{\ominus}(T) + RT\ln\left\{\frac{(1-y)p^{\mathrm{sat}}}{p^{\ominus}}\right\} + \int_{0}^{p^{\mathrm{sat}}}\left\{V_{\mathrm{A,g}}(T,p,x) - \frac{RT}{p}\right\}\mathrm{d}p. \tag{4.103}$$

Equation 4.101 is then

$$\mu_{\mathrm{A,l}}(T,p,x) = \mu_{\mathrm{A,g}}^{\ominus}(T) + RT\ln\left\{\frac{(1-y)p^{\mathrm{sat}}}{p^{\ominus}}\right\} + \int_{0}^{p^{\mathrm{sat}}}\left\{V_{\mathrm{A,g}}(T,p,x) - \frac{RT}{p}\right\}\mathrm{d}p + \int_{p^{\mathrm{sat}}}^{p} V_{\mathrm{A,l}}(T,p,x)\,\mathrm{d}p. \tag{4.104}$$

For substance B the equation is

$$\mu_{\mathrm{B,l}}(T,p,x)=\mu_{\mathrm{B,g}}^{\ominus}(T)+RT\ln\left\{\frac{(1-y)p^{\mathrm{sat}}}{p^{\ominus}}\right\}+\int_0^{p^{\mathrm{sat}}}\left\{V_{\mathrm{B,g}}(T,p,x)-\frac{RT}{p}\right\}\mathrm{d}p$$
$$+\int_{p^{\mathrm{sat}}}^{p}V_{\mathrm{B,l}}(T,p,x)\,\mathrm{d}p. \tag{4.105}$$

The chemical potentials of pure A and B are given by

$$\mu_{\mathrm{A,l}}^{*}(T,p)=RT\ln\{(1-x)p_{\mathrm{A}}^{\mathrm{sat}}\}-\int_0^{p_{\mathrm{A}}^{\mathrm{sat}}}\left\{V_{\mathrm{A,g}}^{*}(T,p)-\frac{RT}{p}\right\}\mathrm{d}p$$
$$-\int_{p^{\mathrm{sat}}}^{p}V_{\mathrm{A,l}}^{*}(T,p)\,\mathrm{d}p-\int_{p_{\mathrm{A}}^{\mathrm{sat}}}^{p^{\mathrm{sat}}}V_{\mathrm{A,l}}^{*}(T,p)\,\mathrm{d}p, \tag{4.106}$$

and

$$\mu_{\mathrm{B,l}}^{*}(T,p)=RT\ln\{xp_{\mathrm{B}}^{\mathrm{sat}}\}-\int_0^{p_{\mathrm{B}}^{\mathrm{sat}}}\left\{V_{\mathrm{B,g}}^{*}(T,p)-\frac{RT}{p}\right\}\mathrm{d}p$$
$$-\int_{p^{\mathrm{sat}}}^{p}V_{\mathrm{B,l}}^{*}(T,p)\,\mathrm{d}p-\int_{p_{\mathrm{B}}^{\mathrm{sat}}}^{p^{\mathrm{sat}}}V_{\mathrm{A,l}}^{*}(T,p)\,\mathrm{d}p. \tag{4.107}$$

In Equations 4.106 and 4.107 the * indicates the pure substance.

For a binary mixture xA + (1 − x)B the excess chemical potential of A is given by

$$\mu_{\mathrm{A,l}}^{\mathrm{E}}(T,p,x)=RT\ln\left\{\frac{(1-y)p^{\mathrm{sat}}}{(1-x)p_{\mathrm{A}}^{\mathrm{sat}}}\right\}+\int_0^{p^{\mathrm{sat}}}\left\{V_{\mathrm{A,g}}(T,p,y)-\frac{RT}{p}\right\}\mathrm{d}p$$
$$+\int_{p^{\mathrm{sat}}}^{p}V_{\mathrm{A,l}}(T,p,x)\,\mathrm{d}p-\int_0^{p_{\mathrm{A}}^{\mathrm{sat}}}\left\{V_{\mathrm{A,g}}^{*}(T,p)-\frac{RT}{p}\right\}\mathrm{d}p$$
$$-\int_{p^{\mathrm{sat}}}^{p}V_{\mathrm{A,l}}^{*}(T,p)\,\mathrm{d}p-\int_{p_{\mathrm{A}}^{\mathrm{sat}}}^{p^{\mathrm{sat}}}V_{\mathrm{A,l}}^{*}(T,p)\,\mathrm{d}p, \tag{4.108}$$

and for substance B

$$\mu_{\mathrm{B,l}}^{\mathrm{E}}(T,p,x)=RT\ln\left\{\frac{yp^{\mathrm{sat}}}{xp_{\mathrm{B}}^{\mathrm{sat}}}\right\}+\int_{0}^{p^{\mathrm{sat}}}\left\{V_{\mathrm{B,g}}(T,p,y)-\frac{RT}{p}\right\}\mathrm{d}p+\int_{p^{\mathrm{sat}}}^{p}V_{\mathrm{B,l}}(T,p,x)\,\mathrm{d}p$$
$$-\int_{0}^{p_{\mathrm{A}}^{\mathrm{sat}}}\left\{V_{\mathrm{B,g}}^{*}(T,p,y)-\frac{RT}{p}\right\}\mathrm{d}p-\int_{p^{\mathrm{sat}}}^{p}V_{\mathrm{B,l}}^{*}(T,p,x)\,\mathrm{d}p-\int_{p_{\mathrm{B}}^{\mathrm{sat}}}^{p^{\mathrm{sat}}}V_{\mathrm{B,l}}^{*}(T,p)\,\mathrm{d}p. \tag{4.109}$$

The $G_{\mathrm{m}}^{\mathrm{E}}$ can be obtained from Equation 4.97 using Equations 4.108 and 4.109.

The activity coefficients f_{A} and f_{B} for a binary mixture $(1-x)\mathrm{A}+x\mathrm{B}$ of liquids (or solids) are defined by

$$RT\ln f_{\mathrm{A,l}}=G_{\mathrm{m}}^{\mathrm{E}}-x\left(\frac{\partial G_{\mathrm{m}}^{\mathrm{E}}}{\partial x}\right)_{T,p}=\mu_{\mathrm{A}}^{\mathrm{E}}=RT\ln\left\{\frac{\lambda_{\mathrm{A}}}{(1-x_{\mathrm{B}})\lambda_{\mathrm{A}}^{*}}\right\}, \tag{4.110}$$

and

$$RT\ln f_{\mathrm{B,l}}=G_{\mathrm{m}}^{\mathrm{E}}-(1-x)\left(\frac{\partial G_{\mathrm{m}}^{\mathrm{E}}}{\partial x}\right)_{T,p}=\mu_{\mathrm{B}}^{\mathrm{E}}=RT\ln\left\{\frac{\lambda_{\mathrm{B}}(1,T,p,x_{\mathrm{C}})}{x_{\mathrm{B}}\lambda_{\mathrm{B}}^{*}(1,T,p)}\right\}, \tag{4.111}$$

where $\mu_{\mathrm{A}}^{\mathrm{E}}$ and $\mu_{\mathrm{A}}^{\mathrm{E}}$ are given by Equations 4.108 and 4.109, respectively. From the definition of the standard chemical potential of B of Equation 1.132 we have

$$\mu_{\mathrm{B,l}}^{\ominus}(T)=\mu_{\mathrm{B,l}}^{*}(T,p)+\int_{p}^{p^{\ominus}}V_{\mathrm{B,l}}^{*}(T,p)\,\mathrm{d}p, \tag{4.112}$$

or in terms of the absolute activity from Equation 1.133

$$\lambda_{\mathrm{B,l}}^{\ominus}(T)=\lambda_{\mathrm{B,l}}^{*}(T,p)+\exp\left\{\frac{1}{RT}\int_{p}^{p^{\ominus}}V_{\mathrm{B,l}}^{*}(T,p)\,\mathrm{d}p\right\}, \tag{4.113}$$

Equation 4.111 can be written for the chemical potential as

$$RT\ln x_{\mathrm{B}}f_{\mathrm{B,l}}(T,p,x_{\mathrm{C}})=\mu_{\mathrm{B,l}}(T,p,x_{\mathrm{C}})-\mu_{\mathrm{B,l}}^{\ominus}(T)+\int_{p}^{p^{\ominus}}V_{\mathrm{B,l}}^{*}(T,p)\,\mathrm{d}p, \tag{4.114}$$

or for the absolute activity

$$RT\ln f_{\mathrm{B,l}}(T,p,x_{\mathrm{C}}) = RT\ln\left(\frac{\lambda_{\mathrm{B,l}}(T,p,x_{\mathrm{C}})}{x_{\mathrm{B}}\left[\lambda_{\mathrm{B,l}}^{\ominus}(T)\exp\left\{(1/RT)\int_{p^{\ominus}}^{p} V_{\mathrm{B,l}}^{*}(T,p)\,\mathrm{d}p\right\}\right]}\right). \tag{4.115}$$

In Equations 4.112 through 4.115 $V_{\mathrm{B,l}}^{*}(T,p)$ is the molar volume of pure liquid B at temperature T and pressure p, and we have used the definition of the standard absolute activity. When the pressure is close to $p^{\ominus}$ ($p \to p^{\ominus}$) the integral makes a negligible contribution to Equations 4.112 through 4.115 and is often taken to be zero so that, Equation 4.115 becomes

$$RT\ln f_{\mathrm{B,l}} \approx RT\ln\left\{\frac{\lambda_{\mathrm{B,l}}(T,p,x_{\mathrm{C}})}{x_{\mathrm{B}}\lambda_{\mathrm{B,l}}^{\ominus}(T)}\right\}, \tag{4.116}$$

and, with Equation 4.50, Equation 4.116 becomes

$$RT\ln\left\{\frac{\lambda_{\mathrm{B,l}}(T,p,x_{\mathrm{C}})}{x_{\mathrm{B}}\lambda_{\mathrm{B,l}}^{\odot}(T)}\right\} = R\ln\left(\frac{a_{\mathrm{B}}}{x_{\mathrm{B}}}\right) = RT\ln f_{\mathrm{B,l}}. \tag{4.117}$$

From Equation 4.117 the definition

$$a_{\mathrm{B}} = f_{\mathrm{B,l}}x_{\mathrm{B}}, \tag{4.118}$$

emerges, which is equivalent to

$$\frac{\lambda_{\mathrm{B,l}}(T,p,x_{\mathrm{C}})}{\lambda_{\mathrm{B,l}}^{*}(T,p)} = \frac{\tilde{p}_{\mathrm{B,l}}(T,p,x_{\mathrm{C}})}{\tilde{p}_{\mathrm{B,l}}^{*}(T,p)} = f_{\mathrm{B,l}}x_{\mathrm{B}}. \tag{4.119}$$

If the $V_{\mathrm{B,l}}(T,p,x) \ll V_{\mathrm{B,g}}(T,p,x)$ so that $V_{\mathrm{B,l}}(T,p,x)$ can be neglected and if we assume that the gas phase is ideal so that $V_{\mathrm{m,g}} = RT/p$ then Equation 4.108 reduces to

$$\mu_{\mathrm{A}}^{\mathrm{E}} = RT\ln\left\{\frac{(1-y)p^{\mathrm{sat}}}{(1-x)p_{\mathrm{A}}^{\mathrm{sat}}}\right\}, \tag{4.120}$$

and similarly Equation 4.109 reduces to

$$\mu_{\mathrm{B}}^{\mathrm{E}} = RT\ln\left(\frac{yp^{\mathrm{sat}}}{xp_{\mathrm{B}}^{\mathrm{sat}}}\right). \tag{4.121}$$

For an ideal mixture $\mu_A^E = 0$ and $\mu_B^E = 0$, Equations 4.120 and 4.121 become

$$(1-y)p^{sat} = (1-x)p_A^{sat}, \tag{4.122}$$

and

$$yp^{sat} = xp_B^{sat}. \tag{4.123}$$

Equations 4.122 and 4.123 are known as Raoult's law and require, in principle, that the gas mixture is perfect and that the molar volumes of the liquid are negligible.

When the interest is in low pressure so that the (p, V_m, T) can be represented adequately by a virial expansion up to the second virial coefficient the properties of the integral of Equation 4.103 can be cast as

$$\int_0^{p^{sat}} \left\{ V_{A,g}(T,p,x) - \frac{RT}{p} \right\} dp = B_{AA} p^{sat} + 2(x_g)^2 \delta_{AB}\, p^{sat} \tag{4.124}$$

and provided the p^{sat} is similar to the pressure p then the integral of Equation 4.104 will be

$$\int_{p^{sat}}^{p} V_{A,l}(T,p,x)dp = V_{A,l}(T,x)(p-p^{sat}). \tag{4.125}$$

In Equation 4.124 δ_{AB} is defined by Equation 4.75 so that Equations 4.108 and 4.109 for μ_A^E and μ_B^E, respectively, can be approximated by

$$\mu_A^E = RT\ln\left\{\frac{(1-y)p^{sat}}{(1-x)p_A^{sat}}\right\} + (B_{AA} - V_{A,l}^*)(p^{sat} - p_A^{sat}) \\ + 2(y)^2\delta_{AB}p^{sat} + \{V_{A,l}(x) - V_{A,l}^*\}(p - p^{sat}), \tag{4.126}$$

and

$$\mu_B^E = RT\ln\left\{\frac{yp^{sat}}{xp_B^{sat}}\right\} + (B_{BB} - V_{B,l}^*)(p^{sat} - p_B^{sat}) \\ + 2(y)^2\delta_{AB}p^{sat} + \{V_{B,l}(x) - V_{B,l}^*\}(p - p^{sat}), \tag{4.127}$$

and from Equation 4.97 the molar-excess Gibbs function is given by

$$G_{\mathrm{m}}^{\mathrm{E}}(T,p,x)=(1-x)RT\ln\left\{\frac{(1-y)p^{\mathrm{sat}}}{(1-x)p_{\mathrm{A}}^{\mathrm{sat}}}\right\}+xRT\ln\left\{\frac{yp^{\mathrm{sat}}}{xp_{\mathrm{B}}^{\mathrm{sat}}}\right\}$$

$$+(1-x)(B_{\mathrm{AA}}-V_{\mathrm{A,l}}^{*})(p^{\mathrm{sat}}-p_{\mathrm{A}}^{\mathrm{sat}})+x(B_{\mathrm{BB}}-V_{\mathrm{B,l}}^{*})(p^{\mathrm{sat}}-p_{\mathrm{B}}^{\mathrm{sat}})$$

$$+\{(1-x)y^{2}x(1-y)^{2}\}2\delta_{\mathrm{AB}}p^{\mathrm{sat}}+V_{\mathrm{m,l}}^{\mathrm{E}}(T,x)(p-p^{\mathrm{sat}}). \quad (4.128)$$

No matter what model is used to describe deviations from ideality, it remains necessary to satisfy the Gibbs–Duhem equation (Equation 4.4).

4.6.1 How Do I Measure the Ratio of Absolute Activities at a Phase Transition?

From Question 3.5.1, Equations 3.43, 3.63, and 3.67 we have

$$\left(\frac{\partial \mu_{\mathrm{B}}}{\partial p}\right)_{T}=V_{\mathrm{B}}, \quad (4.129)$$

$$\left(\frac{\partial \ln \lambda_{\mathrm{B}}}{\partial p}\right)_{T}=\frac{V_{\mathrm{B}}}{RT}, \quad (4.130)$$

and

$$\left(\frac{\partial \ln \lambda_{\mathrm{B}}}{\partial T}\right)_{p}=-\frac{H_{\mathrm{B}}}{RT^{2}}, \quad (4.131)$$

and they can be used to obtain the ratio of absolute activities

$$\frac{\lambda_{\mathrm{B,l}}(T,p,x_{\mathrm{C}})}{\lambda_{\mathrm{B,l}}^{*}(T,p)}. \quad (4.132)$$

To carry this through we will consider a very common situation where one system consists of pure ice, denoted as β^*, coexisting in equilibrium with a liquid solution of a solute (such as sucrose) dissolved in water and denoted as phase α; and a separate system containing pure ice denoted as β^*, coexisting with pure water denoted as α^*. For the case chosen, the phase transition temperatures for the mixture containing B and pure B is denoted by T_{B}; for the pure liquid substance in equilibrium with the pure B the transition temperature is T_{B}^{*}. In this case,

$$\lambda_{\mathrm{B},\alpha^*}^{*}(T_{\mathrm{B}}^{*},p)=\lambda_{\mathrm{B},\beta}^{*}(T_{\mathrm{B}}^{*},p), \quad (4.133)$$

and

$$\lambda_{B,\alpha}(T_B, p, x_C) = \lambda^*_{B,\beta^*}(T_B, p), \tag{4.134}$$

where in each equation x_C denotes the set of mole fractions in α. In view of these definitions Equation 4.132 can be recast as

$$\ln\left\{\frac{\lambda^*_{B,\alpha^*}(T_B, p)}{\lambda_B(\alpha, T_B, p, x_C)}\right\} = \ln\left\{\frac{\lambda^*_{B,\beta^*}(T_B, p)}{\lambda^*_{B,\beta}(T^*_B, p)}\right\} - \ln\left\{\frac{\lambda^*_{B,\alpha^*}(T^*_B, p)}{\lambda^*_{B,\alpha^*}(T_B, p)}\right\}$$
$$= \int_{T_B}^{T^*_B}\left\{\left(\frac{\partial \ln \lambda_{B,\beta^*}}{\partial T}\right)_p - \left(\frac{\partial \ln \lambda_{B,\alpha^*}}{\partial T}\right)_p\right\}, \tag{4.135}$$

and, on using Equations 4.131, becomes

$$\ln\left\{\frac{\lambda^*_B(\alpha^*, T_B, p)}{\lambda_B(\alpha, T_B, p, x_C)}\right\} = \int_{T_B}^{T^*_B}\left\{\frac{(H^{\alpha^*}_B - H^{\beta^*}_B)}{RT^2}\right\}dT = \int_{T_B}^{T^*_B}\left\{\frac{\Delta^\alpha_\beta H^*_B}{RT^2}\right\}dT. \tag{4.136}$$

The molar enthalpy difference $\Delta^\alpha_\beta H^*_B$, in the specified case, is the molar enthalpy of melting $\Delta^l_s H^*_B$. The ratio $\lambda^*_{B,\alpha^*}(T_B, p)/\lambda_{B,\alpha}(T_B, p, x_C)$ is, as a consequence of the inequality $(\partial\mu_B/\partial n_B)_{T,p,n_{C\neq B}} > 0$, always greater than unity.

In the case when α is a liquid and β is a gas (then $\Delta^g_l H^*_B$ is the molar enthalpy of evaporation) it follows from Equation 4.136 that $T_B > T^*_B$ so that the boiling temperature of a solvent is always increased by the addition of an involatile solute. Unfortunately, numerical evaluation of Equation 4.136 can only be achieved using a Taylor series and it requires for convergence that $T_B \approx T^*_B$ so that Equation 4.136 can be written as

$$\ln\left\{\frac{\lambda^*_{B,\alpha^*}(T_B, p)}{\lambda_B(\alpha, T_B, p, x_C)}\right\} \approx \left\{\frac{\Delta^\alpha_\beta H^*_B(T^*_B, p)}{RT^*_B}\right\}\left(1 - \frac{T_B}{T^*_B}\right)$$
$$+\left\{\frac{\Delta^\alpha_\beta H^*_B(T^*_B, p)}{RT^*_B} - \frac{\Delta^\alpha_\beta C^*_{p,B}(T^*_B, p)}{2R}\right\}\left(1 - \frac{T_B}{T^*_B}\right)^2. \tag{4.137}$$

An alternative approach is to determine the ratio of absolute activities at any temperature T and to use the relationship

$$\ln\left\{\frac{\lambda^*_{B,\alpha^*}(T, p)}{\lambda_{B,\alpha}(T, p, x_C)}\right\} = \ln\left\{\frac{\lambda^*_{B,\alpha^*}(T_B, p)}{\lambda_{B,\alpha}(T_B, p, x_C)}\right\}$$
$$-\int_{T_B}^{T^*_B}\left[\frac{\{H^*_{B,\alpha}(T, p) - H^*_{B,\alpha}(T, p, x_C)\}}{RT^2}\right]dT, \tag{4.138}$$

where the enthalpy difference is either the molar enthalpy of mixing or the molar enthalpy of dissolution obtained over a composition range. Combining Equations 4.137 and 4.138 gives

$$\ln\left\{\frac{\lambda^{*}_{\mathrm{B},\alpha^{*}}(T,p)}{\lambda_{\mathrm{B},\alpha}(T,p,x_{\mathrm{C}})}\right\}=\left\{\frac{\Delta^{\alpha}_{\beta}H^{*}_{\mathrm{B}}(T^{*}_{\mathrm{B}},p)}{RT^{*}_{\mathrm{B}}}\right\}\left(1-\frac{T_{\mathrm{B}}}{T^{*}_{\mathrm{B}}}\right)$$

$$+\left\{\frac{\Delta^{\alpha}_{\beta}H^{*}_{\mathrm{B}}(T^{*}_{\mathrm{B}},p)}{RT^{*}_{\mathrm{B}}}-\frac{\Delta^{\alpha}_{\beta}C^{*}_{p,\mathrm{B}}(T^{*}_{\mathrm{B}},p)}{2R}\right\}\left(1-\frac{T_{\mathrm{B}}}{T^{*}_{\mathrm{B}}}\right)$$

$$-\int_{T_{\mathrm{B}}}^{T^{*}_{\mathrm{B}}}\left[\frac{\left\{H^{*}_{\mathrm{B},\alpha}(T,p)-H^{*}_{\mathrm{B},\alpha}(T,p,x_{\mathrm{C}})\right\}}{RT^{2}}\right]\mathrm{d}T. \qquad (4.139)$$

4.6.2 What Is Thermodynamic Consistency?

For a binary mixture Equation 4.28 is

$$0=\left(\frac{H_{\mathrm{m}}}{RT^{2}}\right)\mathrm{d}T-\left(\frac{V_{\mathrm{m}}}{RT}\right)\mathrm{d}p+(1-x)\,\mathrm{d}\ln\lambda_{\mathrm{A}}+x\,\mathrm{d}\ln\lambda_{\mathrm{B}}. \qquad (4.140)$$

At constant T and p this becomes

$$0=(1-x)\,\mathrm{d}\ln\lambda_{\mathrm{A}}+x\,\mathrm{d}\ln\lambda_{\mathrm{B}}, \qquad (4.141)$$

or, with Equations 4.110 and 4.111, it becomes

$$0=(1-x)\,\mathrm{d}\ln f_{\mathrm{A}}+x\,\mathrm{d}\ln f_{\mathrm{B}}. \qquad (4.142)$$

From Equations 4.130 and 4.131 we obtain

$$\left(\frac{\partial \ln f_{\mathrm{B}}}{\partial p}\right)_{T}=\frac{(V_{\mathrm{B}}-V^{*}_{\mathrm{B}})}{RT}, \qquad (4.143)$$

and

$$\left(\frac{\partial \ln f_{\mathrm{B}}}{\partial T}\right)_{p}=-\frac{(H_{\mathrm{B}}-H^{*}_{\mathrm{B}})}{RT^{2}}. \qquad (4.144)$$

For reference, for a multicomponent mixture the generalization of Equation 4.142 is

$$\sum_{\mathrm{B}} x_{\mathrm{B}}\left(\frac{\partial \ln f_{\mathrm{B}}}{\partial x_{\mathrm{B}}}\right)_{T,p,x_{\mathrm{C}\neq\mathrm{B}}} = 0. \qquad (4.145)$$

Integration of Equation 4.142 gives

$$\int_0^1 \ln\left(\frac{f_{\mathrm{B}}}{f_{\mathrm{A}}}\right)\mathrm{d}x = 0. \qquad (4.146)$$

Equation 4.146 is a necessary but not sufficient criterion for thermodynamic consistency of values of f_{A} and f_{B} that may have been measured separately. Equation 4.146 is very often used (or at least should be used) to test the validity measurements of p, x, and y, including isobaric phase equilibria observations obtained for chemical engineering purposes.

Values of the activity coefficient are usually determined from measurements of x, y, and p^{sat}. Because the phase rule yields $F=2$ for a binary mixture only two quantities are required but the measurements must be tested for thermodynamic consistency and this can be done through measurements of the third quantity.

4.6.3 How Do I Use Activity Coefficients Combined with Fugacity to Model Phase Equilibrium?

For a system at constant and uniform temperature and pressure, and of constant amount of substance

$$\left(\frac{\partial G^{\Sigma}}{\partial t}\right)_{T,p,N} < 0. \qquad (4.147)$$

At the equilibrium of two or more phases according to Equation 4.147 the Gibbs function has reached a minimum. The phases are indicated by α, β, γ, $\cdots$, π, and for each substance B of the mixture of C components {A, B, $\cdots$} the following equilibrium conditions in terms of the chemical potential result

$$\mu_{\mathrm{B},\alpha} = \mu_{\mathrm{B},\beta} = \cdots = \mu_{\mathrm{B},\pi}\,. \qquad (4.148)$$

Generalization of Equation 4.78 permits phase equilibrium to be defined in terms of the fugacity $\tilde{p}$ for each substance B of the mixture of C components {A, B, $\cdots$} by

$$\tilde{p}_{\mathrm{B},\alpha} = \tilde{p}_{\mathrm{B},\beta} = \cdots = \tilde{p}_{\mathrm{B}}(\pi). \qquad (4.149)$$

For vapor + liquid equilibrium Equation 4.149 becomes

$$\tilde{p}_{\mathrm{i,g}}(T,p,y)=\tilde{p}_{\mathrm{i,l}}(T,p,x),\quad \mathrm{i}=\mathrm{A},\mathrm{B},\cdots,C. \tag{4.150}$$

From Equations 4.69 and 4.82 with Equation 4.117 (or 4.119) and Equation 4.97 and Equations 4.108 and 4.109 the equilibrium condition of Equation 4.150 becomes, for each substance B of the mixture of C components

$$y_{\mathrm{B}}\phi_{\mathrm{B,g}}(T,p^{\mathrm{sat}},y_{\mathrm{B}})p=x_{\mathrm{B}}f_{\mathrm{B,l}}(T,p,x_{\mathrm{B}})\tilde{p}_{\mathrm{B,l}}(T,p,x_{\mathrm{B}})p_{\mathrm{B}}^{\mathrm{sat}}F_{\mathrm{B}}. \tag{4.151}$$

The equilibrium of the mixture of C components requires C equations of the type of 4.151 one for each component. In Equation 4.151 f is the activity coefficient and $\tilde{p}$ the fugacity. This formalism is known as the gamma-phi approach for calculating vapor-liquid equilibria. The fugacity coefficient $\phi_{\mathrm{B,g}}(T,p,y_{\mathrm{B}})$ that accounts for the nonideality of the vapor phase of each component can be evaluated from an equation of state as can the fugacity $\tilde{p}_{\mathrm{B,l}}(T,p,x_{\mathrm{B}})$, while the activity coefficient $f_{\mathrm{B,l}}(T,p,x_{\mathrm{B}})$ used to describe the nonideal behavior of the liquid phase can be determined from an excess Gibbs function model. For further details the reader should refer to Modell and Reid (1983), Van Ness and Abbott (1982), Poling et al. (2001), Smith et al. (2004), and Prausnitz et al. (2001).

4.6.4 How Do We Obtain Activity Coefficients?

The experimental methods used to acquire values of the activity coefficient have been alluded to in Section 4.6.2. Other methods rely on the use of Equation 4.151. For the majority of cases it is necessary to have experimental values of the activity coefficients for substance B in a binary mixture. A typical experimental determination of the activity coefficient therefore requires measurements of the total pressure p, the mole fractions y_{B} and x_{B} in the vapor and liquid phase, respectively, for a binary mixture at vapor-liquid equilibrium at a temperature.

Measurements as a function of mole fraction for the liquid phase are used to determine the parameters in a suitable activity-coefficient model. As an example, Figure 4.2 shows the measured $(p, x, y)_T$ at $T = 318.15$ K for (nitromethane + tetrachloromethane) while, in Figure 4.3, the corresponding activity coefficients of both components are shown also as a function of liquid composition. The mixture (nitromethane + tetrachloromethane) is not ideal and, as expected, the activity coefficients for both substances are greater than unity. The Poynting factor given by Equation 4.88 is set equal to unity which is a reasonable assumption provided the pressure does not differ significantly from the vapor pressure of the pure components.

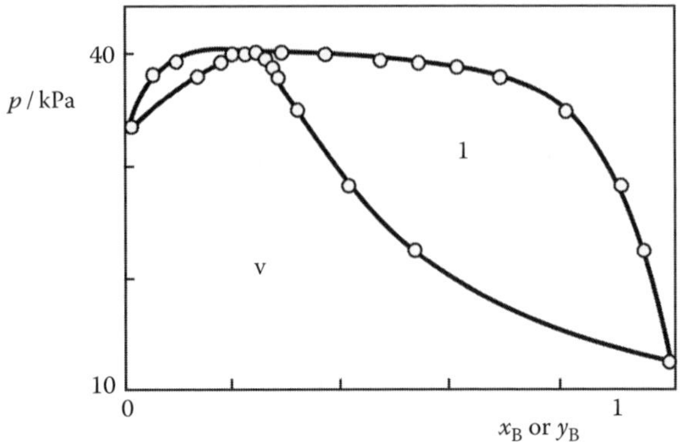

Figure 4.2 $(p, x)_T$ section for {tetrachloromethane(A) + nitromethane(B)} at $T = 318.15$ K. Symbols denote experimental values. Curves represent values calculate using Wilson's equation (Wilson 1964).

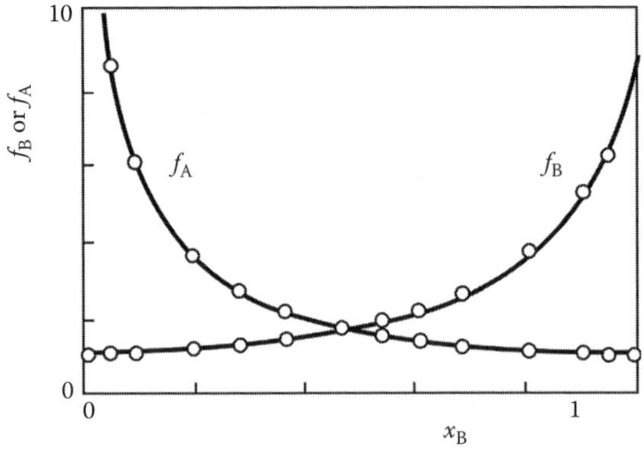

Figure 4.3 Activity coefficients f_A and f_B for {tetrachloromethane(A) + nitromethane(B)} at $T = 318.15$ K as a function of mole fraction x_B. ○: experimental values; ——: values obtained from Wilson's equation (Wilson 1964).

4.6.5 Activity Coefficient Models

The first model of this type was reported by Margules (1895) and represented the logarithm of the activity coefficient by a power series in composition for each component. van Laar (1910 and 1913) proposed a model based on van der Waals' equation of state with two adjustable parameters; predictive capabilities of that scheme have been found to be limited.

Typically, the model requires the measurement of (vapor + liquid) equilibria at a temperature for all possible binary mixtures formed from the components of the fluid. The parameters of the activity coefficient model are then fit to experimental data for binary mixtures. The resulting model can be applied to predict the activity coefficients of a multicomponent mixture over a range of temperature and pressure. For binary mixtures the model is used to extrapolate the measured values with respect to temperature and pressure. For multicomponent mixtures the model also exploits extrapolation of the composition. Examples of this approach are the methods reported by Wilson (1964), T-K-Wilson (Tsuboka and Katayama 1975), the Non-Random Two-Liquid model (NRTL) of Renon (1968 and 1969) and UNIQUAC (Abrams and Prausnitz 1975). Certainly, the most reliable procedure for the determination of parameters in any activity-coefficient model involves a fit to experimental data over a range of liquid compositions. The solution of the model for the parameters which best represent the data is a matter for nonlinear regression analysis. However, the solution found must still conform to the Gibbs–Duhem Equation 4.4. A description of activity coefficient models has been given by Assael et al. (1996).

The requirement to measure the (vapor + liquid) equilibria for all binary mixtures can be rather onerous and it will be no surprise to learn that engineers have created other approximate routes that either reduce or eliminate recourse to specific measurements.

In the absence of sufficient measurements, the model parameters are often estimated from Equations 4.126 to 4.128. In this case, the activity coefficient of component A in a binary mixture $(1-x)\mathrm{A}+x\mathrm{B}$ in the limit as $x \to 0$ is denoted by f_A^∞ and from Equation 4.128 assuming Equation 4.95 $\{\text{i.e., } \Delta_\mathrm{mix}V_\mathrm{m}(\mathrm{id})=0\}$ is then solely a function of temperature at constant pressure given by

$$f_\mathrm{A}^\infty = \frac{p_\mathrm{B}^\mathrm{sat}}{p_\mathrm{A}^\mathrm{sat}}\left[1-\left\{1+p_\mathrm{B}^\mathrm{sat}\left(\frac{B_\mathrm{B}-V_\mathrm{B}^*(\mathrm{l})}{RT}\right)\right\}\frac{\mathrm{d}\ln p_\mathrm{B}^\mathrm{sat}}{\mathrm{d}T}\left(\frac{\partial T}{\partial x_\mathrm{A}}\right)_p^{x_\mathrm{A}\to 0}\right]$$
$$\times\exp\left[\frac{\{B_\mathrm{A}-V_\mathrm{A}^*(\mathrm{l})\}(p_\mathrm{B}^\mathrm{sat}-p_\mathrm{A}^\mathrm{sat})+\delta_\mathrm{AB}p_\mathrm{B}^\mathrm{sat}}{RT}\right]. \quad (4.152)$$

The quantity, f_A^∞, is often incorrectly called the *activity coefficient* at infinite dilution, because that terminology should be reserved for solutions, especially dilute solutions and not for gaseous mixtures. However, Equation 4.152 does provide an alternative approach to model vapor-liquid equilibrium of mixtures because the parameters of the empirical model are simplified. For example, the Wilson method may be implemented from the two infinite-dilution activity coefficients for a binary pair. Other models of this type, have been proposed by Pierotti et al. (1959) for polar mixtures; Helpinstill and van Winkle (1968) proposed an extension of the Scratchard and Hildebrand equations applied to polar systems. More recently, Thomas and Eckert (1984) proposed the modified separation of cohesive energy density (given the acronym MOSCED) model for predicting infinite-dilution activity coefficients from pure-component parameters only.

In the absence of specific measurements, the parameters of the activity-coefficient model can be estimated using a group-contribution method, which assumes that groups of atoms within a molecule contribute in an additive manner to the overall thermodynamic property for the entire molecule. Thus a methyl group may make one kind of contribution, while a hydroxyl group makes another contribution. Once the contributions to the property from each group of the molecule have been determined the activity coefficient of the molecule can be obtained from the contributions of the groups it contains. Schemes of this type ultimately rely on (vapor + liquid) equilibria measurements that are used with definitions of the groups within molecules to determine the parameters of a model for the molecular group by regression. Examples of this approach are the Analytical Solution of Groups (ASOG) (Wilson and Deal 1962; Wilson 1964; Kojima and Toshigi 1979) and the Universal Functional Group Activity Coefficients (UNIFAC) (Fredenslund et al. 1975; 1977) models; the UNIFAC method is widely used.

4.6.6 How Can I Estimate the Equilibrium Mole Fractions of a Component in a Phase?

To complete the description of phase equilibrium, a means of determining the distribution of the substance B between the liquid and gas phases. This can be done by analogy with the methods used for chemical equilibrium given by Equation 1.30 in terms of the standard equilibrium constant given by Equation 1.116; for (vapor + liquid) equilibrium the components of the mixture are unchanged by the vaporization and condensation. The equilibrium constant therefore describes the distribution of the components between the various phases. When the (vapor + liquid) equilibrium can be represented by fugacity coefficients the distribution is determined for each species from the ratio of the fugacity coefficients for the liquid $\phi_{\mathrm{B,l}}(T,p,x_{\mathrm{C}})$ to that of the gas $\phi_{\mathrm{B,g}}(T,p,y_{\mathrm{C}})$ (given by Equations 4.83 and 4.69, where the fugacity of the liquid $\tilde{p}_{\mathrm{B,l}}(T,p,x_{\mathrm{C}})$ and gas $\tilde{p}_{\mathrm{B,g}}(T,p,y_{\mathrm{C}})$ are given by Equations 4.81 and 4.68) by

$$K_{\mathrm{p}} = \prod_{\mathrm{B}} \frac{\phi_{\mathrm{B,l}}(T,p,x_{\mathrm{C}})}{\phi_{\mathrm{B,g}}(T,p,y_{\mathrm{C}})} = \prod_{\mathrm{B}} \frac{\tilde{p}_{\mathrm{B,l}}(T,p,x_{\mathrm{C}})y_{\mathrm{B}}}{\tilde{p}_{\mathrm{B,g}}(T,p,y_{\mathrm{C}})x_{\mathrm{B}}} = \prod_{\mathrm{B}} \frac{y_{\mathrm{B}}}{x_{\mathrm{B}}}, \tag{4.153}$$

because at equilibrium $\tilde{p}_{\mathrm{B,l}}(T,p,x_{\mathrm{C}})=\tilde{p}_{\mathrm{B,g}}(T,p,y_{\mathrm{C}})$. In Equation 4.153 y_{B} and x_{B} are the mole fraction of the gas and liquid, respectively, of substance B.

For (vapor + liquid) equilibrium that requires the use of activity coefficients Equation 4.151 can be used so that for each substance B

$$K_{\mathrm{B}}=\frac{y_{\mathrm{B}}}{x_{\mathrm{B}}}=\frac{f_{\mathrm{B,l}}(T,p,x_{\mathrm{B}})\tilde{p}_{\mathrm{B,l}}(T,p,x_{\mathrm{B}})p_{\mathrm{B}}^{\mathrm{sat}}F_{\mathrm{B}}}{\phi_{\mathrm{B,g}}(T,p^{\mathrm{sat}},y_{\mathrm{B}})p}, \tag{4.154}$$

and thus,

$$K_{\mathrm{p}}=\prod_{\mathrm{B}}\frac{f_{\mathrm{B,l}}(T,p,x_{\mathrm{B}})\tilde{p}_{\mathrm{B,l}}(T,p,x_{\mathrm{B}})p_{\mathrm{B}}^{\mathrm{sat}}F_{\mathrm{B}}}{\phi_{\mathrm{B,g}}(T,p^{\mathrm{sat}},y_{\mathrm{B}})p}=\prod_{\mathrm{B}}\frac{y_{\mathrm{B}}}{x_{\mathrm{B}}}. \tag{4.155}$$

4.7 HOW DO I CALCULATE VAPOR + LIQUID EQUILIBRIUM?

The coexisting phases of liquid and gas of a pure component are of considerable importance in both chemistry and engineering applications, so we devote here some space to particular aspects of the behavior of these two-phase systems. For the initial examples in Question 4.7.1 water and air are used because of their considerable importance in practical applications. However, the issues raised in Question 4.7.1 have relevance to every system.

The reader interested specifically in the computation of phase boundaries for nonpolar and polar fluid mixtures should consult Questions 7.5.4 and 7.5.5, respectively, as well as Question 7.5.6.

4.7.1 Is There a Difference between a Gas and a Vapor?

When water is boiled one observes water above the liquid in a form that is commonly referred to as "vapor" or "steam." Thermodynamically, this nomenclature is incorrect. Steam refers to gaseous water that is a clear colorless substance invisible to the human eye. The observer actually observes a mist of water droplets formed from condensed steam and they are thus liquid water. Before continuing to address the question posed by this section heading we digress to consider evaporation.

Figure 4.4 illustrates the concept of the vaporization of a liquid of fixed amount of substance and initial mass m at constant pressure achieved by a piston and added force given by a mass and local acceleration of free fall. The corresponding points on a $p(v_{\mathrm{c}})$ section are shown in Figure 4.5. When energy is provided to the liquid it expands from points 1 to 2 as shown in Figure 4.4 and

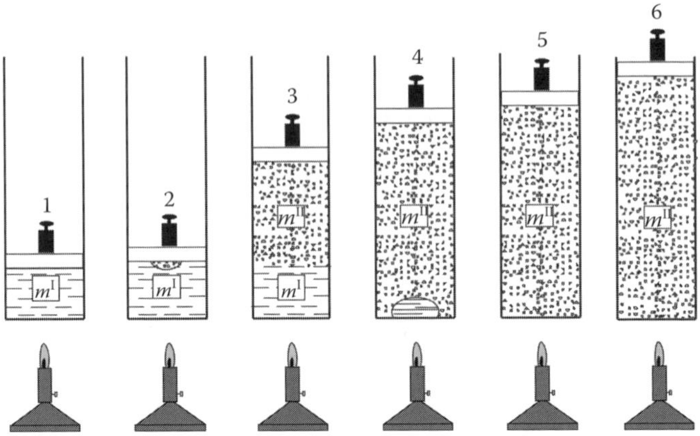

Figure 4.4 Vaporization of a liquid at constant pressure.

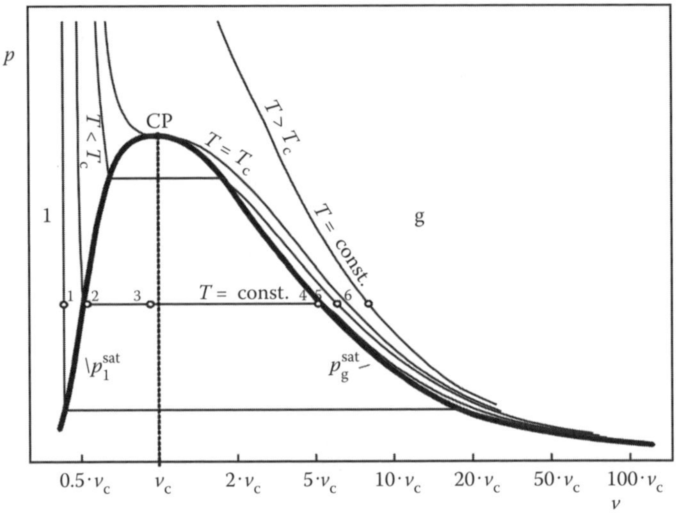

Figure 4.5 $p(v_c)$ section for a isobaric vaporization process, where v_c is the specific critical volume. The saturated liquid (bubble curve) and saturated vapor (dew curve) are shown along with items 1 through 6 of Figure 4.4.

Figure 4.5. When the vapor forms at the boiling temperature for the pressure the saturation line shown in Figure 4.1 has been reached and a vapor bubble forms as shown at step 2 of Figures 4.4 and 4.5.

At step 3 of Figures 4.4 and 4.5 the vessel contains a mixture of saturated liquid of mass m' and a mass of saturated vapor designated m''. During evaporation the volume occupied by the fluid increases because the vapor phase requires a much larger volume than the liquid phase. The mass m' decreases while m'' increases as illustrated in step 4 of Figure 4.4 and Figure 4.5. This process continues until $m' = 0$ and all the liquid has evaporated (just after point 4 of Figures 4.4 and 4.5). Addition of energy to a purely gaseous phase results in an increase in temperature of the phase and also the volume occupied as illustrated in steps 5 and 6 of Figures 4.4 and 4.5. The temperature of steps 2 through 4 of Figures 4.4 and 4.5 is constant and equal for both gas and liquid owing to the absorption of heat equivalent to the specific enthalpy of evaporation $\Delta_l^g h$.

In points 2 to 4 of Figures 4.4 and 4.5 the temperature and pressure are insufficient to unambiguously determine the state of the system as it is possible for the states to be in either one-phase region. To specify the state of the two-phase system requires introduction of the quality x given by

$$x = \frac{m''}{m}, \tag{4.156}$$

which is the ratio of the mass of the vapor phase to the total mass of fluid and has a value between 0 and 1 for the saturated liquid and vapor, respectively. Extensive properties Z are related to the specific values z through

$$z = \frac{Z}{m} = \frac{Z' + Z''}{m' + m''} \tag{4.157}$$

that combine the properties for the liquid and the vapor phases and may be expressed with the quality x and the tabulated values for the saturated states $'$ and $''$ using

$$z = (1 - x)z' + x z'' = z' + (z'' - z'). \tag{4.158}$$

This relation is routinely used for the specific volume v, specific internal energy u, specific enthalpy h, and specific entropy s. We will now return to address the question posed regarding the difference between vapor and gas.

In common understanding, the term vapor implies that it has emerged from the evaporation of a liquid. But one can also vaporize liquid nitrogen and would hardly speak about air containing nitrogen vapor. We can get closer to an answer if we reverse the vaporization process and compress to liquefy a vapor. Compression of gases is usually performed isothermally as discussed in Chapter 1. If we start at point 5 in the $p(v_c)$ diagram of Figure 4.5 and compress the vapor isothermally the system reaches the saturation line and the vapor begins to condense. If the compression commenced at point 6 of Figure 4.5 the isotherm would follow the line to infinite pressure without crossing the saturation line and forming liquid. From Figure 4.5 the resulting difference between the compression starting at point 5 or point 6 arises because the starting temperature 5 is below the critical temperature T_c, while point 6 is above the critical temperature. The word vapor may be defined as a gas at a temperature below its critical temperature, and steam is therefore simply water vapor.

Thus, we conclude that moist air is a mixture of air and water given by a gaseous phase (air and water vapor) and a condensed phase liquid. The condensed phase consists essentially of pure water in either liquid or solid form; if the system temperature $T > T(H_2O, s+l+g) = 273.16$ K liquid water is the phase while if $T < T(H_2O, s+l+g)$ the condensed phase is ice.

For many technical applications and especially for air conditioning the gaseous phase may be approximated by a mixture of two components that behave as ideal gases. These are dry air, which here will be given the subscript a and will be treated as a pure component, and water vapor, given the subscript v, which because of the low partial pressure relative to atmospheric pressure of about 0.1 MPa for air can also be considered an ideal gas. In the ideal mixture the total pressure p of the gas phase is simply the sum of the partial pressures of the two constituents given by $p = p_a + p_v$. Condensation of water occurs when the water content of the moist air increases to saturation that is when the partial pressure of water vapor (hypothetically) exceeds the maximum permissible value $p_{v,max}$ that is equal to the vapor pressure of pure water at the specified temperature $p^{sat}_{H_2O}(T)$. The reasoning behind this statement is that each component in an ideal gas mixture behaves as if it existed alone. As the vapor pressure, which may be taken from steam tables (see Chapter 7) or be calculated from Equation 4.20 (the Antoine equation of Question 4.2), depends on temperature, the temperature affects the capacity of air to maintain water vapor before it condenses as illustrated in Figures 4.1 and 4.6.

Moist air, as Figure 4.6 shows, is characterized by a partial pressure p_v of water vapor in air. Isothermal addition of water is shown in Figure 4.6 by a vertical line connecting p_v to $p_v^{sat}(T)$, while isobaric cooling of moist air is shown in Figure 4.6 by a horizontal line connecting p_v to $p_v^{sat}(T_d)$ at the dew-point temperature T_d. Water condenses when the saturation line is reached. Dehumidification of moist air is achieved by cooling to a temperature below the dew-point temperature.

Condensation of water vapor occurs in every day life when the temperature of the system is lowered below the saturation temperature corresponding to the partial pressure of the water in atmospheric air. For example, condensation happens when a person wearing spectacles enters a heated room from the external environment in winter. Because the lenses of the spectacles are cold the chilled air near the surface cannot hold the same amount of water as the air in the heated room, and small water droplets start to form on the lenses. The same phenomenon may occur at the inner surface of the windows of a house in winter, or when the windscreen in your car fogs up from the water vapor

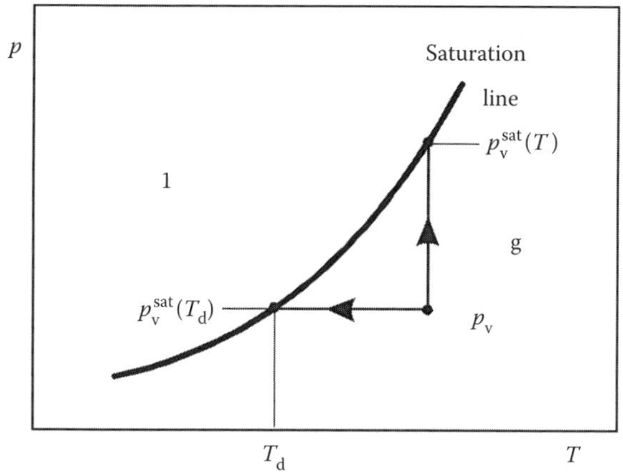

Figure 4.6 Schematic of the $p(T)$ section for the evaporation of water.

content of your warm breath. You will often find dew on the lawn after a cool night in summer or on the outer surface of a container holding a chilled drink. It is also possible then to see that a similar process happens in our initial example with water boiling in a kettle: hot steam at a temperature of about 100 °C exits the kettle, and the air in the room at a temperature below 100 °C is locally supersaturated, so that small water droplets form, which are observed as fog or mist.

For completeness, two important variables characterizing moist air are introduced. The first is the relative humidity ϕ (or sometimes φ), a property you can read from a device called a *hygrometer* and which is expressed as the ratio of the actual partial pressure of water vapor in air p_v at a particular temperature to its saturation value $p_v^{sat}(T)$ at the same temperature T defined by

$$\phi = \frac{p_v(T)}{p_v^{sat}(T)}. \tag{4.159}$$

In IUPAC nomenclature (Quack et al. 2007) Equation 4.159 would be cast as follows:

$$\phi = \frac{p_{H_2O}(T)}{p_{H_2O}^{sat}(T)} = \frac{\rho_g\{p_{H_2O}(T),T\}}{\rho_g\{p_{H_2O}^{sat}(T),T\}}, \tag{4.160}$$

where $p_{H_2O}(T)$ is the partial pressure of water in (air + water), $p_{H_2O}^{sat}(T)$ is the vapor pressure of water, $\rho_g\{p_{H_2O}(T),T\}$ is the mass density of the gas determined at the pressure $p_{H_2O}(T)$ and $\rho_g\{p_{H_2O}^{sat}(T),T\}$ is the mass density at the pressure $p_{H_2O}^{sat}(T)$, all at temperature T.

The quantity ϕ in Equations 4.159 and 4.160 varies between 0 and 1. Cooling moist air increases ϕ up to unity when liquid water (or ice) forms. Another quantity, which relates the mass of vapor m_v to the mass of dry air m_a, has been given several names, including the absolute or specific humidity, the humidity ratio or the moisture content, usually with symbol ω or X (sometimes—and very unfortunately—also x, which may be easily confused with the quality defined by Equation 4.156), thus,

$$\omega = \frac{m_v}{m_a}. \tag{4.161}$$

Equation 4.161 is used with the mass of dry air because in air conditioning the mass of dry air often remains constant, while the total mass of humid air varies. In some cases, the moisture content is extended to include all water, now designated by a subscript w, and typically given the symbol X defined by

$$X = \frac{m_w}{m_a}, \tag{4.162}$$

where the moisture content is also given for supersaturated air or pure water (where $X \rightarrow \infty$). When moist air is heated or cooled absolute humidity is not altered but in contrast relative humidity is.

The values of the absolute and relative humidity can be interrelated. For an ideal gas this is given by

$$\omega = \frac{m_v}{m_a} = \frac{p_v V M_v/(RT)}{p_a V M_a/(RT)} = \frac{p_v \cdot M_v}{p_a \cdot M_a} = \frac{18.02}{28.96}\frac{p_v}{p_a} = 0.622\frac{p_v}{p_a}, \tag{4.163}$$

where m_v is the mass of water vapor, m_a is the mass of air, p_v is the pressure of the water vapor, p_a is the air pressure, R is the gas constant, V is the volume occupied, and M_v and M_a are the molar mass of water vapor and air, respectively. Combining Equation 4.163 with Equation 4.159 we obtain

$$\omega = 0.622\frac{p_v}{p_a} = 0.622\frac{p_v}{p - p_v} = 0.622\frac{\phi \cdot p_v^{sat}}{p - \phi \cdot p_v^{sat}} = 0.622\frac{p_v^{sat}}{p/\phi - p_v^{sat}}. \tag{4.164}$$

Cooling moist air below the dew-point temperature is of vital importance in the air conditioning process where, as well as the maintenance of a specific temperature, control of humidity is required. In engineering terms, it is a relatively easy task to add water but the reverse process is more challenging.

Dehumidification is required to remove the moisture generated by human beings in a room, and can also be employed to defog the windscreen in your car on a cold winter day using the A/C rather than the heater. While it is of course possible to avoid moisture, for example, in the packaging of electronic equipment, by adding some hygroscopic material, this approach is not practicable for a continuous process, because the material would have to be removed and dried in some batch process for reuse.

As a consequence, in A/C applications moist air is drawn out of a room into a machine, cooled below its dew point, the condensate removed, and the air with lower moisture is reheated to the desired temperature before being ejected back into the room. To reduce energy consumption the heating is, or should be, achieved using a heat exchanger between the two air streams.

4.7.2 Which Equations of State Should Be Used in Engineering VLE Calculations?

Equations of state are used in engineering to predict thermodynamic properties in particular the phase behavior of pure substances and mixtures. However, since there is neither an exact statistical-mechanical solution relating the properties of dense fluids to their intermolecular potentials, nor detailed information available on intermolecular potential functions, all equations of state are, at least partially, empirical in nature. The equations of state in common use within both industry and academia can be arbitrarily classified as follows: (1) cubic equations such as that of van der Waals that are described by Economou (2010); (2) those based on the virial equation discussed by Trusler (2010) and Chapter 2 of this volume; (3) equations based on general results obtained from statistical mechanics and computer simulations mentioned, including the many forms of statistical associating fluid theory known by the acronym SAFT as described by McCabe and Galindo (2010); and (4) those obtained by selecting, based on statistical means, terms that best represent the available measurements obtained from a broad range of experiments as outlined by Lemmon and Span (2010). Forms of item 3 are particularly advantageous when one of the phases includes water.

The development of an equation of state typically commences with the representation of the thermodynamic properties of pure fluids and the functions are then extended to provide estimates of the properties of mixtures by the introduction of mixing and combining rules.

Mixing rules are used to obtain numerical estimates for the parameters in an equation of state for a specified mixture from the same parameters when the same equation of state is used to represent the properties of the pure substance. However, in the description of a mixture, parameters appear that result from the interactions between unlike species, for example, the second virial coefficient B_{AB} used in Equation 4.73. These parameters are obtained using combining rules. By using mixing and combining rules, measurements are only required for the pure substances and not the very large number of mixtures that it is possible to make. When these mixing and combining rules are used with $p(V_m, T)$ equations of state they provide the link between the microscopic and the macroscopic. The certainty with which the predictions result from the use of an equation of state with its mixing and combining rules can be evaluated using experimental data and additional adjustable parameters are added when there is sufficient experimental data. Therefore, the development of an equation of state for mixtures is largely reduced to the establishment of the mixing and combining rules to describe the thermodynamic properties, especially the phase boundaries.

The plethora of both equations of state and of mixing and combining rules means there is a multitude of options available and that some adopted are purely empirical. Consequently, the task of providing a comprehensive list of all equations of state, mixing and combining rules is rather daunting. The basis for the inclusion of those selected herein were their frequent appearance in the archival literature, which does not necessarily imply that the rules are optimal or even correct. The reader requiring a rather more extensive review of equations of state should consult Goodwin and Sandler (2010) and the recent work of Kontogeorgis and Folas (2010) for mixing and combining rules.

The methods most frequently used to predict the properties of mixtures for over 100 years have inevitably undergone only minor additions and corrections to, it is claimed, improve the representation of experimental data for specific categories of substances. It is, however, possible that completely different alternatives to these traditional approaches are required, particularly for a method to be both predictive and applicable over a wide range of fluids and conditions (Heideman and Fredenslund 1989). Such methods might arise from future research and methods based on statistical mechanics and quantum-mechanical calculations (Leonhard et al. 2007; Singh et al. 2007) are ultimately sought rather than empiricism.

For the purpose of elucidating calculations in the remainder of this section we will consider the cubic equation of state of the form of Equation 4.28 with Equations 4.29 and 4.30; however, we wish to emphasize that our analysis is much more general in reality. We employ the van der Waals one-fluid theory for mixtures. This assumes that the properties of a mixture can be represented by a hypothetical pure fluid. Thus the thermodynamic behavior of a mixture of constant composition is assumed to be isomorphic to that of a one-component fluid; this assumption is not true near the critical point where the thermodynamic behavior of a mixture at constant thermodynamic potential is most definitely not isomorphic with that of a one-component fluid.

The van der Waals one-fluid theory gives the following for the mixing rules for the van der Waals equation of state:

$$a(x) = \sum_{i=0}^{C} \sum_{j=0}^{C} x_i x_j a_{ij}, \quad (4.165)$$

and

$$b(x) = \sum_{i=0}^{C} \sum_{j=0}^{C} x_i x_j b_{ij}. \quad (4.166)$$

Equations 4.165 and 4.166 are quadratic in mole fraction x for the parameters a and b of Equations 4.29 and 4.30 of substances i and j. Equation 4.166 is often approximated by

$$b(x) = \sum_{i=0}^{C} x_i b_i. \quad (4.167)$$

Before the introduction of combining rules we digress to return to intermolecular potentials and, in particular, the Lennard-Jones intermolecular potential (Lennard-Jones 1931), which accounts for the repulsive and attractive forces. For the interaction of spherical substances A and B in (A + B), $\phi_{AB}(r)$ is given by

$$\phi_{AB}(r) = 4\varepsilon_{AB} \left\{ \left(\frac{\sigma_{AB}}{r_{AB}} \right)^{12} - \left(\frac{\sigma_{AB}}{r_{AB}} \right)^{6} \right\}, \quad (4.168)$$

and is frequently used in computer simulation. For a ternary mixture of spherical molecules, it is assumed that $\phi(r_{AB}, r_{BC}, r_{CA})$ is given by the sum of three pair-interaction energies $\{\phi(r_{AB}) + \phi(r_{BC}) + \phi(r_{CA})\}$ of which the first term in the summation is given by Equation 4.168. The parameter ε_{AB} of Equation 4.168 defines the depth of the potential well and σ_{AB} is the separation distance at the potential minimum. Combining rules at the molecular level are required to determine ε_{AB} and σ_{AB} from the pure-component values, and it is the discussion of these that we now turn to because they provide background information for this and other sections of this chapter.

The parameter σ_{AB} for unlike interactions between molecules A and B is most often determined from the rule proposed by Lorentz (1881), which is based on the collision of hard spheres; the result is that σ_{AB} is given by the arithmetic mean of the pure-component values with

$$\sigma_{AB} = \frac{\sigma_A + \sigma_B}{2}. \quad (4.169)$$

The parameter ε_{AB} is obtained from the expression of Berthelot (1889) for the geometric mean of the pure-component parameters of

$$\varepsilon_{AB} = (\varepsilon_A \varepsilon_B)^{1/2}. \tag{4.170}$$

Equation 4.170 arises from consideration of the London theory (1937) of dispersion (Hirschfelder et al. 1954; Rowlinson 1969; and Henderson and Leonard 1971; Maitland et al. 1981).

Equations 4.169 and 4.170 are collectively known as the Lorentz-Berthelot combining rules; they are known to fail particularly in the case of highly nonideal mixtures (Reed 1955a and 1955b; Delhommelle and Millié 2001; Unferer et al 2004; Haslam et al. 2008; Goodwin and Sandler 2010).

Because the core volume b of Equation 4.28 is proportional to σ^3 of Equation 4.169 and a is proportional to the depth of the potential well given by Equation 4.170, Equations 4.169 and 4.170 can be recast as

$$b_{AB} = \frac{(b_A^{1/3} + b_A^{1/3})^3}{8}, \tag{4.171}$$

and

$$a_{AB} = (a_A a_B)^{1/2}, \tag{4.172}$$

respectively. Equations 4.171 and 4.172 provide the means to estimate both a_{AB} and b_{AB}. Molecules are not hard spheres so that Equation 4.171 is corrected, particularly to estimate phase boundaries, by the addition of a parameter β_{AB}. Equation 4.172 is also modified by a parameter k_{AB} for the same reason. These modifications lead to the actual forms of Equations 4.171 and 4.172 that are routinely used in engineering calculations:

$$b_{AB} = (1 - \beta_{AB}) \frac{(b_A^{1/3} + b_A^{1/3})^3}{8}, \tag{4.173}$$

and

$$a_{AB} = (1 - k_{AB})(a_A a_B)^{1/2}. \tag{4.174}$$

The parameters β_{AB} of Equation 4.173 and k_{AB} of Equation 4.174 are frequently called *interaction parameters*. Equation 4.173 is often cast as

$$b_{AB} = 0.5(1 - \beta_{AB})(b_A + b_B). \tag{4.175}$$

Because, in this form, the combined equation of state, mixing and combining rules provide estimates of the properties of the mixture that differ less from the experimental measurements than when Equation 4.173 is used. The

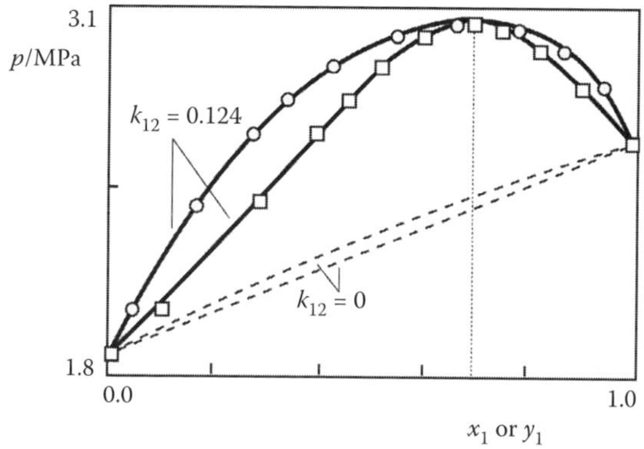

Figure 4.7 $p(x)_T$ section for the vapor + liquid equilibrium of $\{CO_2(1) + C_2H_6(2)\}$ as a function of mole fraction x of the liquid and y of the gas phases. O: liquid phase measured bubble pressure (Fredenslund and Mollerup 1974); □: measured dew pressure; ——, dew pressures (Fredenslund and Mollerup 1974) estimated from the Peng-Robinson equation of state with $k_{12} = 0.124$; - - - - -, dew pressure estimated from the Peng-Robinson equation of state with $k_{12} = 0$; vertical ········, indicates the azeotropic mixture at $x = 0.7$.

importance of the binary interaction parameter k_{AB} of Equation 4.174 in the estimation of phase equilibria can be illustrated by the system $xCO_2 + (1 - x) C_2H_6$ for which the $p(x)_T$ section has been estimated with $k_{ij} = 0$ and $k_{ij} = 0.124$ as shown in Figure 4.7 where the data are compared with the measured values. The system $xCO_2 + (1 - x)C_2H_6$ exhibits azeotropic behavior that will be discussed in Question 4.11.4. As a general rule, increasing the molecular complexity increases the sensitivity of the calculation to the interaction parameter. Hence, in complicated mixtures, the availability of the binary interaction parameters for a particular equation of state might be the overwhelming criterion for choosing a particular functional form for the equation of state.

The cubic equations of state of Peng–Robinson (Peng and Robinson 1976) and Redlich–Kwong-Soave (Soave 1972) are the most commonly used in these calculations. However, other equations that might be categorized as virial equations or as hard sphere approximations with up to 53 adjustable parameters, such as the modified Benedict–Webb–Rubin equation as originally proposed by Strobridge (1962) have also been employed.

4.7.3 What Is a Bubble-Point or Dew-Point Calculation and Why Is It Important?

A specific example of a dew temperature was provided in Question 4.7.1 for gaseous water and in air. This concept will be generalized herein to vapor + liquid equilibrium and to also include the bubble pressure. We recall that the dew point is the point of a thermodynamic surface at which liquid first forms and by analogy the bubble point is the point at which vapor first forms in a system.

The basic "engine" of most phase-equilibrium calculations is an algorithm to calculate the dew or bubble pressure for a mixture of specified composition and temperature. The kind of calculation to be made (dew or bubble) may be specified by giving the *vapor fraction* β, which is defined as the *amount of substance* in the vapor phase divided by the total amount of substance. It follows that this quantity is unity at a dew point and zero at a bubble point. The phase rule (defined in Question 4.1.1) then requires specification of either the temperature or the pressure in addition to the composition of the bulk phase. It is then our task to calculate the remaining variables; these are either p (for specified T) or T (for specified p) and the composition of the coexisting phase at the dew or bubble point. This problem should have either one solution, when two phases are possible under the specified conditions or no solution when they are not. Whether this is the case with a particular thermodynamic model remains to be proven because the model may or may not accord with reality.

The calculation commences with Equation 4.148 or more often for engineers with Equation 4.150, with one for each of the C substances in the mixture to give C simultaneous equations to be solved to determine equilibrium; the simultaneous equations can also be cast for the equality of product of the fugacity coefficients and mole fraction of the gas and liquid phases given by Equations 4.69 and 4.83. At a specified temperature and pressure the fraction of vapor for a component B is given by one element of the continued product of Equation 4.153; at equilibrium Equation 4.153 can also be cast as the ratio of the activity coefficient of a liquid to that of the gas.

We can now proceed to describe a basic algorithm for determining the bubble-point of a fluid mixture, on the basis of Equations 4.69 and 4.83 and Equations 4.70 and 4.81 employing an equation of state for both phases. An equivalent algorithm can be employed in the case of an activity-coefficient model for the liquid phase and an equation of state for the vapor (Assael et al. 1996); Equation 4.151 is used for $f_{\mathrm{B,l}}(T, p, x_{\mathrm{B}})$ determined from the activity-coefficient model and $\phi_{\mathrm{i,g}}(T, p, y_{\mathrm{C}})$ from Equation 4.69, $\tilde{p}_{\mathrm{B,l}}(T, p, x_{\mathrm{B}})$ from Equation 4.82 and F_{B} from Equation 4.86 determined from the equation of state. There are many ways in which one might set about solving the phase-equilibrium problem but the strategy outlined in Figure 4.8. is a simple and reliable approach to the problem and involves the following steps:

1. The liquid composition x_{i} ($\mathrm{i} = 1, 2, \cdots, n$) and either the pressure p or the temperature T must be specified.
2. An initial value is assumed for the unknown bubble-point temperature or pressure; often, Raoult's law (Equations 4.122 and 4.123) is employed for this purpose.

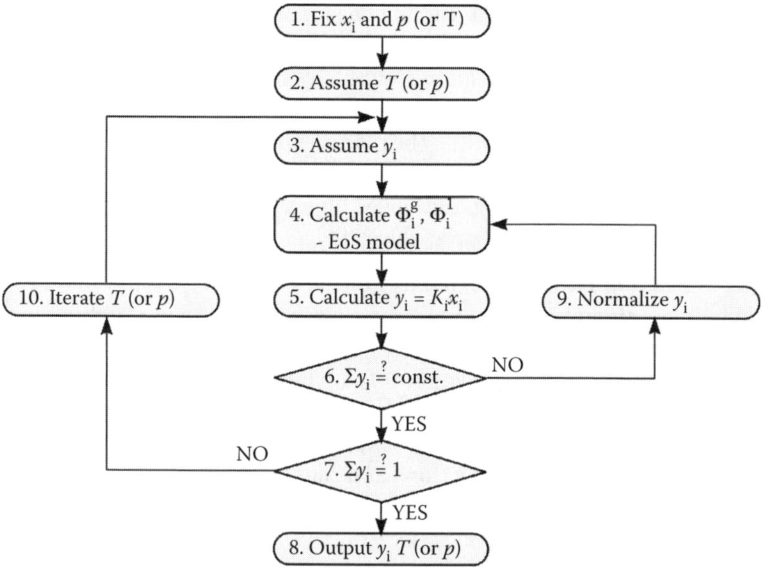

Figure 4.8 Bubble-point algorithm using an equation of state for both phases.

3. Initial values for vapor composition y_i (i = 1, 2,⋯, n) are assumed. Unless the system is known to exhibit nearly ideal behavior, one often sets $y_i = x_i$. The sum $s = \Sigma y_i$ should be initialized at this stage.
4. Next, the fugacity coefficients $\phi_i(\text{g}, T, p, y_C)$ and $\phi_i(\text{l}, T, p, x_C)$ of Equations 4.69 and 4.83, respectively, of each component i in the vapor and liquid phases are calculated at the assumed temperature, pressure, and phase compositions. To do so requires the equation of state for the molar volume of each component in each phase as provided by Equations 4.70 and 4.81 to obtain $\phi_i(\text{g}, T, p, y_C)$ and $\phi_i(\text{l}, T, p, x_C)$, respectively.
5. New approximations to the vapor mole fractions are estimated from one product of Equation 4.153 using $y_i = x_i K_i$ with $K_i = \phi_{i,l}(T, p, x_C)/\phi_{i,g}(T, p, y_C)$.
6. The new sum $s = \Sigma y_i$ is calculated. If this is equal to that for the previous iteration then proceed to step 7; otherwise, go to step 9.
7. Once a constant value of s is obtained subject to the presently assumed estimate of the unknown bubble-point temperature or pressure, test to see if $s = 1$. If this condition is satisfied then proceed to step 8; otherwise to step 10.
8 A solution has been found, which satisfies the thermodynamic requirements for thermal, hydrostatic, and phase equilibrium.

9. Normalized values of the vapor-phase mole fractions are calculated, $y_i' = y_i/s$, and used in another iteration starting at step 4.
10. A new estimate of the unknown bubble-point temperature or pressure must be made. If $s > 1$ then the assumed temperature (pressure) is too high (low) while, if $s < 1$ then the reverse applies. The simplest method for updating the unknown T or p is by means of a bisection algorithm; this requires that upper and lower limits of the unknown be established at the start of the procedure.

The interaction parameters, the k_{ij}'s in the equation of state mixing rules, are usually obtained by regression to measurements of dew and bubble pressures for the binary subsystems.

The determination of the dew-point temperature or pressure and the composition of the coexisting liquid is almost identical to that for the bubble-point problem. In this case, the vapor composition is specified, and iterations are performed over the liquid mole fractions and the unknown temperature or pressure. The algorithms shown in Figure 4.8 may be used after obvious changes. It might be interesting to note that, since a bubble-point routine returns the composition of the coexisting vapor, it may be used as it stands to generate points on the dew-point surface (although not at predetermined vapor compositions).

4.7.4 What Is a Flash Calculation?

The modeling of flash processes is probably the single most important application of chemical engineering thermodynamics. A flash process is one in which a fluid stream of known overall composition and flow rate passes through a throttle, turbine, or compressor and into a vessel (flash drum) where liquid and vapor phase are separated before each passes through the appropriate outlet. Such a process may be operated under many different sets of conditions, including the following: (1) constant temperature and pressure (isothermal flash); (2) constant enthalpy and pressure (isenthalpic flash); and (3) constant entropy and pressure (isentropic flash). The thermodynamic modeling of these processes requires, in each case, determination of the vapor fraction and the vaporization equilibrium ratio for the components in the system. It is also important in general to determine the thermal power (heat duty) absorbed or liberated in the flash process, although this is zero by definition in an isenthalpic or isentropic flash. In performing VLE calculations, we may choose to employ an equation of state for both phases or, where necessary, an activity-coefficient model for the liquid and an equation of state for the vapor.

4.7.4.1 What Is an Isothermal Flash?

The isothermal flash (constant temperature and pressures), illustrated schematically in Figure 4.9, is one of the most common features encountered in

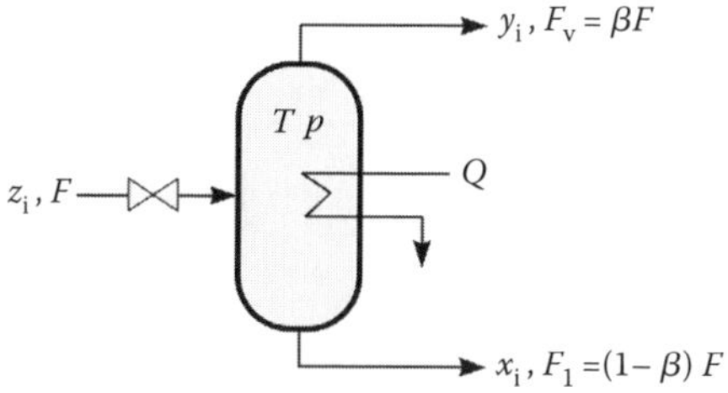

Figure 4.9 Isothermal flash unit.

chemical engineering. The feed, at temperature T_F and pressure p_F, passes through a throttle and enters the flash vessel, where liquid and vapor phases may separate. The operating pressure p of the unit is controlled in some way and heat is supplied or removed at rate Q though a heat exchanger so as to maintain isothermal conditions at temperature T. The molar flow rate F of the feed to the unit is specified, together with the overall composition (mole fractions z_i) and the temperature and pressure at which the unit operates. The objectives of the calculation are to determine the compositions (y_i and x_i) and the molar flow rates (F_v and F_l) of the vapor and liquid streams leaving the unit. From the known composition of the mixture a material balance is used for each of the n components to distribute the substance between the phases:

$$Fz_i = F_l x_i + F_v y_i, \tag{4.176}$$

with

$$y_i = K_i x_i. \tag{4.177}$$

Combining Equation 4.176 with Equation 4.177 and eliminating the flow rates in favor of the vapor fraction $\beta = F_v/F$, the so-called flash condition may be written as

$$f(\beta) = \sum_{i=1}^{n} x_i - 1 = \sum_{i=1}^{n} \frac{z_i}{1+\beta(K_i-1)} - 1 = 0. \tag{4.178}$$

Equation 4.178 may be solved for β with a Newton–Raphson algorithm that gives for successive iterations

$$\beta_{k+1} = \beta_k + \left[\sum_{i=1}^{n}\left(\frac{z_i}{1+\beta_k(K_i-1)}\right)-1\right]\left[\sum_{i=1}^{n}\left(\frac{(K_i-1)z_i}{[1+\beta_k(K_i-1)]^2}\right)\right]^{-1}. \tag{4.179}$$

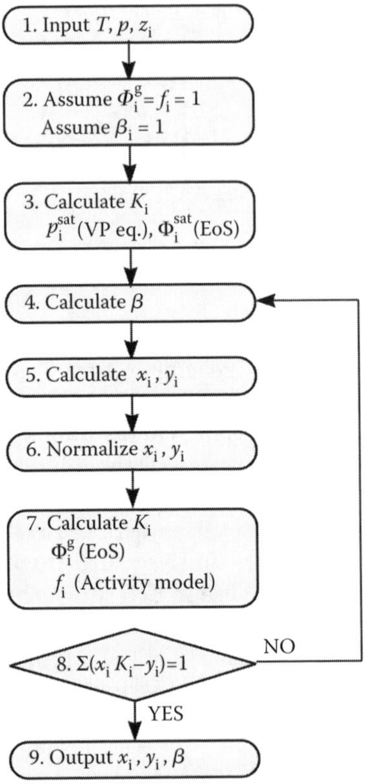

Figure 4.10 Isothermal flash algorithm using an activity-coefficient model for the liquid phase.

Typically, commencing with $\beta_1 = 1$ the convergence is rapid. The phase compositions are then given by

$$x_i = \frac{z_i}{1+\beta(K_i - 1)} \quad \text{and} \quad y_i = K_i x_i. \tag{4.180}$$

Of course both β and the K_i's are unknown and the latter are therefore evaluated during each cycle of Equation 4.179. An algorithm for solving this flash problem is shown in Figure 4.10 for the case in which an activity-coefficient model is applied for the liquid phase.

The isothermal flash algorithm involves the following steps:

1. The temperature, pressure, and overall mixture composition are specified.

2. Initial values of unity are assumed for the vapor-phase fugacity coefficient and liquid-phase activity coefficient of each component. β is initialized with the value unity.
3. A first approximation to the K_i is calculated for each component from Equation 4.153 using $y_i = x_i K_i$ with $K_i = \phi_{i,l}(T, p, x_C)/\phi_{i,g}(T, p, y_C)$ with p_i^{sat} determined from a suitable representation of the vapor pressure and $\phi_{i,l}(T, p, x_C)$ calculated from an equation of state.
4. A new value of β is determined from a single iteration of Equation 4.179.
5. The compositions of each phases are determined from Equations 4.180.
6. The mole fractions are normalized so that $\Sigma x_i = \Sigma y_i = 1$.
7. New vaporization equilibrium ratio's are calculated from Equation 4.151 with $f_{B,l}(T, p, x_B)$ determined from the activity coefficient model and $\phi_{i,g}(T, p, y_C)$ from Equation 4.69, $\tilde{p}_{B,l}(T, p, x_B)$ from Equation 4.82, and F_B from Equation 4.86 are determined from the equation of state.
8. We now test to see if the new vapor composition differs from that of the previous iteration. If it does, begin a new iteration at step 4; otherwise, go to step 9.
9. A solution to the problem has been found.

One rather obvious point that should not be forgotten is that two phases will only form when the specified pressure lies between the dew point and the bubble point for the given temperature and feed composition. Usually the heat duty Q on the flash unit is also required. Q (which is positive for heat supplied to the unit) may be determined from the molar flow rates and the molar enthalpy of the feed and product streams.

The method is a good deal simpler if an equation of state model is applied consistently to both phases during the entire calculation.

4.7.4.2 What Is an Isenthalpic Flash?

In Figure 4.11, an isenthalpic flash (constant enthalpy H and pressure p) is illustrated schematically. The unit is operated under adiabatic conditions ($Q = 0$) and, because no work is done on the fluid, the process is isenthalpic.

The objective of the flash calculation is to find the temperature, vapor fraction, and product compositions for the case in which the operating pressure and the temperature, pressure, and composition of the feed are specified.

4.7.4.3 What Is an Isentropic Flash?

If, instead of expanding through a throttle, the feed is compressed or expanded adiabatically and reversibly before entering the adiabatic flash vessel then the process is an isentropic flash (constant entropy S and pressure p). An isentropic flash unit is illustrated schematically in Figure. 4.12.

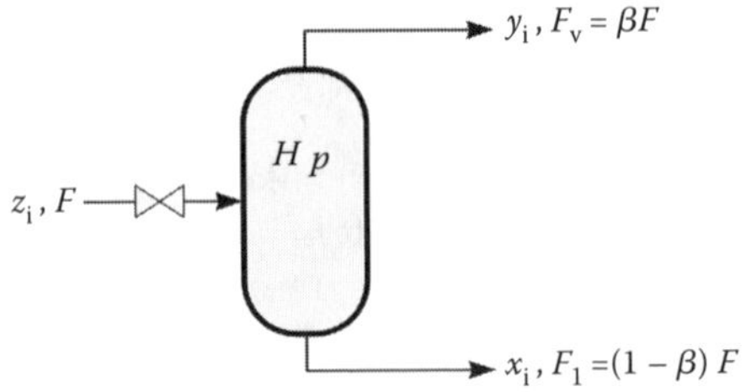

Figure 4.11 Isenthalpic flash unit.

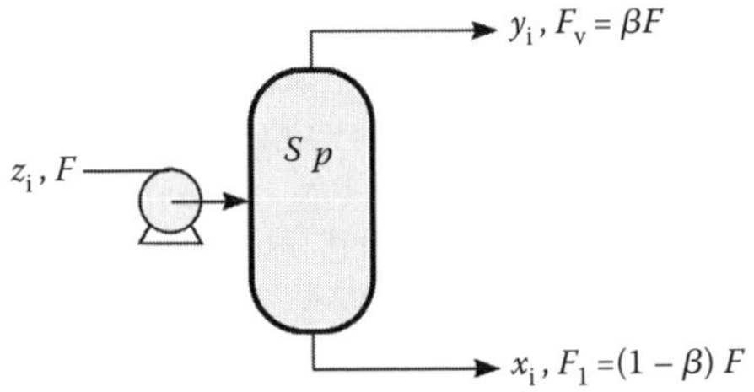

Figure 4.12 Isentropic flash.

The objective of the flash calculation is to find the temperature, vapor fraction, and product compositions for the case in which the operating pressure and the temperature, pressure, and composition of the feed are specified. Both an isenthalpic and an isentropic flash can be solved with methods analogous to Figure. 4.10 and details are given in Assael et al. (1996).

4.8 WOULD PRACTICAL EXAMPLES HELP?

4.8.1 What Is the Minimum Work Required to Separate Air into Its Constituents?

To tackle the problem it might seem straightforward to look for one or more processes that promise to separate air—or more generally a mixture of gases—and then seek to find the optimal conditions for each process under which they require the minimum amount of work. In general it may be difficult to find any such process, and one can never be sure that the result obtained is actually the optimal choice; it may merely be the best from those selected. Thus, it is best to consider the problem from the other end: what is the amount of useful energy that is destroyed by the mixing of gases, or what is the exergy loss E_l in such a process (see Question 3.9).

To simplify the analysis without losing the major thrust of the argument we consider air in the first instance as a mixture of only nitrogen and oxygen ($y_{N_2} = 0.79$, $y_{O_2} = 0.21$) and expand the problem to a more general case later. We further restrict the problem to treating dry air and neglect the varying humidity. When nitrogen and oxygen are mixed at standard conditions ($T = 298.15$ K, $p = 10^5$ Pa) these constituents may be treated as ideal gases. Thus, there is no enthalpy of mixing (nor a change in internal energy), and the mixing at constant pressure and temperature occurs in an adiabatic manner. If we imagine that the two gases are held separately in a single rigid vessel and that we then remove the partition (as shown in Figure 4.13), the system undergoes a diffusion process toward a new equilibrium. This diffusion process is irreversible and is accompanied by a rise in entropy

$$\Delta_{irr} S = -nR \sum_i y_i \ln y_i, \tag{4.181}$$

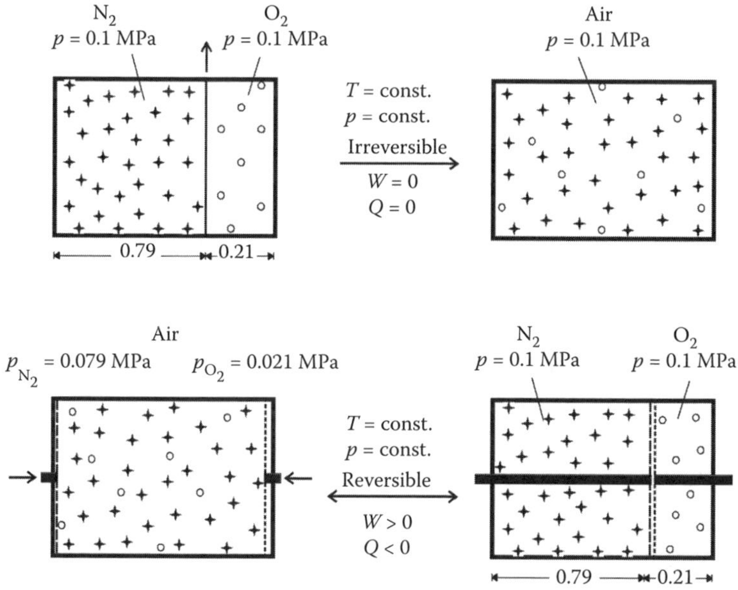

Figure 4.13 Mixing and separation of nitrogen and oxygen as the constituents of air at isothermal and isobaric conditions. Top: When removing the partition N_2 and O_2 mix irreversibly; there is no transfer of work or heat across the system boundaries. Bottom: In a hypothetical process the separation of air may be performed in a reversible manner: When work is applied to two semipermeable pistons the components are compressed from their respective partial pressures to system pressure under discharge of heat.

where n is the total amount of subtance and y represents the mole fractions of species in the gas phase. The amount of useful energy destroyed by such a process, or in other words the exergy loss, may be generally described by $E_l = T_0\Delta_{irr}S$ where T_0 is the temperature of the surroundings to give

$$E_l = -nT_0 R \sum_i y_i \ln y_i. \tag{4.182}$$

Because of the irreversibility of the equilibration processes there is no direct inverse of this process. However, it is obvious that the minimum work required to restore the initial state cannot be less in magnitude than the exergy loss in the mixing process: $W_{min} \geq E_l$.

Now we can ask what such a separation process might look like? As the constituents are to be present at the original temperature the restoration process should obviously be performed in an isothermal way. From the 1st law of thermodynamics, $\delta W + \delta Q = dU$, and $dU = n \cdot C_{V,m}\, dT = 0$. For a mixture of ideal gases it follows that the amount of work applied to the system must be balanced by the same amount of heat rejected from the system. If we suppose that in an idealized circumstance the heat is rejected at constant temperature T_0 for both the system and the surroundings then the heat transfer δQ is connected with the change of entropy dS via $\delta Q = T_0\, dS$. The total amount of entropy discharged during that "demixing" process equals, in magnitude, the entropy generated during the irreversible mixing process. So finally:

$$W_{min} = -Q = -T_0\left(S_{final} - S_{initial}\right) = -T_0\left(-\Delta_{irr}S\right) = -nT_0 R \sum_i y_i \ln y_i = E_l \tag{4.183}$$

Therefore, the minimal work required to separate air into its constituents has the same magnitude as the exergy destroyed during the mixing process.

It is particularly valuable to use this result to point out that it does not contradict the statement that the mixing process itself is irreversible. "Reversibility" (see Chapter 1) always implies that a process is reversed without any effect on the surroundings. While there is no energy flux whatsoever across system boundaries during the mixing process, the separation process requires the input of work and the discharge of heat across the system boundary.

It is also interesting to note that the work required to separate the mixture in the case that it is ideal is

$$W_{min} = -nT_0 R \sum_i y_i \ln y_i = -T_0 R \sum_i n_i \ln y_i = -T_0 R \sum_i n_i \ln\left(\frac{p_i}{p}\right), \tag{4.184}$$

which is equivalent to the total work required to compress the constituents from their partial pressures p_i to the system pressure p in an isothermal process (because $y_i = p_i/p$).

This analysis suggests that a hypothetical separation process might be as follows. The vessel containing the air (gas mixture) possesses two pistons, one on each side as shown in Figure 4.13. The piston on the left consists of a semi-permeable membrane where only nitrogen molecules may pass through and oxygen molecules are withheld; for the piston on the right conditions are interchanged, so oxygen may pass and nitrogen is blocked. When we move the two pistons simultaneously in a way to achieve a final position where the left piston has travelled 79 % of the total way (and accordingly the right one 21 %), all the nitrogen is enclosed in the left compartment and all the oxygen in the right one. This process exactly corresponds to the compression of each component from its partial pressure in the mixture to a final pressure of 10^5 Pa, which then equals the system pressure.

Let us finally illustrate the process with a numerical example and then examine the effect of the real composition of air on the results. As an example, 1 m^3 of air at standard conditions contains about 40 mol of an ideal gas mixture irrespective of composition. From Equation 4.184

$$W_{\text{min}} = -40.34\ \text{mol} \cdot 298.15\ \text{K} \cdot 8.314\ \text{J} \cdot \text{mol}^{-1} \cdot \text{K}^{-1} \sum_{\text{i}} y_{\text{i}} \ln y_{\text{i}}$$

$$\approx -100.0\ \text{kJ} \sum_{\text{i}} y_{\text{i}} \ln y_{\text{i}} \tag{4.185}$$

and we see that the actual work to separate the mixture depends only on the relative composition and not on the nature of the individual constituents. When we return to our simplest model of a binary mixture $y_{\text{N}_2} = 0.79$, $y_{\text{O}_2} = 0.21$ we obtain

$$W_{\text{min}} = -100.0\ \text{kJ}\left[0.79(-0.236) + 0.21(-1.56)\right]$$

$$= 100.0\ \text{kJ}\left[0.186 + 0.328\right] = 51\ \text{kJ}. \tag{4.186}$$

In a next approximation we also consider argon as a constituent of air, now with a composition of $y_{\text{N}_2} = 0.781$, $y_{\text{O}_2} = 0.210$, $y_{\text{Ar}} = 0.009$, resulting in

$$W_{\text{min}} = -100.0\ \text{kJ}\left[0.781(-0.2472) + 0.210(-1.561) + 0.009(-4.711)\right]$$

$$= 100.0\ \text{kJ}\left[0.1930 + 0.3277 + 0.0424\right] = 56.3\ \text{kJ}, \tag{4.187}$$

This procedure may be expanded in a straightforward manner to include other components of air such as carbon dioxide, neon, and so on. The main point about the numerical example, however, is to make clear that rather small or even spurious amounts of further components considerably increase the work required to separate the gas mixture. The reason behind this increase is the strong rise in the entropy of mixing with increasing dilution or—in other words—the comparatively large amount of work required to compress a volume containing a component at small partial pressure to system pressure. It should be obvious that the actual work for gas separation in a process is a multiple of the minimum work obtained from the idealized calculation.

Finally, the hypothetical separation process might—admittedly only theoretically—be inverted to obtain a mixing process producing work. The process is similar to the extraction of work from the isothermal expansion of a volume of a pure gas with a heat supply. In the case of the separation the expansion is allowed through the movement of the two semipermeable pistons lowering the pressure from the system pressure p to the respective partial pressures $y_i \cdot p$ for each component.

4.8.2 How Does a Cooling Tower Work?

A consequence of the 2nd law of thermodynamics is that a large power plant inevitably has to discharge energy of the order of 1 GW, because only a portion of the energy provided can be used as work to generate electricity. In a steam-powered electricity generation, the steam exiting the turbine must be condensed with cooling water and owing to the volume required the water is often extracted from lakes or rivers, passed through heat exchangers and discharged to the source of the water at a temperature greater than the source; this action in principle has an environmental consequence that will not be considered further here. To limit the temperature increment wet cooling towers are used, which rely on the enthalpy of vaporization to cool the water. In the case of water the enthalpy of vaporization is relatively high and requires a relatively low mass to evaporate as a function of time to decrease the water temperature. Cooling towers are used in other industrial applications or with large air conditioning systems where single phase energy exchangers utilizing air or water are insufficient to dissipate the energy.

Cooling towers may operate with either forced or natural convection, and a schematic of the latter is shown in Figure 4.14. In this case, a stream of warm water is sprayed onto a solid surface labeled as inserts in Figure 4.14 that ensure that the droplets are broken up and that there is intense mixing of them with atmospheric air, which is drawn in from below and takes up moisture as water vapor leaving at the top of the tower with a higher humidity. When this exhaust air mixes with colder ambient air, a plume of fog may become visible, in a similar fashion to that discussed in Question 4.7.1.

The energy required for the evaporation of some of the water is mostly taken from the warm water, which leaves with a lower temperature at the bottom of the tower and may be returned to the coolant stream. Because some cooling water is evaporated as a part of this process it is necessary to add water to

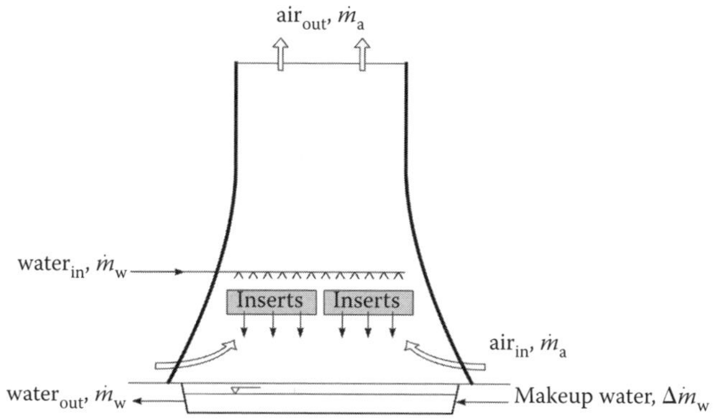

Figure. 4.14 Scheme of natural-draught cooling tower: Incoming water is cooled by mixing with ambient air and partial evaporation; only a small portion of the water has to be replaced by additional so-called make-up water.

replace that which is lost, but because water has a relatively large enthalpy of evaporation the amount of water required is, as we will now show, rather small by comparison. To illustrate this point we will consider the mass flow rate of water required for a steam-powered electricity generating plant that must dissipate a heat flux of $\dot{Q} = 1$ GW. In the first case, we assume that cooling occurs solely by water obtained from a river, which enters the cooling system at a temperature of 10 °C (283 K) and then exits at a temperature of 35 °C (308 K) where the specific heat capacity of water is approximated as $4.2{\cdot}10^3\ \mathrm{J{\cdot}kg^{-1}\cdot K^{-1}}$. From

$$\dot{Q} = \Delta\dot{H} = \dot{m}_w c_p \Delta T, \tag{4.188}$$

we have

$$\dot{m}_w = \frac{\dot{Q}}{c_p \Delta T} = \frac{10^9\ \mathrm{W}}{4.2\cdot 10^3\ \mathrm{J\cdot kg^{-1}\cdot K^{-1}} \times (35-10)\mathrm{K}}\ 9.5\cdot 10^3\ \mathrm{kg\cdot s^{-1}}. \tag{4.189}$$

Alternatively, the cooling water may be circulated and chilled with a cooling tower. If we assume that the water is only cooled down to a temperature of 20 °C in that circle, providing a temperature difference of only 15 K instead of 25 K, a higher mass flow rate of $\dot{m}_w = 16 \cdot 10^3\ \mathrm{kg \cdot s^{-1}}$ results. We now introduce a cooling tower where ambient air enters at a temperature of $t = 10$ °C and a relative humidity $\phi = 0.7$ and where the air exits the cooling tower saturated with water vapor at a temperature of 25 °C. The moisture content ω of the air, defined by Equation 4.161, can be calculated from Equation 4.164 with p_v^{sat} (10 °C) = 1.23 kPa and p_v^{sat} (25 °C) = 3.17 kPa to give $\omega_{in} = 0.0054$ and $\omega_{out} = 0.020$. Using

IUPAC nomenclature (Quack et al. 2007) and Equation 4.160 combined with an equation of state gives a mass ratio of water to air of the input stream of

$$\frac{\rho_g\{p_{H_2O}(283\text{ K}),283\text{ K}\}}{\rho_g\{p^{\text{sat}}_{H_2O}(283\text{ K}),283\text{ K}\}}=0.0054, \tag{4.190}$$

and of the output stream ratio of

$$\frac{\rho_g\{p_{H_2O}(298\text{ K}),298\text{ K}\}}{\rho_g\{p^{\text{sat}}_{H_2O}(298\text{ K}),298\text{ K}\}}=0.020, \tag{4.191}$$

when the vapor pressure $p^{\text{sat}}_{H_2O}(T)$ may be obtained from Equation 4.27. From an energy balance over the whole cooling tower a mass flow of dry air $\dot{m}_a = 19 \cdot 10^3\text{ kg}\cdot\text{s}^{-1}$ is obtained. Because water evaporates during the cooling process, it must be replaced by what is termed *additional make-up water* at an assumed temperature of 10 °C. The mass flow rate of the additional water $\Delta\dot{m}_w$ required is obtained from

$$\Delta\dot{m}_w = \dot{m}_a(\omega_{out} - \omega_{in}) = 0.28\cdot 10^3\text{ kg}\cdot\text{s}^{-1}, \tag{4.192}$$

which is less than 2 % of the total mass flow. In IUPAC nomenclature Equation 4.192 for the mass flow rate of water $\Delta\dot{m}(H_2O)$ is given by

$$\Delta\dot{m}(H_2O)=\dot{m}(\text{air})\left[\frac{\rho\{g,p(298\text{ K},H_2O),298\text{ K}\}}{\rho\{g,p^{\text{sat}}(298\text{ K},H_2O),298\text{ K}\}}-\frac{\rho\{g,p(283\text{ K},H_2O),283\text{ K}\}}{\rho\{g,p^{\text{sat}}(283\text{ K},H_2O),283\text{ K}\}}\right]=0.28\cdot 10^3\text{ kg}\cdot\text{s}^{-1}. \tag{4.193}$$

4.9 WHAT IS THE TEMPERATURE CHANGE OF DILUTION?

This example is intended to illustrate in a simple manner the nature of the calculations that the preceding material makes possible.

The concepts required to describe the properties of a mixture of two liquids have been introduced in Question 4.5, and these include ideal mixtures and the definition of the excess properties given by Equations 4.97 through 4.100. For an ideal mixture the molar volume of mixing and the molar enthalpy of mixing are zero as given by Equations 4.95 and 4.94, respectively. The excess molar enthalpy and excess molar volume are given in Equations 4.99 and 4.100. Normally, $\Delta_{mix}H_m$ and $\Delta_{mix}V_m$ are nonzero.

Atkins (1987) has a description of a "corrupt barman" and the argument will be used here as an example. The barman mixes at a temperature of 298.15 K a volume of 100 cm^3 of substance he intends to sell as pseudowhiskey. This barman uses 40 cm^3 of ethanol and 60 cm^3 of water to make the drink. The negative volume of mixing $\Delta_{mix}V_m$ results in a volume of 96 cm^3 of ethanol + water. We note here parenthetically that, based on the densities of the two pure substances at a temperature of 298.15 K and a pressure of 0.1 MPa, an amount of substance of ethanol $n(C_2H_5OH) = 0.71$ mol and an amount of substance of water of $n(H_2O) = 3.45$ mol would actually be required to provide 100 cm^3 of (ethanol + water).

If a further volume of 50 cm^3 of water is then added to the original mixture the volume will change again and, if the dilution is prepared adiabatically, so will the temperature of the resulting mixture. Adiabatic conditions can be approximated adequately for our purposes by rapid mixing or by the use of a Styrofoam cup.

The temperature change can be determined from the 1st law of thermodynamics for a closed system in the absence of external work from stirring and energy transfer from the surroundings, when the internal energy U of the system remains unaltered. However, for mixtures it is more common to consider the enthalpy $H = U + pV$ where in the case of liquids practically no difference arises. Dilution of the pseudowhiskey results in a volume of mixing $\Delta_{mix}V_m$ that is less then 1 cm^3; the corresponding change of enthalpy at atmospheric pressure of about 0.1 MPa (10^5 Pa) is $p\Delta V = 10^5\ \mathrm{Pa} \cdot 10^{-6}\ \mathrm{m}^3 = 0.1\ \mathrm{J}$ and this is negligible compared to the other energies involved in the mixing.

Because the enthalpies of mixing (equivalent to the excess enthalpies) are defined and measured for constant temperature (and pressure) and because we expect a change in temperature during our mixing process, we notionally split up the process into two steps:

1. First we perform the dilution step at constant temperature of 298.15 K by rejecting exactly an amount of heat Q_D to render the temperature unaltered. (We anticipate that heat is released during the dilution step but note that the sign of Q_D does not affect the following calculations.)
2. Then we use exactly this heat to increase the temperature of the resulting mixture to a final temperature.

First of all, we calculate the mole fractions of ethanol in the respective mixtures before the dilution (initial, i) and after dilution (final, f) and obtain $x_i = 0.17$ and $x_f = 0.10$ (with an amount of substance $n_a = 2.77$ mol of water added).

Using the formal IUPAC nomenclature adopted by chemists (Quack et al. 2007), the energy balance for the first step is given by

$$
\begin{aligned}
& n(C_2H_5OH)H_m(C_2H_5OH)+\{n(H_2O)+n_a(H_2O)\}H_m(H_2O) \\
& \quad +\{n(C_2H_5OH)+n(H_2O)+n_a(H_2O)\}H_m^E(x_f, 298.15\ K) - \\
& \left[\begin{array}{l} n(C_2H_5OH)H_m(C_2H_5OH)+n(H_2O)H_m(H_2O)+ \\ \{n(C_2H_5OH)+n(H_2O)\}H_m^E(x_i, 298.15\ K)+n_a(H_2O)H_m(H_2O) \end{array}\right] = Q_D, \quad (4.194)
\end{aligned}
$$

where $H_m^E(x)$ is the excess molar enthalpy at the respective concentration, and H_m are the molar enthalpy for each pure component. However, as we recognize throughout this text, the language of the chemist is neither familiar nor practical for all and so we adopt a simplification to the notation used in Equation 4.194. In particular, we substitute n_E for $n(C_2H_5OH)$, $H_{m,E}$ for $H_m(C_2H_5OH)$, n_W for $n(H_2O)$, and $H_{m,W}$ for $H_m(H_2O)$ so that Equation 4.194 reads

$$
\begin{aligned}
& n_E H_{m,E} + n_W H_{m,W} + n_a H_{m,W} + (n_E + n_W + n_a) H_m^E(x_f, 298.15\ K) - \\
& \quad \left[n_E H_{m,E} + n_W H_{m,W} + (n_E + n_W) H_m^E(x_i, 298.15\ K) + n_a H_{m,W} \right] = Q_D, \quad (4.195)
\end{aligned}
$$

In either form of Equation 4.194 or 4.195 the enthalpies of the pure components cancel so that Equation 4.194 simplifies to

$$
\begin{aligned}
& \{n(C_2H_5OH) + n(H_2O) + n_a(H_2O)\} H_m^E(x_f, 298.15\ K) \\
& \quad - \{n(C_2H_5OH) + n(H_2O)\} H_m^E(x_i, 298.15\ K) = Q_D, \quad (4.196)
\end{aligned}
$$

or

$$
(n_E + n_W + n_a) H_m^E(x_f, 298.15\ K) - (n_E + n_W) H_m^E(x_i, 298.15\ K) = Q_D. \quad (4.197)
$$

Figure 4.15 shows the variation of the excess molar enthalpy for ethanol + water as a function of composition as determined experimentally. By careful interpolation in the data that support Figure 4.15 we can obtain H_m^E ($x_f = 0.10$, $T_i = 298$ K) = −711 J · mol^{-1} and H_m^E ($x_f = 0.17$, $T_i = 298$ K) = −784 J · mol^{-1}. Thus

$$
\begin{aligned}
Q_D &= (0.71+3.45+2.77)(-711\ J) - (0.71+3.45)(-784\ J) \\
&= -4.93\ kJ + 3.26\ kJ = -1.67\ kJ, \quad (4.198)
\end{aligned}
$$

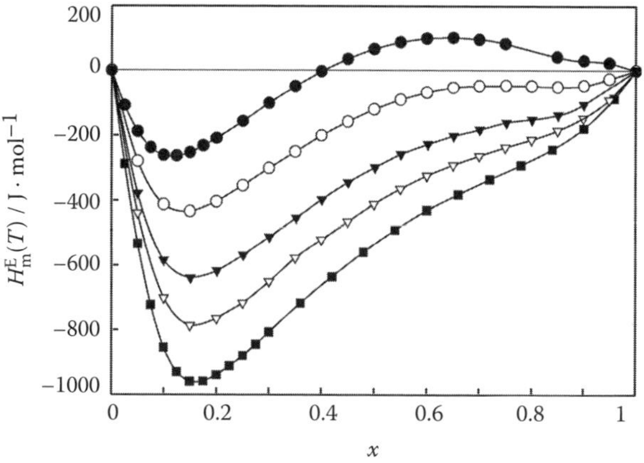

Figure 4.15 Molar excess enthalpy $H_m^E(T)$ for (ethanol + water) as a function of mole fraction of ethanol x and temperature T. ●: $T = 338.15$ K; ○: $T = 323.15$ K; ▼: $T = 308.15$ K; ▽: $T = 298.15$ K; and ■: $T = 285.65$ K. Data from Friese et al. (1998; 1999).

which is negative; this means that heat must be discarded to hold the temperature constant.

In the second step we add this heat to increase the temperature of the mixture to the final temperature T_f obtained from

$$H_m(x_f,T_f) - H_m(x_f,298.15\,\text{K}) = \\ \{n(C_2H_5OH) + n(H_2O) + n_a(H_2O)\}C_{p,m}(T_f - 298.15\,\text{K}) = -Q_D, \quad (4.199)$$

or the alternative form

$$H_m(x_f,T_f) - H_m(x_f,298.15\,\text{K}) = (n_E + n_W + n_a)C_{p,m}(T_f - 298.15\,\text{K}) = -Q_D. \quad (4.200)$$

where $C_{p,m}$ is the molar heat capacity at constant pressure for the mixture. In this step, however, we must proceed with some caution. From Figure 4.15 we recognize that H_m^E is a function of both temperature and composition and, from the definition of the heat capacity at constant pressure, we have

$$C_{p,m} = \left(\frac{\partial H_m}{\partial T}\right)_{p,x}. \quad (4.201)$$

We can now split up the heat capacity into two parts

$$C_{p,\mathrm{m}} = \left(\frac{\partial \left[\{ n(\mathrm{C_2H_5OH}) H_\mathrm{m}(\mathrm{C_2H_5OH}) + \{ n(\mathrm{H_2O}) + n_\mathrm{a}(\mathrm{H_2O}) \} H_\mathrm{m}(\mathrm{H_2O}) \} / n + H_\mathrm{m}^\mathrm{E} \right]}{\partial T} \right)_{p,x}$$

$$= \left(\frac{\partial \left[\{ n(\mathrm{C_2H_5OH}) H_\mathrm{m}(\mathrm{C_2H_5OH}) + \{ n(\mathrm{H_2O}) + n_\mathrm{a}(\mathrm{H_2O}) \} H_\mathrm{m}(\mathrm{H_2O}) \} / n \right]}{\partial T} \right)_{p,x}$$

$$+ \left(\frac{\partial H_\mathrm{m}^\mathrm{E}}{\partial T} \right)_{p,x} = C_{p,\mathrm{m}}^\mathrm{id} + C_{p,\mathrm{m}}^\mathrm{E}, \tag{4.202}$$

or in the alternative form

$$C_{p,\mathrm{m}} = \left[\frac{\partial \left\{ \left(n_\mathrm{E} H_\mathrm{m,E} + n_\mathrm{W} H_\mathrm{m,W} + n_\mathrm{a} H_\mathrm{m,W} \right) / n + H_\mathrm{m}^\mathrm{E} \right\}}{\partial T} \right]_{p,x}$$

$$= \left[\frac{\partial \left\{ \left(n_\mathrm{E} H_\mathrm{m,E} + n_\mathrm{W} H_\mathrm{m,W} + n_\mathrm{a} H_\mathrm{m,W} \right) / n \right\}}{\partial T} \right]_{p,x} + \left(\frac{\partial H_\mathrm{m}^\mathrm{E}}{\partial T} \right)_{p,x},$$

$$= C_{p,\mathrm{m}}^\mathrm{id} + C_{p,\mathrm{m}}^\mathrm{E} \tag{4.203}$$

where $n = n(\mathrm{C_2H_5OH}) + n(\mathrm{H_2O}) + n_\mathrm{a}(\mathrm{H_2O}) = n_\mathrm{E} + n_\mathrm{W} + n_\mathrm{a}$ is the total amount of substance. Here, $C_{p,\mathrm{m}}^\mathrm{id}$ is the ideal part of the heat capacity, which can be easily obtained by summing up the heat capacities of the individual components weighted with their respective mole fractions, and $C_{p,\mathrm{m}}^\mathrm{E}$ is the excess part of the heat capacity, which takes the nonideality of the solution into account. From the tabulated heat capacities of the pure substances, we obtain $C_{p,\mathrm{m}}^\mathrm{id} = 79.1\ \mathrm{J \cdot mol^{-1} \cdot K^{-1}}$, which is assumed to be constant over the small temperature range of interest. On the other hand $C_{p,\mathrm{m}}^\mathrm{E}$ is estimated from Figure. 4.15. If we consider the excess enthalpies for $x = 0.10$ and temperatures of 285.65 K, 298.15 K, and 308.15 K we see that H_m^E is almost linear with temperature in this range and from a fit through these three points we obtain for the gradient $C_{p,\mathrm{m}}^\mathrm{E} = 12.2\ \mathrm{J \cdot mol^{-1} \cdot K^{-1}}$.

From Equation 4.202 it then follows that

$$(T_\mathrm{f} - 298.15\ \mathrm{K}) = \frac{1.67\ \mathrm{kJ}}{6.93\ \mathrm{mol}(79.1 + 12.2)\ \mathrm{J \cdot mol^{-1} \cdot K^{-1}}} = 2.6\ \mathrm{K}. \tag{4.204}$$

In this case of dilution we therefore observe a moderate, yet easily measurable temperature increase of the mixture of 2.6 K.

The question may arise what would happen if we added ethanol to the pseudowhiskey. In general, when we start with a mixture of pure substances (water and ethanol), a negative excess enthalpy means that we have to discharge heat to keep the solution at constant temperature, and the temperature would rise if the system was adiabatic. At a temperature of $T = 338.15$ K and depending on the final composition the opposite effect is also possible.

When we dilute our original mixture with water at $T = 298.15$ K the corresponding point for the final concentration on the connecting line (between the original state and that for pure water) is above the curve for $H^E_m(T_i = 298.15\,\text{K})$, thus the temperature will rise for adiabatic mixing as shown in Figure 4.16. On the other hand one recognizes that when connecting this initial point with the point of pure ethanol there are portions of the connecting line {roughly in a range $0.27 < x(C_2H_5OH) < 0.75$} that are below the curve H^E_m, which means that the mixture would cool down during mixing. We note, however, that neither the cooling effect nor the high alcohol content of the drink would encourage consumption.

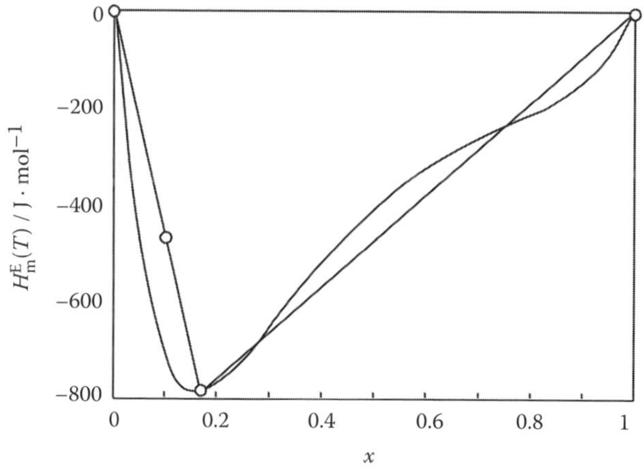

Figure 4.16 Molar excess enthalpy $H^E_m(T)$ for (ethanol + water) as a function of mole fraction of ethanol x at $T = 298.15$ K and illustration of the effect of adding water or ethanol, respectively, to a mixture with an initial concentration $x(C_2H_5OH) = 0.17$. The temperature of the mixture will rise (fall) if the point on the connecting line for the final concentration is above (below) the curve $H^E_m(298.15\,\text{K})$.

4.10 WHAT ABOUT LIQUID + LIQUID AND SOLID + LIQUID EQUILIBRIA?

We now return to the discussion of liquids and in particular some issues regarding liquid + liquid equilibrium and solid + liquid equilibrium. These are certainly important industrially and to our way of life.

4.10.1 What Are Conformal Mixtures?

The assumption that the pair-interaction energy ϕ_{AB} between substances A and B of a mixture is solely a function of the intermolecular separation r and is given by

$$\phi_{AB}(r) = \varepsilon_{AB}\Phi\left(\frac{r}{r^*_{AB}}\right), \tag{4.205}$$

where ε_{AB} is the well depth at the equilibrium r, and r^*_{AB} is a characteristic separation, and Φ is the same function for A and B. Strictly, the dependence of ϕ_{AB} solely on r means that the theory is limited to mixtures of spherical molecules and the requirement of Φ requires that the molecules conform to the principle of corresponding states described in Chapter 2.

There are many routes that can be followed from Equation 4.205 that depend on the method used for a mixture. The most common is the one-fluid theory as applied to the van der Waals equation as discussed earlier. In the one-fluid theory the liquid mixture is assumed to be represented by a hypothetical pure fluid that also conforms to Equation 4.205. To complete the theory we require a definition of Φ, typically from an equation of state such as Carnahan and Starling (1972), and a selection of mixing and combining rules. The mixing rules for ε and r^* are analogous to those obtained for van der Waals one-fluid approximation and are given by Equations 4.165 and 4.166. The combining rules for r^*_{AB} and ε_{AB} are obtained from expressions analogous to Equations 4.169 and 4.170, including a disposable parameter often called an *interaction parameter*. This set of equations can then be used to determine G^E_m, H^E_m, and V^E_m from the critical properties of the pure substances A and B.

4.10.2 What Are Simple Mixtures?

For nonelectrolytes a simple mixture can be defined by the excess molar Gibbs function that can be written as

$$G^E_m = x(1-x)Lw, \tag{4.206}$$

where L is the Avogadro constant, and w depends on temperature and pressure only. From Equations 4.110 and 4.111 the activity coefficients of a simple mixture are given by

$$\ln f_{\rm A} = \frac{x^2 w}{kT}, \tag{4.207}$$

and

$$\ln f_{\rm B} = \frac{(1-x)^2 w}{kT}. \tag{4.208}$$

Hildebrand's theory of solubility with parameter $\delta_{\rm i}$ of a substance i given by

$$\delta_{\rm i} = \left\{ \frac{\Delta_{\rm l}^{\rm g} H_{\rm i}^* - RT}{V_{\rm i}^*} \right\}^{1/2}, \tag{4.209}$$

and it can be used to estimate w from the properties of pure substances. For a binary mixture $(1 - x)$A + xB the expression is

$$w = \frac{V_{\rm A}^* V_{\rm B}^* (\delta_{\rm A} - \delta_{\rm B})^2}{L\{(1-x)V_{\rm A}^* + xV_{\rm B}^*\}}. \tag{4.210}$$

4.10.3 What Are Partially Miscible Liquid Mixtures?

Liquid mixture can separate into two-liquid phases. The two phases appear at a temperature below what is called an *upper critical solution temperature* (UCST) or above a *lower critical solution temperature* (LCST). Mixtures with a LCST can also have a USCT at higher temperature and exhibit what is called *closed loop miscibility.* It is possible to have a LCST at a temperature greater than the UCST so that at temperature between the liquids are miscible, for example, for $(1 - x)H_2O + xCH_2(OH)CH_2OC_4H_9$. Examples of UCST and LCST are shown in Figure 4.17.

For a simple mixture, as defined by Equation 4.206, the chemical potentials of substances A and B of a binary liquid mixture are given by

$$\mu_{\rm A} = \mu_{\rm A}^* + RT\ln(1-x) + x^2 Lw, \tag{4.211}$$

and

$$\mu_{\rm B} = \mu_{\rm B}^* + RT\ln x + (1-x)^2 Lw. \tag{4.212}$$

So that the condition for coexisting phases α and β is given by

$$\ln\left\{\frac{(1-x^{\alpha})}{(1-x^{\beta})}\right\} = \frac{(x^{\beta} - x^{\alpha})(x^{\beta} + x^{\alpha})w}{kT}, \tag{4.213}$$

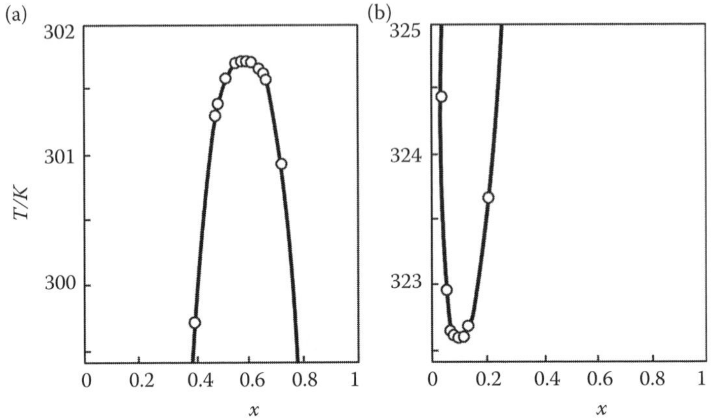

Figure 4.17 $p(x)$ section illustrating the partial miscibility. (a): $(1 - x)C_6H_{12} + xCH_2I_2$ showing a UCST. o: experimental values; ——: estimated with a critical exponent $\beta = 0.347$. (b): $(1 - x)H_2O + xCH_3(C_2H_5)_2N$ illustrating the LCST. o: experimental values; ——, estimated with a critical exponent $\beta = 0.34$.

and

$$\ln\left(\frac{x^{\alpha}}{x^{\beta}}\right) = \frac{(x^{\beta} - x^{\alpha})(x^{\beta} + x^{\alpha} - 2)w}{kT}. \quad (4.214)$$

Equations 4.213 and 4.214 do not represent well the measured properties of mixtures but they nevertheless provide estimates that are qualitatively correct.

4.10.4 What Are Critical Points in Liquid Mixtures?

The criteria for a critical point of a mixture is the same for both (vapor + liquid) equilibria and solutions. Thus, the critical temperature T_c is defined by

$$\left(\frac{\partial^2 G_m}{\partial x^2}\right)_{T,p} = 0, \quad (4.215)$$

$$\left(\frac{\partial^3 G_m}{\partial x^3}\right)_{T,p} = 0, \quad (4.216)$$

and

$$\left(\frac{\partial^4 G_m}{\partial x^4}\right)_{T,p} > 0. \quad (4.217)$$

For the variables T, x, and V_m the Helmholtz function is the appropriate thermodynamic energy, and Equations 4.215 to 4.217 can be recast as

$$\left(\frac{\partial^2 A_m}{\partial x^2}\right)_{T,V_m} - 2\frac{(\partial^2 A_m/\partial V_m\partial x)_T}{(\partial^2 A_m/\partial V_m^2)_{T,x}}\left(\frac{\partial^2 A_m}{\partial V_m \partial x}\right)_T + \frac{(\partial^2 A_m/\partial V_m\partial x)_T}{(\partial^2 A_m/\partial V_m^2)_{T,x}}\left(\frac{\partial^2 A_m}{\partial V_m^2}\right)_{T,x} = 0, \tag{4.218}$$

$$\left(\frac{\partial^3 A_m}{\partial x^3}\right)_{T,V_m} - 3\frac{(\partial^2 A_m/\partial V_m\partial x)_T}{(\partial^2 A_m/\partial V_m^2)_{T,x}}\left(\frac{\partial^3 A_m}{\partial V_m \partial x^2}\right)_T + 3\left\{\frac{(\partial^2 A_m/\partial V_m\partial x)_T}{(\partial^2 A_m/\partial V_m^2)_{T,x}}\right\}^2\left(\frac{\partial^3 A_m}{\partial V_m^2 \partial x}\right)_T$$
$$- \left\{\frac{(\partial^2 A_m/\partial V_m\partial x)_T}{(\partial^2 A_m/\partial V_m^2)_{T,x}}\right\}^3\left(\frac{\partial^3 A_m}{\partial x^3}\right)_{T,V_m} = 0, \tag{4.219}$$

and

$$\left(\frac{\partial^4 A_m}{\partial x^4}\right)_{T,V_m} - 4\frac{(\partial^2 A_m/\partial V_m\partial x)_T}{(\partial^2 A_m/\partial V_m^2)_{T,x}}\left(\frac{\partial^4 A_m}{\partial V_m \partial x^3}\right)_T + 6\left\{\frac{(\partial^2 A_m/\partial V_m\partial x)_T}{(\partial^2 A_m/\partial V_m^2)_{T,x}}\right\}^2\left(\frac{\partial^4 A_m}{\partial V_m^2 \partial x^2}\right)_T$$
$$- 4\left\{\frac{(\partial^2 A_m/\partial V_m\partial x)_T}{(\partial^2 A_m/\partial V_m^2)_{T,x}}\right\}^3\left(\frac{\partial^4 A_m}{\partial V_m^3 \partial x}\right)_T + \left\{\frac{(\partial^2 A_m/\partial V_m\partial x)_T}{(\partial^2 A_m/\partial V_m^2)_{T,x}}\right\}^4\left(\frac{\partial^4 A_m}{\partial V_m^4}\right)_{T,x} > 0. \tag{4.220}$$

(Liquid + liquid) and (liquid + gas) critical points can be indistinguishable.

For liquid mixtures with UCST and LCST at low pressure Equations 4.215 and 4.216 can be cast in terms of the excess molar Gibbs function as

$$\left(\frac{\partial^2 G_m^E}{\partial x^2}\right)_{T_c} = -\frac{RT_c}{x_c(1-x_c)}, \tag{4.221}$$

and

$$\left(\frac{\partial^3 G_m^E}{\partial x^3}\right)_{T_c} = \frac{RT_c(1-2x_c)}{(x_c)^2(1-x_c)^2}. \tag{4.222}$$

For the simple mixture defined by Equations 4.206, 4.221, and 4.222

$$-2w = -\frac{RT_c}{x_c(1-x_c)}, \tag{4.223}$$

and

$$0 = \frac{RT_c(1-2x_c)}{(x_c)^2(1-x_c)^2}. \qquad (4.224)$$

From Equation 4.224 $x_c = 0.5$ and thus from Equation 4.223

$$w = 2RT_c, \qquad (4.225)$$

so that T_c exists only for $w > 0$ and it is a UCST for $(w - T\mathrm{d}w/\mathrm{d}T) > 0$ and an LCST for $(w - T\mathrm{d}w/\mathrm{d}T) < 0$. These are useful approximations for estimating the conditions under which UCST and LCST will occur. For associating liquid and fluid mixtures, that is, those mixtures that form compounds through, for example, hydrogen bonding, the reader should refer to the methods of SAFT and the so-called Cubic Plus Association equation of state (Economou 2010).

4.10.5 What about the Equilibrium of Liquid Mixtures and Pure Solids?

For a mixture of liquids A and B that form a solid the $T(x)_p$ sections of the phase diagrams are similar to the two examples given in Figure 4.18. In Figure 4.18 the point labeled E is the intersection of, for the case of Figure 4.18a, the curves for the melting of solid A and the melting of solid AB. Its coordinates are called the *eutectic temperature* and the *eutectic composition*. The eutectic temperature is the lowest temperature where liquid A and B can exist and at this temperature there are three coexisting phases consisting of liquid mixture, solid A, and solid AB. Figure 4.18a also shows the congruent melting temperature $T_C \approx 297$ K of $C_6H_6 \cdot C_6F_6$(s) in $(1 - x)C_6H_6 + xC_6F_6$, while Figure 4.18b gives the peritectic temperature $T_P \approx 237$ K of $C_5H_5N \cdot C_6F_6$(s) that decomposes into a liquid of mole fraction x_P and pure solid B from $(1 - x)C_5H_5N + xC_6F_6$.

4.11 WHAT PARTICULAR FEATURES DO PHASE EQUILIBRIA HAVE?

Excess molar functions as given, for example, in the simplified forms of Equations 4.126 through 4.128, are useful for mixtures of liquids of similar volatility at pressures that are about $p^{\ominus} = 0.101325$ MPa ≈ 0.1 MPa and do not exceed $2p^{\ominus}$. For a mixture of liquids of similar volatility at higher pressure p, however, a virial expansion is inadequate. Unfortunately in just those cases insufficient information is usually available to determine an equation of state for the coexisting gas phase. This means it is difficult to use Equation 4.103 and

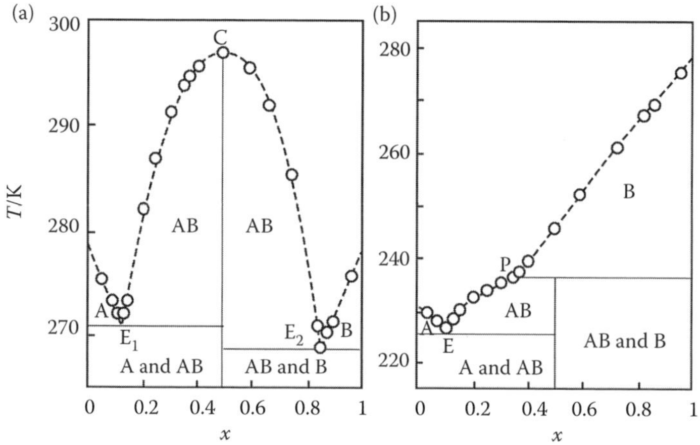

Figure 4.18 $T(x)$ section for two-liquid mixtures that form a solid compound. (a): $(1-x)C_6H_6 + xC_6F_6$ that forms solid compound $C_6H_6{\cdot}C_6F_6(s)$ that melts at a congruent melting temperature $T_C \approx 297$ K. (b): $(1-x)C_5H_5N + xC_6F_6$ that forms $C_5H_5N{\cdot}C_6F_6(s)$ that decomposes into a liquid of mole fraction x_P and pure solid B at an incongruent melting temperature or peritectic temperature $T_P \approx 237$ K.

the circumstances severely limit the temperature range over which G_m^E can be determined. Furthermore, the activity coefficient includes the absolute activity of each pure substance and that requires it to be a liquid at the relevant temperature and pressure. Thus, for mixtures of substances of very different volatility, it is possible that one component at the relevant temperature and pressure may be either a gas or a solid. Evidently in such cases the approach of activity coefficients is rather difficult to apply; for example at $T = 300$ K for $(1-x)C_6H_6 + xN_2$ the nitrogen is a gas and for $(1-x)C_6H_6 + xC_{14}H_{10}$ the anthracene is a solid.

4.11.1 What Is a Simple Phase Diagram?

Phase diagrams for mixtures are at least three dimensional (p, T, x). These are usually shown as two dimensional projections of $p(T)$. In this case the pressure as a function of temperature $p(T)_x$ at constant composition (these are called *isopleths*) would reveal the dew and bubble pressure that meet on the critical line $p(x)_T$ that are isotherms and $T(x)_p$ isobars. For a $p(x)$ diagram it is possible to show several x and the critical line is then the locus of the maxima of $p(x)$ isothermal sections. For a particular temperature the lines joining the mole fractions of the coexisting fluid phases are called *tie lines* and, in very simple mixtures, define two curves one for the gas the other for the liquid. Here we will restrict comments to those dealing with fluid phases and exclude the formation of solids.

For a binary mixture the vapor + liquid phase equilibria is simple when the critical points of the two pure substances are joined by a continuous curve and there is neither azeotropy (Question 4.11.4) nor three fluid phases.

4.11.2 What Is Retrograde Condensation (or Evaporation)?

Typically, retrograde condensation occurs when the dew (or for that matter the bubble) curve is intersected twice for an isothermal section by a pathway of constant composition as shown in Figure 4.19 for $(1-x)$Ar + xKr at $T = 177.38$ K. Figure 4.19 also shows the relative volumes of the more dense phase that is formed when the pressure increases. At $x = 0.39$ from a pressure below that of dew formation the gas is compressed and a more dense phase forms at the dew pressure $p^{\mathrm{d}} \approx 5.4$ MPa. The volume of the more dense phase varies with increasing pressure as the quality lines within the two phase region are intersected; in the case shown in Figure 4.19 increasing pressure initially increases the volume of the more dense phase. Further increase in pressure result in a decrease in the volume of the more dense phase (liquid) as the pressure tends toward a second intersection with the dew pressure $p^{\mathrm{d}} \approx 5.9$ MPa. Further increase of pressure above the dew pressure results in the disappearance of the more dense phase (liquid is no longer present). For $x = 0.42$ a more dense phase forms and the volume of this phase increases until the pressure is greater than the bubble pressure $p^{\mathrm{b}} \approx 5.95$ MPa. The pathway at $x = 0.39$ appears to defy the concept that at a given temperature increasing pressure must decrease the volume occupied by a fixed amount of substance.

4.11.3 What Is the Barotropic Effect?

This occurs when a mixture of two substances is at a temperature and pressure such that the molar mass of the pure substances and the molar volumes of the coexisting phases give nearly equal densities for the phases. This means that

$$\frac{(1-x^{\alpha})M_{\mathrm{A}}+x^{\alpha}M_{\mathrm{B}}}{V_{\mathrm{m}}^{\alpha}} \approx \frac{(1-x^{\beta})M_{\mathrm{A}}+x^{\beta}M_{\mathrm{B}}}{V_{\mathrm{m}}^{\beta}}. \tag{4.226}$$

In this case, within a gravitational field, a change in pressure or temperature can cause the two phases to invert so that what was the more dense becomes the less dense. The question that can be asked then is which is the gas phase

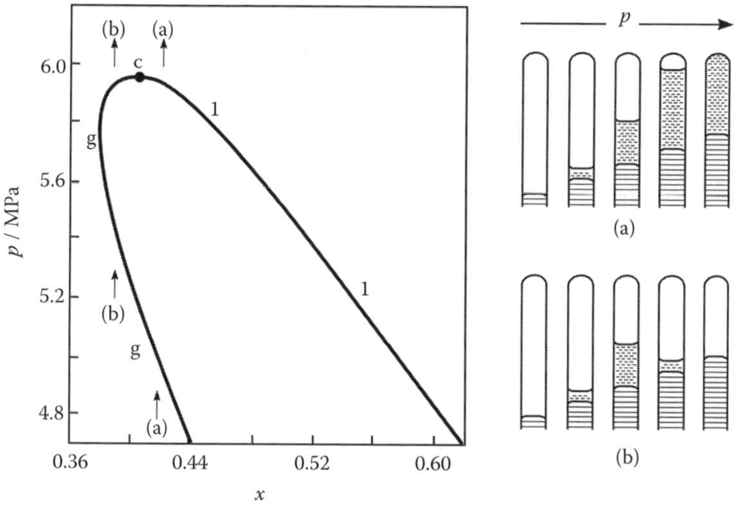

Figure 4.19 Left: The $p(x)_T$ section at $T = 177.38$ K for $(1 - x)$Ar + xKr illustrates retrograde condensation. C denotes the critical point; 1, is the bubble curve at $x > x_c$; and g, labels the dew curve. Right: Illustrates the relative volumes of liquid and gas obtained for changing pressure with a mercury piston, indicated by horizontal lines, at constant composition and temperature. For $x = 0.42$ and illustrated in schema (a), the gas is compressed to condense (in this case we will assume to a liquid) a phase of greater density indicated by a dashed horizontal line forms at the dew line $p \approx 4.95$ MPa; continual compression results in a tube filled with a more dense phase (in this case a liquid) at a pressure greater than the bubble pressure $p^b \approx 5.95$ MPa. For $x = 0.39$ and illustrated by schema (b) the gas is compressed and forms a more dense phase (we will call liquid) at the dew pressure $p^d \approx 5.4$ MPa; the volume of the more dense phase varies as the quality lines within the two-phase region are intersected with increasing pressure, in this case increasing pressure increases the volume of the more dense phase. Further pressure increases result in a decrease in the volume of the more dense phase (liquid) as the pressure tends toward a second intersection with the dew pressure $p^d \approx 5.9$ MPa. Further increase of pressure above the dew pressure results in the more dense phase disappearing so that liquid is no longer present.

and which is the liquid? The question is merely a semantic one based on common experience and it is certainly best to regard both as fluid phases.

4.11.4 What Is Azeotropy?

The $(p, x)_T$ section for the vapor + liquid equilibrium of $\{CO_2(1) + C_2H_6(2)\}$ is shown in Figure 4.7 and, at $x = 0.7$, this mixture exhibits an azeotrope. We see from Figure. 4.7 that for an azeotrope $x^{\alpha} = x^{\beta} = x^{az}$ but $V_m^{\alpha} \neq V_m^{\beta}$. Equations 4.44, 4.45, 4.46, and 4.47 define the conditions for an azeotrope. The fluid mixture at the azeotropic composition behaves as if it were a pure fluid and has a unique vapor pressure. Figure 4.7 shows a positive azeotrope, for which there is a maximum in the vapor pressure of the system at a given temperature, while a negative azeotropy has a minimum vapor pressure at a temperature and is relatively uncommon. The diagram of Figure 4.7 will be repeated at other temperatures and thus an azeotropic line (the line joining azeotropic points) can persist to the critical line. However, this is not always the case and when it does not the maximum of the curve occurs at a mole fraction that attains $x_B = 1$ at a temperature below the critical temperature of pure substance B.

4.12 WHAT ARE SOLUTIONS?

Chapter 1 defines a solution as a mixture for which it is convenient to distinguish between the solvent and the solutes. The amount of substance of solvent is often much greater than that of the solutes and this is called a *dilute solution*. It is usual in solutions to use molality m_B of a solute B rather than mole fraction x_B in a solvent A of molar mass M_A, where the molality is given by

$$x_B = \frac{M_A m_B}{1 + M_A \sum_B m_B}, \qquad (4.227)$$

and

$$m_B = \frac{x_B}{M_A \left(1 - \sum_B x_B\right)}. \qquad (4.228)$$

4.12.1 What Is the Activity Coefficient at Infinite Dilution?

The activity coefficient γ_B of a solute B in a solution (especially a dilute liquid solution that follows Henry's law) containing molalities m_B, m_C, . . . of solutes B, C, . . . in a solvent A is defined by

$$\lambda_B = m_B \gamma_B \left(\frac{\lambda_B}{m_B}\right)^{\infty}, \qquad (4.229)$$

where the superscript ∞ implies infinite dilution or $\Sigma_B m_B \to 0$. The activity coefficient γ_B of Equation 4.229 can also be defined by the chemical potential μ_B through

$$RT\ln\left(\frac{m_B\gamma_B}{m^{\ominus}}\right) = \mu_B - \left\{\mu_B - RT\ln\left(\frac{m_B}{m^{\ominus}}\right)\right\}^{\infty}, \tag{4.230}$$

where $m^{\ominus}$ is the standard molality and is typically taken to be $m^{\ominus} = 1\ \text{mol}\cdot\text{kg}^{-1}$.

4.12.2 What Is the Osmotic Coefficient of the Solvent?

The osmotic coefficient ϕ of the solvent is defined by

$$\ln\left(\frac{\lambda_A}{\lambda_A^*}\right) = -\phi M_A \sum_B m_B, \tag{4.231}$$

and is related to the activity coefficient of Equation 4.229 by the Gibbs–Duhem equation at constant temperature and pressure that can be written as

$$\mathrm{d}\left\{(1-\phi)\sum_B m_B\right\} + \sum_B m_B\ \mathrm{d}\ln\gamma_B = 0. \tag{4.232}$$

For a single solute B Equation 4.232 reduces to

$$\mathrm{d}\left\{(1-\phi)m_B\right\} + m_B\ \mathrm{d}\ln\gamma_B = 0, \tag{4.233}$$

and f_B can be determined from measurements of ϕ as a function of composition using

$$-\ln\gamma_B = (1-\phi) + \int_0^{m_B}(1-\phi)\,\mathrm{d}\ln\left(\frac{m_B}{m_B^{\ominus}}\right), \tag{4.234}$$

For an ideal and dilute solution $\phi = 1$ and $\gamma_B = 1$ for each solute B so that from Equation 4.231 we have

$$\lambda_B = \left(\frac{\lambda_B}{m_B}\right)^{\infty} m_B, \tag{4.235}$$

and is commonly known as Henry's law.

4.13 REFERENCES

Abbott M.M., and Nass K.K., 1986, *Equations of State and Classical Solution Thermodynamics: Survey of the Connection*, in *Equations of State: Theories and Applications*, eds. Chao K.C., and Robinson R.L., ACS Symposium Series 300, American Chemical Society, Washington DC.

Abrams D.S., and Prausnitz J.M., 1975, "Statistical thermodynamics of liquid-mixtures—new expression for excess Gibbs energy of partly or completely miscible systems," *AIChE J.* **21**:116–128.

Assael M.J., Trusler J.P.M., and Tsolakis Th., 1996, *Thermophysical Properties of Fluids. An Introduction to Their Prediction*, Imperial College Press, London.

Atkins P.W., 1987, *Physical Chemistry*, Oxford University Press, Oxford.

Behnejard H., Sengers J.V., and Anisimov M.A., 2010, *Thermodynamic Behavior of Fluids Near Critical Points, Ch. 10*, in *Experimental Thermodynamics Volume VIII: Applied Thermodynamics of Fluids*, eds. Goodwin A.R.H., Sengers J.V., and Peters C.J., for IUPAC, Royal Society of Chemistry, Cambridge, UK.

Berthelot D., 1889, "Sur le Mélange des Gaz," *C. R. Acad. Sci.* (Paris) **126**:1703.

Carnahan N.F., and Starling K.E., 1972, "Intermolecular repulsions and the equation of state for fluids," *AIChE J.* **18**:1184–1189.

Colbeck S.C., 1995, "Pressure melting and ice skating," *Am. J. Phys.* **63**:888–890.

Curl Jr. R.F., and Pitzer K.S., 1956, "Volumetric and thermodynamic properties of fluids—enthalpy, free energy and entropy," *Ind. Eng. Chem.* **48**:265–274.

Dash J.G., Rempel A.W., and Wettlaufer J.S., "The physics of premelted ice and its geophysical consequences," 2006, *Rev. Mod. Phys.* **78**:695–741.

Delhommelle J., and Millié P., 2001, "Inadequacy of the Lorentz-Berthelot combining rules for accurate predictions of equilibrium properties by molecular simulation," *Mol. Phys.* **99**:619–625.

Economou I.G., 2010, *Cubic and Generalized van der Waals Equations of State*, Chapter 4, *Experimental Thermodynamics Volume VIII: Applied Thermodynamics of Fluids*, eds. Goodwin A.R.H., Sengers J.V., and Peters C.J., for IUPAC, Royal Society of Chemistry, Cambridge, UK.

Feistel R., and Wagner W., 2006, "A new equation of state for H_2O ice Ih," *J. Phys. Chem. Ref. Data* **35**:1021–1047.

Fredenslund Aa., Gmehling J., and Rasmussen P., 1977, *Vapor-Liquid Equilibria Using UNIFAC*, Elsevier, Amsterdam.

Fredenslund Aa., Jones R.L., and Prausnitz J.M., 1975, "Group-contribution estimation of activity-coefficients in nonideal liquid mixtures," *AIChE J.* **21**:1086–1099.

Fredenslund A., and Mollerup J., 1974, "Measurement and prediction of equilibrium ratios for C_2H_6+CO_2 system," *J. Chem. Soc. Faraday Trans. I* **70**:1653–1660.

Friese T., Ulbig P., Schulz S., and Wagner K., 1998, "Effect of NaCl on the excess enthalpies of binary liquid systems," *Thermochim. Acta* **310**:87–94.

Friese T., Ulbig P., Schulz S., and Wagner K., 1999, "Effect of NaCl or KCl on the excess enthalpies of alkanol plus water mixtures at various temperatures and salt concentrations," *J. Chem. Eng. Data* **44**:701–714.

Goodwin A.R.H., and Ambrose D., 2005, *Vapor Pressure*, in *Encyclopedia of Physics, Volume 2*, 3rd ed., eds. Lerner R.G., and Trigg G.L., and Wiley-VCH, Berlin, pp. 2846–2848.

Goodwin A.R.H., and Sandler S.I., 2010, *Mixing and Combining Rules*, Ch. 5, in *Experimental Thermodynamics Volume VIII: Applied Thermodynamics of Fluids*, eds. Goodwin A.R.H., Sengers J.V., and Peters C.J., for IUPAC, Royal Society of Chemistry, Cambridge, UK.

Goodwin A.R.H., Sengers J.V., and Peters C.J., 2010, *Experimental Thermodynamics Volume VIII: Applied Thermodynamics of Fluids*, for IUPAC, Royal Society of Chemistry, Cambridge, UK.

Guggenheim E.A., 1959, *Thermodynamics: An Advanced Treatment for Chemists and Physicists*, 4th ed., North-Holland Publishing Company, Amsterdam.

Haslam A.J., Galindo A., and Jackson G., 2008, "Prediction of binary intermolecular potential parameters for use in modelling fluid mixtures," *Fluid Phase Equilib.* **266**:105–128.

Heideman R., and Fredenslund Aa., 1989, "Vapor-liquid equilibria in complex mixtures," *Chem. Eng. Res. Des.* **67**:145–158.

Helpinstill J.G., and van Winkle M., 1968, "Prediction of infinite dilution activity coefficients for polar-polar binary systems," *Ind. Eng. Chem. Proc. Des. Dev.* **7**:213–220.

Henderson D., and Leonard P.J., 1971, *Liquid Mixtures*, in *Physical Chemistry and Advanced Treatise, Volume 8B, The Liquid State*, eds. Eyring H., Hederson D., and Jost W., Academic Press, New York.

Hirschfelder J.O., Curtis C.F., and Bird R.B., 1954, *Molecular Theory of Gases and Liquids*, Wiley, New York.

Kojima K., and Toshigi K., 1979, *Prediction of Vapor-Liquid Equilibria by the ASOG Method, Physical Sciences Data 3*, Elsevier Publishing Company, Tokyo.

Kontogeorgis G., and Folas G., 2010, *Thermodynamic Models for Industrial Applications: From Classical and Advanced Mixing Rules to Association Theories*, Wiley, Chichester.

Lemmon E., and Span R., 2010, *Multiparameter Equations of State for Pure Fluids and Mixtures*, Chapter 12, *Experimental Thermodynamics Volume VIII: Applied Thermodynamics of Fluids*, eds. Goodwin A.R.H., Sengers J.V., and Peters C.J., for IUPAC, Royal Society of Chemistry, Cambridge, UK.

Lennard-Jones J.E., 1931, "Cohesion," *Proc. Phys. Soc.* **43**:461–482.

Leonhard K., Nguyen V.N., and Lucas K., 2007, "Making equation of state models predictive—Part 2: An improved PCP-SAFT equation of state," *Fluid Phase Equilib.* **258**:41–50.

London F., 1937, "The general theory of molecular forces," *Trans. Faraday. Soc.* **33**:8–26.

Lorentz H.A., 1881, "Über die Anwendung des Satzes vom Virial in der kinetischen Theorie der Gase," *Ann. Phys.* **12**:127–136.

Maitland G.C., Rigby M., Smith E.B., and Wakeham W.A., 1981, *Intermolecular Forces: Their Origin and Determination*, Clarendon Press, Oxford.

Margules S., 1895, "On the composition of the saturated vapours of mixtures," *Math. Naturw. Akad. Wiss.* (Vienna) **104**:1243–1249.

McCabe C., and Galindo A., 2010, *SAFT Associating Fluids and Fluid Mixtures*, Ch. 8, *Experimental Thermodynamics Volume VIII: Applied Thermodynamics of Fluids*, eds. Goodwin A.R.H., Sengers J.V., and Peters C.J., for IUPAC, Royal Society of Chemistry, Cambridge, UK.

McGlashan M.J., 1979, *Chemical Thermodynamics*, Academic Press, London.

Modell M., and Reid R.C., 1983, *Thermodynamics and Its Applications*, 2nd ed., Prentice Hall, New York.

Peng D.Y., and Robinson D.B., 1976, "New 2-constant equation of state," *Ind. Eng. Chem. Fundam.* **15**:59–64.

Pierotti G.J., Deal C.H., and Derr E.L., 1959, "Activity coefficients and molecular structure," *Ind. Eng. Chem.* **51**:95–102.

Pitzer K.S., 1955, "The volumetric and thermodynamic properties of fluids 1. Theoretical basis and virial coefficients," *J. Am. Chem. Soc.* **77**:3427–3433.

Pitzer K.S., Lippmann D.J., Curl Jr. R.F., and Huggins C.M., 1955, "The volumetric and thermodynamic properties of fluids 2. Compressibility factor, vapor pressure and entropy of vaporization," *J. Am. Chem. Soc.* **77**:3433–3440.

Poling B., Prausnitz J.M., and O'Connell J.P., 2001, *The Properties of Gases and Liquids*, 5th ed., McGraw-Hill, New York.

Prausnitz J.M., Lichtenthaler R.N., and Gomes de Azevedo E., 1986, *Molecular Thermodynamics of Fluid-Phase Equilibria*, 2nd ed., Prentice Hall, Englewood Cliffs.

Quack M., Stohner J., Strauss H.L., Takami M., Thor A.J., Cohen E.R., Cvitas T., Frey J.G., Holström B., Kuchitsu K., Marquardt R., Mills I., and Pavese F., 2007, *Quantities, Units and Symbols in Physical Chemistry*, 3rd ed., RSC Publishing, Cambridge.

Reed III T.M., 1955a, "The theoretical energies of mixing for fluorocarbon-hydrocarbon mixtures," *J. Phys. Chem.* **59**:425–428.

Reed III T.M., 1955b, "The ionization potential and the polarizability of molecules," *J. Phys. Chem.* **59**:428–433.

Renon H., and Prausnitz J.M., 1968, "Local compositions in thermodynamic excess functions for liquid mixtures," *AIChE J.* **14**:135–144.

Renon H., and Prausnitz J.M., 1969, "Estimation of parameters for NRTL equation for excess Gibbs energies of strongly nonideal liquid mixtures," *Ind. Eng. Chem. Process Des. Dev.* **8**:413–419.

Rosenberg R., 2005, "Why is ice slippery?," *Phys. Today* **58(12)**:50–55.

Rowlinson J.S., and Swinton F.L., 1982, *Liquids and Liquid Mixtures*, 3rd ed., Butterworth Publishers, London.

Singh M., Leonhard K., and Lucas K., 2007, "Making equation of state models predictive — Part 1: Quantum chemical computation of molecular properties," *Fluid Phase Equilib.* **258**:16–28.

Smith J.M., van Ness H.C., and Abbott M.M., 2004, *Introduction to Chemical Engineering Thermodynamics*, 6th ed., McGraw-Hill, New York.

Soave G., 1972, "Equilibrium constants from a modified Redlich-Kwong equation of state," *Chem. Eng. Sci.* **27**:1197–1203.

Strobridge T.R., 1962, "The thermodynamic properties of nitrogen from 64 to 300 K between 0.1 and 200 Atmospheres," NBS Technical Note No. 129.

Thomas E.R., and Eckert C.A., 1984, "Prediction of limiting activity coefficients by a modified separation of cohesive energy density model and UNIFAC," *Ind. Eng. Chem. Process Des. Dev.* **23**:194–209.

Trusler J.P.M., 2010, *The Virial Equation of State*, Chapter 3, *Experimental Thermodynamics Volume VIII: Applied Thermodynamics of Fluids*, eds. Goodwin A.R.H., Sengers J.V., and Peters C.J., for IUPAC, Royal Society of Chemistry, Cambridge, UK.

Tsuboka T., and Katayama T., 1975, "Modified Wilson equation for vapor-liquid and liquid-liquid equilibria," *J. Chem. Eng. Japan* **8**:181–187.

Unferer P., Wender A., Demoulin G., Bourasseau E., and Mougin P., 2004, "Application of Gibbs ensemble and NPT Monte Carlo simulation to the development of improved processes for H2S-rich gases," *Mol. Simul.* **30**:631–640.

van Laar J.J., 1910, "Über Dampfspannungen von binären Gemischen," *Z. Physik Chem.* **72**:723–751.

van Laar J.J., 1913, "Zur Theorie der Dampfspannungen von binären Gemischen," *Z. Physik Chem.* **83**:599–609.

van Ness H.C., and Abbott M.M., 1982, *Classical Thermodynamics of Nonelectrolyte Solutions*, McGraw-Hill, New York.

Walas S.M., 1985, *Phase Equilibrium in Chemical Engineering*, Butterworth, Boston.

Wettlaufer J.S., and Grae Worster M., 2006, "Premelting dynamics," *Ann. Rev. Fluid Mech.* **38**:427–452.

Wilson G.M., 1964, "Vapor-liquid equilibrium. 11. New expression for excess free energy of mixing," *J. Am. Chem. Soc.* **86**:127–130.

Wilson G.M., and Deal C.H., 1962, "Activity coefficients and molecular structure—activity coefficients in changing environments. Solutions of Groups," *Ind. Eng. Chem. Fundam.* **1**:20–23.

Chapter 5

Reactions, Electrolytes, and Nonequilibrium

5.1 INTRODUCTION

In this chapter we consider a number of aspects of chemically reacting systems at both equilibrium and as they approach equilibrium. The role of the equilibrium constant and how it is affected by temperature is discussed. Examples of enthalpy of reactions are given. A discussion for reacting systems not in equilibrium is provided, as well as a number of examples to illustrate the calculation of the variation of substance concentration with time. We also consider briefly the language and purpose of irreversible thermodynamics (Kjelstrup and Bedeaux 2010; de Groot and Mazur 1984) and electrolyte solutions (Robinson and Stokes 2002) although in both cases our treatment is not intended to provide more than an opportunity to refer to the substantive literature on the topics that interested readers may wish to consult.

5.2 WHAT IS CHEMICAL EQUILIBRIUM?

On either the laboratory or the industrial chemical engineering scales many processes involve chemical reactions as well as flow, work, and heat transfer. It is therefore important for us to consider thermodynamic principles and practice as they apply to systems in which there are chemical reactions. In a reaction vessel a number of chemical components (reactants) are mixed together and a chemical reaction or chemical reactions take place that produce different chemical species (products). In general, there is an incomplete conversion of reactants to products, but after some time a point is reached where there is no change with respect to time of the amount of substance of either the reactants or the products. A fixed amount of each substance from the reactants exists simultaneously with the fixed amount of substance of the products. This state is called *chemical equilibrium* and was discussed formally in Section 1.3.18.

For a closed system in equilibrium at constant temperature and pressure as we saw with Equation 1.30 a chemical reaction from reagents R to products P can be written as

$$\sum_{R}(-\nu_R)R = \sum_{P}\nu_P P, \tag{5.1}$$

where ν is the stoichiometric number and is, by convention, negative for reactants and positive for products. A general chemical equation can be written as

$$0 = \sum_{B}\nu_B B. \tag{5.2}$$

The energy change for Equation 5.2 is $\Delta_r U$ and can be converted to the enthalpy change $\Delta_r H$ for Equation 5.2 with $\Delta_r(pV)$. For liquids and solids $\Delta_r(pV)$ is given by the difference

$$\Delta_r(pV) = \sum_{B}\{n_B(\text{fin})p(\text{fin})V_B(\text{fin}) - n_B(\text{int})p(\text{int})V_B(\text{int})\}, \tag{5.3}$$

between the final and initial amount of substance B, pressure p, and partial molar volume V_B, for the condensed states this change is usually small and often negligible. For gases $\Delta_r(pV)$ is given by

$$\Delta_r(pV) = \xi RT\sum_{B}\nu_B, \tag{5.4}$$

where ξ is the extent of reaction defined in Section 1.3.18 and ν_B is the stoichiometric number for the gaseous substances in the reaction.

5.2.1 What Are Enthalpies of Reaction?

We begin with an example. The standard enthalpy of formation is a particular example of a standard enthalpy of reaction (see Question 1.10, Equation 1.118) in which a compound is formed from its elements, where each is in their stable state. We consider first the two oxides of carbon for which $\Delta_f H_m^{\ominus}$ are

$$C(s) + O_2(g) = CO_2(g), \quad \Delta_f H_m^{\ominus} = -393.5\ \text{kJ}\cdot\text{mol}^{-1}, \tag{5.5}$$

and

$$C(s) + \frac{1}{2}O_2(g) = CO(g), \quad \Delta_f H_m^{\ominus} = -100.5\ \text{kJ}\cdot\text{mol}^{-1}. \tag{5.6}$$

When Equation 5.6 is subtracted from Equation 5.5 we obtain the enthalpy of reaction

$$CO(g) + \frac{1}{2}O_2(g) = CO_2(g), \quad \text{of} \quad \Delta_r H_m^{\ominus} = -283\ \text{kJ}\cdot\text{mol}^{-1}. \tag{5.7}$$

Only two of the three reactions given by Equations 5.5, 5.6, and 5.7 are independent and the third can be determined by subtraction (or addition). This demonstrates the requirement to determine only the standard enthalpy of formation for substances, and the enthalpy of reaction of the same components can then be determined algebraically, affording a considerable saving on the number of measurements.

Values of $\Delta_f H_m^{\ominus}$ are generally obtained from $\Delta_r H_m^{\ominus}$, which can be determined from reactions carried out within calorimeters. For example, the $\Delta_f H_m^{\ominus}$ for $C_6H_5CO_2H(s)$ cannot be measured directly from the reaction

$$7C(\text{graphite}) + 3H_2(g) + O_2(g) = C_6H_5CO_2H(s), \tag{5.8}$$

which has for reactants each of the components in their most stable state. Instead, $\Delta_f H_m^{\ominus}$ is obtained indirectly from $\Delta_c H_m^{\ominus}$ for the combustion reactions

$$C_6H_5CO_2H(s) + \frac{15}{2}O_2(g) = 7CO_2(g) + 3H_2O(l), \tag{5.9}$$

and

$$C(\text{graphite}) + O_2(g) = CO_2(g), \tag{5.10}$$

with the $\Delta_f H_m^{\ominus}$ for the formation reaction

$$H_2(g) + \frac{1}{2}O_2(g) = H_2O(l). \tag{5.11}$$

The algebraic manipulation of Equations 5.9 through 5.11, and for that matter Equations 5.5 through 5.7 are examples of the application of Hess's law that applies only to standard enthalpy changes. Further examples can be found in text books such as Atkins and de Paula (2006). Each of the enthalpy changes will depend on composition, temperature, and pressure.

The methods of determining enthalpies of combustion for the cases when the substance contains carbon, hydrogen, oxygen, and nitrogen, as well as metals that form oxides make use of an adiabatic bomb calorimeter as described in Chapter 1. Here the volume is constant so that $\Delta_c U$ is actually determined and corrected to give $\Delta_c U^{\ominus}$.

It is the standard (defined in Section 1.8) molar enthalpy of formation denoted by $\Delta_f H_m^{\ominus}$ that is tabulated for each substance at a temperature of 298.15 K (TRC Tables NSRDS-NIST-74 and NSRDS-NIST-75). Of course measurements are neither performed at $T = 298.15$ K nor at $p^\ominus$ so that corrections, which we will address shortly, must be applied. At this time it is sufficient to state that the correction of $\Delta_f H_m$ to $\Delta_f H_m^\ominus$ or $\Delta_r H_m$ to $\Delta_r H_m^\ominus$ is small; on the other hand, the differences between $\Delta_r G_m$ and $\Delta_r G_m^\ominus$ as well as $\Delta_r S_m$ and $\Delta_r S_m^\ominus$ are typically not small. It is to the corrections of the determination of the $\Delta_r H_m^\ominus$ from the calorimetrically determined $\Delta_r H_m$ that we now turn to.

We consider the reaction given by Equation 5.2 conducted solely in the gas phase with an initial amount of substance n_B^{int} of substance B that reacts in a thermally insulated calorimeter at an initial temperature T_1 and pressure p_1. After the reaction of extent ξ (see Question 1.3.8) the calorimeter may be at a temperature T_2 and pressure p_2 and the energy change is given by

$$U(T_2, p_2, \xi) - U(T_1, p_1) = W^{int}, \tag{5.12}$$

where W^{int} is the work done to initiate the reaction. The calorimeter with products is now returned from temperature T_2 to temperature T_1 and the pressure becomes p_3. The calorimeter temperature is then increased from T_1 to T_2 using electrical work and the pressure returns to p_2 and the resulting energy change is given by

$$U(T_2, p_2, \xi) - U(T_1, p_3, \xi) = W^{elec}. \tag{5.13}$$

Subtracting Equation 5.13 from Equation 5.12 gives

$$U(T_1, p_3, \xi) - U(T_1, p_1) = -W^{elec} + W^{int}, \tag{5.14}$$

or

$$H(T_1, p_3, \xi) - H(T_1, p_1) = -W^{elec} + W^{int} + \{p_3 V(T_1, p_3, \xi) - p_1 V(T_1, p_1)\}. \tag{5.15}$$

Equations 5.14 and 5.15 assume the energy (heat) losses from the insulated calorimeter are calculable and that any change in the energy content of the calorimeter during each part of the experiment is exactly the same and cancels in Equations 5.14 and 5.15 so that

$$H(T, p, \xi) = \sum_B (n_B^{int} - \nu_B \xi) H_B(T, p, \xi), \tag{5.16}$$

and

$$V(T, p, \xi) = \sum_B (n_B^{int} - \nu_B \xi) V_B(T, p, \xi). \tag{5.17}$$

In Equations 5.16 and 5.17 H_B and V_B are the partial molar enthalpy and partial molar volume, respectively, of substance B.

Equation 1.85 defines $H^{\ominus}_{\mathrm{B,g}}(T)$ as

$$H^{\ominus}_{\mathrm{B,g}}(T) = H_{\mathrm{B,g}}(T,p,x) - \int_0^p \left\{ V_{\mathrm{B,g}}(T,p,x) - \left(\frac{\partial V_{\mathrm{B,g}}(T,p,x)}{\partial T} \right)_p \right\} \mathrm{d}p, \tag{5.18}$$

that can be used to rewrite Equation 5.16 as

$$\begin{aligned} H(T,p,\xi) &= \sum_{\mathrm{B}} \left(n_{\mathrm{B}}^{\mathrm{int}} - \nu_{\mathrm{B}}\xi \right) H^{\ominus}_{\mathrm{B}}(T) \\ &\quad + \sum_{\mathrm{B}} \left(n_{\mathrm{B}}^{\mathrm{int}} - \nu_{\mathrm{B}}\xi \right) \int_0^p \left[V_{\mathrm{B}}(T,p,\xi) - T \left\{ \frac{\partial V_{\mathrm{B}}(T,p,\xi)}{\partial T} \right\}_p \right] \mathrm{d}p, \end{aligned} \tag{5.19}$$

where for the sake of simplicity the superscript g has been dropped. Substituting Equation 5.19 into Equation 5.15 we obtain

$$\begin{aligned} \xi \sum_{\mathrm{B}} \nu_{\mathrm{B}} H^{\ominus}_{\mathrm{B}}(T_1) &= \xi \Delta H^{\ominus}_{\mathrm{m}}(T_1) \\ &= -W^{\mathrm{elec}} - W^{\mathrm{int}} \\ &\quad - \sum_{\mathrm{B}} \left\{ n_{\mathrm{B}}^{\mathrm{int}} - \nu_{\mathrm{B}}\xi \right\} \int_0^{p_3} \left[V_{\mathrm{B}}(T_1,p,\xi) - T_1 \left\{ \frac{\partial V_{\mathrm{B}}(T,p,\xi)}{\partial T} \right\}_{p,T=T_1} \right] \mathrm{d}p \\ &\quad + \sum_{\mathrm{B}} n_{\mathrm{B}}^{\mathrm{int}} \int_0^{p_1} \left[V_{\mathrm{B}}(T_1,p,\xi) - T_1 \left\{ \frac{\partial V_{\mathrm{B}}(T,p,\xi)}{\partial T} \right\}_{p,T=T_1} \right] \mathrm{d}p \\ &\quad + \sum_{\mathrm{B}} \left\{ n_{\mathrm{B}}^{\mathrm{int}} - \nu_{\mathrm{B}}\xi \right\} p_3 V_{\mathrm{B}}(T_1,p_3,\xi) - \sum_{\mathrm{B}} n_{\mathrm{B}}^{\mathrm{int}} p_3 V_{\mathrm{B}}(T_1,p_1). \end{aligned} \tag{5.20}$$

If the pressure is sufficiently low so that the gas can be considered perfect (see Question 2.6) then Equation 5.20 becomes

$$\Delta H^{\ominus}_{\mathrm{m}}(T_1) = -\frac{W^{\mathrm{elec}}}{\xi} + \frac{W^{\mathrm{int}}}{\xi} + \left(\sum_{\mathrm{B}} \nu_{\mathrm{B}} \right) RT_1. \tag{5.21}$$

To correct a value determined thermodynamically in this manner a thermodynamic path is selected where the reactants are first heated or cooled to the temperature of 298.15 K, then the reaction takes place at 298.15 K, and finally the products are cooled or heated to the final temperature (compare with Question 4.9).

The total enthalpy change for the three steps is given as

$$\Delta_r H_m^\ominus(T) = \Delta_r H_m^\ominus(298.15\,\mathrm{K}) + \sum_R n_R \int_T^{T=298.15\,\mathrm{K}} C_{p,m}^\ominus(\mathrm{R})\,\mathrm{d}T + \sum_P n_P \int_{T=298.15\,\mathrm{K}}^{T} C_{p,m}^\ominus(\mathrm{P}) \quad (5.22)$$

It is this standard molar enthalpy of formation denoted by $\Delta_f H_m^\ominus$ that is tabulated for each substance at a temperature of 298.15 K (TRC Tables NSRDS-NIST-74 and NSRDS-NIST-75).

5.3 WHAT ARE EQUILIBRIUM CONSTANTS?

For a chemical reaction given by Equation 5.2 the standard equilibrium constant $K^\ominus$, according to Question 1.10, is defined by Equation 1.116, that is,

$$K^\ominus(T) \overset{\text{def}}{=} \exp\left\{\frac{-\sum_B \nu_B \mu_B^\ominus(T)}{RT}\right\}, \quad (5.23)$$

or

$$K^\ominus(T) = \prod_B \left\{\lambda_B^\ominus(T)\right\}^{-\nu_B}. \quad (5.24)$$

For a chemical reaction $0 = \Sigma_B \nu_B B$ of a liquid (or solid) mixture

$$K^\ominus(T) = \prod_B \left\{\lambda_B^\ominus(\mathrm{s\ or\ l}, T)\right\}^{-\nu_B} = \prod_B \left\{\lambda_B^*(\mathrm{s\ or\ l}, T, p^\ominus)\right\}^{-\nu_B}, \quad (5.25)$$

and at equilibrium

$$K^\ominus(T) \approx \prod_B \left\{x_B f_B\right\}^{\nu_B}. \quad (5.26)$$

Equation 5.26 was obtained from Equation 4.116 that was itself obtained by from Equation 4.117 by omission of the integral. In view of Equations 4.118 and 4.119, Equation 5.26 can be cast as

$$K^{\ominus}(T) = \prod_{\mathrm{B}} \{a_{\mathrm{B}}\}^{\nu_{\mathrm{B}}} = \prod_{\mathrm{B}} \left\{ \frac{\tilde{p}_{\mathrm{B}}(1, T, p^{\ominus}, x_{\mathrm{C}})}{\tilde{p}^{*}_{\mathrm{B}}(1, T, p^{\ominus})} \right\}^{\nu_{\mathrm{B}}}. \quad (5.27)$$

The equilibrium constant is from Equations 5.27 and 5.26 given by the relative activities or the fugacity ratio and activity coefficients of the substances in the mixture. These quantities can be obtained from either an equation of state or an activity coefficient model as discussed in Questions 4.4.1, 4.6, and 4.7.

5.3.1 What Is the Temperature Dependence of the Equilibrium Constant?

Equation 1.120 from Chapter 1, Section 1.8 is

$$\frac{\mathrm{d}\ln K^{\ominus}}{\mathrm{d}T} = \frac{\Delta_{\mathrm{r}} H^{\ominus}_{\mathrm{m}}}{RT^2}, \quad (5.28)$$

which provides the temperature dependence of the equilibrium constant. The evaluation of Equation 5.28 is discussed in Chapter 1 with Equations 1.120 through 1.123. In particular, when $\Delta_{\mathrm{r}} H^{\ominus}_{\mathrm{m}}$ is known at a temperature T_3 and it is assumed independent of temperature and values of the standard molar heat capacities at constant pressure $C^{\ominus}_{p,\mathrm{B}}$ are available over a range of temperature, Equation 5.28 becomes

$$\ln\{K^{\ominus}(T_2)\} = \ln\{K^{\ominus}(T_1)\} + \frac{\Delta_r H^{\ominus}_{m}(T_3)(T_2 - T_1)}{RT_1T_2}$$
$$+ \int_{T_1}^{T_2} \left\{ \int_{T_3}^{T} \sum_{B} \nu_B C^{\ominus}_{p,\mathrm{B}}(T)\, dT \right\} \left(\frac{1}{RT^2} \right) dT. \quad (5.29)$$

If the $C^{\ominus}_{p,\mathrm{B}}$ are not known over a temperature range then they too can be assumed independent of T to obtain an approximate temperature dependence of the equilibrium constant.

An alternative derivation of Equation 5.28 can be obtained from Equation 1.113 for a perfect gas at the standard pressure because

$$H^{\ominus}_{\mathrm{B}} = G^{\ominus}_{\mathrm{B}} + TS^{\ominus}_{\mathrm{B}} = \mu^{\ominus}_{\mathrm{B}} - T\frac{\mathrm{d}\mu^{\ominus}_{\mathrm{B}}}{\mathrm{d}T} = -RT^2 \frac{\mathrm{d}\ln \lambda^{\ominus}_{\mathrm{B}}}{\mathrm{d}T}. \quad (5.30)$$

Equation 1.74 can be cast in terms of Equation 5.1 as

$$\ln K^{\ominus}(T) = \sum_{\mathrm{R}} \nu_{\mathrm{R}} \ln \lambda_{\mathrm{R}}^{\ominus}(T) - \sum_{\mathrm{P}} \nu_{\mathrm{P}} \ln \lambda_{\mathrm{P}}^{\ominus}(T). \tag{5.31}$$

Differentiating Equation 5.31 with respect to T and substituting Equation 5.30 gives

$$\frac{\mathrm{d}\ln K^{\ominus}}{\mathrm{d}T} = \frac{\sum_{\mathrm{R}} \nu_{\mathrm{R}} H_{\mathrm{R}}^{\ominus}(T) - \sum_{\mathrm{P}} \nu_{\mathrm{P}} H_{\mathrm{P}}^{\ominus}(T)}{RT^2} = \frac{\Delta_{\mathrm{r}} H_{\mathrm{m}}^{\ominus}}{RT^2}, \tag{5.32}$$

which is Equation 5.28. This equation will be of considerable use in the determination of equilibrium constants.

5.3.2 What Is the Equilibrium Constant for a Reacting Gas Mixture?

For the reaction given by Equation 5.2 in a gas mixture the standard chemical potential is defined by Equation 1.126 and, together with the definition of absolute activity given by Equation 1.111, can be rearranged to yield

$$\lambda_{\mathrm{B,g}}(T,p,y_{\mathrm{C}}) = \lambda_{\mathrm{B,g}}^{\ominus}(T)\left(\frac{y_{\mathrm{B}}p}{p^{\ominus}}\right)\exp\left\{\int_0^p \left\{\frac{V_{\mathrm{B,g}}(T,p,y_{\mathrm{C}})}{RT} - \frac{1}{p}\right\}\mathrm{d}p\right\}, \tag{5.33}$$

which is Equation 4.67. Thus, Equation 5.24 can be written as

$$K^{\ominus}(T) = \prod_{\mathrm{B}}\left(\frac{y_{\mathrm{B}}p}{p^{\ominus}}\right)^{\nu_{\mathrm{B}}} \exp\left[\sum_{\mathrm{B}} \nu_{\mathrm{B}} \int_0^p \left\{\frac{V_{\mathrm{B,g}}(T,p,y_{\mathrm{c}}^{\mathrm{eq}})}{RT} - p^{-1}\right\}\mathrm{d}p\right]. \tag{5.34}$$

For a perfect gas mixture Equation 5.34 becomes

$$K^{\ominus} = \prod_{\mathrm{B}}\left(\frac{y_{\mathrm{B}}p}{p^{\ominus}}\right)^{\nu_{\mathrm{B}}}, \tag{5.35}$$

or for a real gas for which $(p \to 0)$ it becomes

$$K^{\ominus} = \lim_{p\to 0}\left\{\prod_{\mathrm{B}}\left(\frac{y_{\mathrm{B}}p}{p^{\ominus}}\right)^{\nu_{\mathrm{B}}}\right\}. \tag{5.36}$$

Omitting $p^{\ominus}$ from Equation 5.35 gives

$$K_p = \prod_B (y_B p)^{\nu_B}, \tag{5.37}$$

where K_p has dimensions of $(\text{pressure})^{\nu_B}$. The mole fractions in Equations 5.35 through 5.37 are at equilibrium.

At pressures that are sufficiently low to permit the use of the approximation for the pressure, explicit virial expansion of

$$V_m = \frac{RT}{p} + B, \tag{5.38}$$

and assuming Equation 4.76 applies, then Equation 5.34 becomes

$$K^{\ominus}(g,T) \approx \prod_B \left(\frac{y_B^{eq} p}{p^{\ominus}}\right)^{\nu_B} \exp\left(\frac{\sum_B \nu_B B_B p}{RT}\right). \tag{5.39}$$

From the definition of fugacity given by Equation 4.68 of

$$\ln \phi_B = \ln\left(\frac{\tilde{p}_{B,g}}{y_B p}\right) = \int_0^p \left\{\frac{V_{B,g}(T,p,y_C)}{RT} - \frac{1}{p}\right\} dp, \tag{5.40}$$

Equation 5.34 can be cast as

$$K^{\ominus}(T) = \prod_B \left(\frac{y_B p}{p^{\ominus}}\right)^{\nu_B} \exp\left[\sum_B \nu_B \ln \phi_B\right]. \tag{5.41}$$

An example of the calculation of $K^{\ominus}(T)$ for a gaseous reaction is given for the reaction

$$CO(g) + H_2O(g) \overset{T=1{,}000\,K}{=} CO_2(g) + H_2(g). \tag{5.42}$$

For Equation 5.42 the $K^{\ominus}(T)$ is from Equation 5.24 given by either

$$K^{\ominus}(T) = \frac{\lambda_B^{\ominus}(CO,T)\lambda_B^{\ominus}(H_2O,T)}{\lambda_B^{\ominus}(CO_2,T)\lambda_B^{\ominus}(H_2,T)}, \tag{5.43}$$

or

$$\ln K^{\ominus}(T) = \ln \lambda_B^{\ominus}(CO,T) + \ln \lambda_B^{\ominus}(H_2O,T) - \ln \lambda_B^{\ominus}(CO_2,T) - \ln \lambda_B^{\ominus}(H_2,T). \tag{5.44}$$

The temperature of 1,000 K was chosen to ensure that all the constituents were gases, and Equation 5.43 can be evaluated from statistical mechanical results with spectroscopic data combined with a calorimetric determination of $\Delta_r H_m^\ominus$ and the molar mass of the reactants and products as discussed in Question 2.4.4, for example, by Equation 2.66 for a reaction of diatomic gases.

5.3.3 What Is the Equilibrium Constant for Reacting Liquid or Solid Mixtures?

The standard equilibrium constant for reaction (Equation 5.2) in a liquid or solid mixture (or for that matter a liquid and solid mixture) is obtained from Equation 5.23 as

$$K^\ominus(T) = \exp\left\{-\frac{\sum_B \nu_B \mu_B^\ominus(T)}{RT}\right\}$$

$$= \prod_B \left\{\lambda_B^\ominus(\text{l or s}, T)\right\}^{-\nu_B} = \prod_B \left\{\lambda_B^*(\text{l or s}, T)\right\}^{-\nu_B}. \qquad (5.45)$$

The definitions of the standard chemical potential of a liquid and solid given by Equations 1.130 and 1.131 have been used to obtain the right hand side of Equation 5.45. For liquid or solid mixtures the standard chemical potential is given by Equation 1.136. In the light of the definition of activity coefficient $f_{B,l}$ given by Equation 4.114 the standard chemical potential of a liquid (or solid by change of l to s) substance B is given by

$$\mu_{B,l}^\ominus(T) = \mu_{B,l}(T,p,x) - RT\ln(x_B f_{B,l}) + \int_p^{p^\ominus} V_{B,l}^*(T,p)\,\mathrm{d}p. \qquad (5.46)$$

The standard equilibrium constant for reaction (Equation 5.2) in a liquid or solid mixture (or for that matter a liquid and solid mixture) is obtained from Equation 5.45 by the insertion of Equation 5.46 for each B. If the difference between p and $p^\ominus$ is neglected then the integral in Equation 5.46 can be neglected and Equation 5.45 becomes approximately

$$K^\ominus(T) \approx \prod_B \left\{x_B f_{B,l}\right\}^{\nu_B}. \qquad (5.47)$$

For any pure substance $x_B = 1$ and $f_B = 1$, and so we see that pure solid or liquid substances have no effect upon the determination of the equilibrium constant.

5.3.4 What Is the Equilibrium Constant for Reacting Solutes in Solution?

For solutes B, C, . . . in a solvent A, the chemical reaction (Equation 5.2) is recast as

$$0 = \nu_A + \sum_B \nu_B B, \tag{5.48}$$

because in solutions the solvent is conventionally treated differently from the solutes. The equilibrium constant of Equation 5.24 (or Equation 1.116) is in this question given by

$$K^\ominus(T) = \{\lambda_A^\ominus(T)\}^{-\nu_A} \prod_B \{\lambda_B^\ominus(T)\}^{-\nu_B}. \tag{5.49}$$

The chemical potential (absolute activity) of the solvent is given by Equation 1.137. Thus introducing molality in place of mole fractions, because we are working with solutions, and using the osmotic coefficient defined by Equation 4.231 we find that Equation 1.136 becomes

$$\mu_{A,l}^\ominus(T) = \mu_{A,l}(T, p, m_c) + RT\phi M_A \sum_B m_B + \int_p^{p^\ominus} V_{A,l}^*(T, p)\,\mathrm{d}p. \tag{5.50}$$

For the solute (sol) B the chemical potential is given by Equation 1.138 and this can be rearranged using the earlier results so that

$$\mu_{B,sol}^\ominus(T) = \mu_{B,sol}(T, p^\ominus, m_C) - RT\ln\left(\frac{m_B}{m^\ominus}\right) + \int_p^{p^\ominus} V_{B,sol}^*(T, p)\,\mathrm{d}p$$
$$+ \left[\begin{array}{l} \left\{\mu_{B,sol}(T, p, m_C) - RT\ln\left(\frac{m_B}{m^\ominus}\right)\right\}^\infty \\ -\left\{\mu_{B,sol}(T, p, m_C) - RT\ln\left(\frac{m_B}{m^\ominus}\right)\right\} \end{array}\right]. \tag{5.51}$$

The last term of Equation 5.51 in square brackets can be measured, but it is also interesting to proceed in a slightly different way. That is, for an ideal dilute solution the term in square brackets disappears (indeed it is known to vanish for real solutions, which are sufficiently dilute) so that in these circumstances Equation 5.51 can be cast as

$$\mu_{B,sol}^\ominus(T) = \mu_{B,sol}(T, p^\ominus, m_C) - RT\ln\left(\frac{m_B}{m^\ominus}\right) + \int_p^{p^\ominus} V_{B,sol}^*(T, p)\,\mathrm{d}p. \tag{5.52}$$

Assuming the molality, activity coefficient, pressure, and osmotic coefficient are the values at equilibrium of m_B^{eq}, γ_B^{eq}, p^{eq}, and ϕ^{eq}, respectively, then the equilibrium constant of Equation 5.49 can with Equations 5.50 and 5.52 be written as

$$K^{\ominus}(T) = \exp\left(-\nu_A RT\phi^{eq} M_A \sum_B m_B\right) \prod_B \left(\frac{m_B^{eq}\gamma_B^{eq}}{m^{\ominus}}\right)^{\nu_B} \times \exp\left\{\int_{p^{\ominus}}^{p^{eq}} \left(\frac{\nu_A V_A^* + \sum_B \nu_B V_B^{\infty}}{RT}\right) dp\right\}. \quad (5.53)$$

In Equation 5.53 the activity coefficient γ_B^{eq} of a solute B in a solution has been introduced from Equation 4.229. When $p^{eq} = p^{\ominus}$ Equation 5.53 reduces to

$$K^{\ominus}(T) = \exp\left(-\nu_A RT\phi^{eq} M_A \sum_B m_B^{eq}\right) \prod_B \left(\frac{m_B^{eq}\gamma_B^{eq}}{m^{\ominus}}\right)^{\nu_B}. \quad (5.54)$$

For an ideal and dilute solution, where $\phi^{eq} = 1$ for the solvent and $\gamma_B^{eq} = 1$ for each solute B, Equation 5.54 becomes

$$K^{\ominus}(T) = \exp\left(-\nu_A M_A \sum_B m_B^{eq}\right) \prod_B \left(\frac{m_B^{eq}}{m^{\ominus}}\right)^{\nu_B}. \quad (5.55)$$

This can be interpreted as the result for a real solution by writing Equation 5.55 in the form

$$K^{\ominus}(T) = \lim_{m_B^{eq} \to 0} \prod_B \left(\frac{m_B^{eq}}{m^{\ominus}}\right)^{\nu_B}, \quad (5.56)$$

where the term $\exp(-\nu_A M_A \Sigma_B m_B^{eq})$ can be set to unity because the argument of the exponential is usually very small. The $m^{\ominus}$ in the term $\Pi_B (m_B^{eq}/m^{\ominus})^{\nu_B}$ can be eliminated in view of its definition, and Equation 5.56 can then be written as

$$K_m = \lim_{m_B^{eq} \to 0} \prod_B (m_B^{eq})^{\nu_B}, \quad (5.57)$$

where the K_m has units of $(\text{molality})^{\Sigma_B \nu_B}$.

5.3.5 What Are the Enthalpy Changes in Mixtures with Chemical Reactions?

For a change of extent of reaction from ξ' to ξ'' for the reaction Equation 5.2 the enthalpy difference is given by

$$\Delta_r H = H(T, p, \xi'') - H(T, p, \xi') = \int_{\xi'}^{\xi''} \left(\frac{\partial H}{\partial \xi}\right)_{T,p} d\xi, \tag{5.58}$$

and can be obtained from a calorimeter when the extent of reaction ξ'' is equal to the extent of reaction obtained when the reaction has reached equilibrium of ξ^{eq}.

The difference in Gibbs function $\Delta_r G$ as a result of a change in chemical composition arising from a chemical reaction is given by

$$\Delta_r G = G(T, p, \xi'') - G(T, p, \xi') = \int_{\xi'}^{\xi''} \left(\frac{\partial G}{\partial \xi}\right)_{T,p} d\xi. \tag{5.59}$$

The $(\partial G/\partial \xi)_{T,p}$ along with Equations 3.34 and 3.39 can be written as

$$\left(\frac{\partial G}{\partial \xi}\right)_{T,p} = -A = \sum_{B} \nu_B \mu_B \tag{5.60}$$

so that Equation 5.59 becomes

$$\Delta_r G = \int_{\xi'}^{\xi''} \left\{\sum_{B} \nu_B \mu_B (T, p, \xi)\right\} d\xi \cdot \tag{5.61}$$

From Question 3.4 equilibrium occurs when

$$\sum_{B} \nu_B \mu_B (T, p, \xi^{eq}) = 0, \tag{5.62}$$

where ξ^{eq} is the extent of reaction at equilibrium. In view of Equation 5.62 we can recast Equation 5.61 as

$$\Delta_r G = \int_{\xi'}^{\xi''} \left[\sum_{B} \nu_B \left\{\mu_B(T, p, \xi) - \mu_B(T, p, \xi^{eq})\right\}\right] d\xi. \tag{5.63}$$

The difference $\mu_B(T,p,\xi) - \mu_B(T,p,\xi^{eq})$ in Equation 5.63 can be measured; this is the subject of Question 5.3.6. For further details the reader should consult, for example, Guggenheim (1967).

5.3.6 What Is the difference between $\Delta_r G_m$ and $\Delta_r G_m^{\ominus}$?

Unfortunately, it is quite common for experimentalists to report $\Delta_r G_m^{\ominus}$ when what has actually been determined is $\Delta_r G_m$. The question that now arises is whether the difference between $\Delta_r G_m^{\ominus}$ and $\Delta_r G_m$ is significant. It is to address this question that we now turn.

For a reaction according to Equation 5.2 in a perfect gas mixture for an extent of reaction ξ the amount of substance $n_B(\xi)$ of substance B is given by

$$n_B(\xi) = n_B(\xi_1 = 0) + \nu_B \xi, \tag{5.64}$$

where ν_B is the stoichiometric number. The standard molar change of Gibbs function $\Delta_r G_m^{\ominus}$ is given by Equation 1.119, that is,

$$\Delta_r G_m^{\ominus} = \sum_B \nu_B \mu_B^{\ominus} = \sum_B \nu_B G_B^{\ominus} = RT \ln K^{\ominus}. \tag{5.65}$$

The Gibbs function of the mixture for the extent of reaction ξ is from Equation 5.52 given by

$$\begin{aligned} G^{pg}(T,p,\xi) = &\sum_B n_B(\xi)\mu_B^{\ominus}(g,T) \\ &+ RT\sum_B n_B(\xi)\ln\left\{\frac{n_B(\xi)}{\sum_B n_B(\xi_1=0) + \xi\sum_B \nu_B}\right\} \\ &+ RT\sum_B n_B(\xi)\ln\left\{\frac{p}{p^{\ominus}}\right\}. \end{aligned} \tag{5.66}$$

The change in Gibbs function for a change in the extent of reaction from ξ_1 to ξ_2 is given by

$$\frac{\Delta_r G^{pg}(T,p)}{\Delta\xi} = \sum_B \nu_B \mu_B^{\ominus}(g,T)$$

$$+\frac{RT}{\Delta\xi}\sum_B n_B(\xi_2)\ln\left\{\frac{n_B(\xi_2)}{\sum_B n_B(\xi_1=0)+\xi_2\sum_B \nu_B}\right\}$$

$$-\frac{RT}{\Delta\xi}\sum_B n_B(\xi_1)\ln\left\{\frac{n_B(\xi_1)}{\sum_B n_B(\xi_1=0)+\xi_1\sum_B \nu_B}\right\}$$

$$+RT\left(\sum_B \nu_B\right)\ln\left\{\frac{p}{p^{\ominus}}\right\}. \tag{5.67}$$

In Equation 5.67, $\Delta\xi = \xi_2 - \xi_1$. The ratio $\Delta_r G^{pg}(T, p)/\Delta\xi$ given by Equation 5.67 has the same dimensions as the molar change of Gibbs function denoted by $\Delta_r G_m$. However, $\Delta_r G_m$ is obtained from Equation 1.4 by taking the ratio of the change of G to the total amount of substance $\Sigma_B n_B$, while Equation 5.67 is the quotient with $\Delta\xi$. Despite this difference in terminology the $\Delta_r G^{pg}(T, p)/\Delta\xi$ of Equation 5.67 is often called the *molar change* of Gibbs function and is given the symbol $\Delta_r G_m$.

Let us consider as an example the values of $\Delta_r G^{pg}(T, p)/\Delta\xi$, $\Delta_r G_m^{\ominus}$, and for completeness $(\partial G/\partial\xi)_{T,p}$ of Equation 5.60 for the reaction

$$N_2(g) + 3H_2(g) \overset{T=492\ K,\,p=0.5\ MPa}{=} NH_3(g), \tag{5.68}$$

for which the standard equilibrium constant $K^{\ominus}=0.15$. At $\xi_1=0$ the mole fractions are assumed to be $x(N_2) = 0.25$, $x(H_2) = 0.75$, and $x(NH_3) = 0$, and the equilibrium extent of reaction ξ^{eq} is at a mole fraction of ammonia given by $x(NH_3) = 0.31$ that we denote as ξ_2, which is also equal to ξ^{eq}. For $\xi_1 = 0$ $\Delta_r G^{pg}(T, p)/\Delta\xi = -\infty$ and $(\partial G/\partial\xi)_{T,p} = -\infty$. For $\xi_2 = \xi^{eq}$ $\Delta_r G^{pg}(T, p)/\Delta\xi = -7.4\ kJ \cdot mol^{-1}$ and $(\partial G/\partial\xi)_{T,p} = 0$ and these two quantities are different. However, the standard molar change of Gibbs function is $\Delta_r G_m^{\ominus} = 7.8\ kJ \cdot mol^{-1}$ and is constant. At mole fractions, $x(NH_3)$, other than equilibrium the values of $\Delta_r G^{pg}(T, p)/\Delta\xi$ and $(\partial G/\partial\xi)_{T,p}$ differ from those obtained at ξ^{eq} and indeed these values can be equal.

5.4 WHAT IS IRREVERSIBLE THERMODYNAMICS?

When a system is not in a state of thermodynamic equilibrium, so-called transport processes will act, in an isolated system, to move the system toward equilibrium. It is also possible for a system with external action to maintain a gradient of one thermodynamic variable or another so that the transport of some quantity is a continuous nonequilibrium process. The most familiar process of either kind is probably that which is associated with conductive heat transport down an imposed temperature gradient in a process of heat conduction. In a mixture of components this heat transport process is also always combined with a process called *thermal diffusion*, which leads to a partial separation of the components so that the composition of the system is inhomogeneous. If the driving force of the temperature gradient is removed the isolated system will tend to equilibrium through a process of thermal relaxation and diffusion. This is an example of a set of coupled processes that are inevitably linked and their study leads to the subject of the thermodynamics of irreversible processes. A further example is a thermocouple, where an applied temperature difference between two junctions of dissimilar conductors generates a potential difference, while a potential difference between them can generate a temperature difference.

In this brief treatment we will consider the entropy of the system and its surroundings. Equation 2.85 states that the entropy is given by

$$S = kT\left(\frac{\partial \ln Q_{\mathrm{PF}}}{\partial T}\right)_{N,V} + k\ln Q_{\mathrm{PF}}, \tag{5.69}$$

where Q_{PF} is the *canonical partition function* and k is the Boltzmann's constant. From Equation 2.137 the change of entropy ΔS is given in terms of the heat Q_{H} transferred by

$$\Delta S = \int T^{-1}\, dQ_{\mathrm{H}}, \tag{5.70}$$

and is taken to mean the energy lost to dissipation (Clausius 1865) as was discussed in Question 2.9. That loss occurs by interaction with the surroundings and we denote that entropy change by S_{e} (with index e for external). We also know from Equation 3.2 that if any measurable quantity changes perceptibly (if anything changes) in an isolated system, (which is one of constant energy U, volume V, and material content Σn_{B}, without regard to chemical state or any state of aggregation) the entropy of the system S, must increase

$$\left(\frac{\partial S}{\partial t}\right)_{U,V,\Sigma n_{\mathrm{B}}} \geq 0\cdot \tag{5.71}$$

Further, Equation 3.3 states that for an isolated system, in which there are no changes in $T, p, V, U, \Sigma n_B$ there is nothing happening so that

$$\left(\frac{\partial S}{\partial t}\right)_{U,V,\Sigma n_B} = 0, \tag{5.72}$$

and the system is in equilibrium. The entropy introduced in Equation 5.69 and 5.72 is called the *internal entropy* denoted by S_i. The derivative of S_i with respect to time refers to the rate of internal entropy production.

The overall change of entropy in a time t is therefore given by

$$dS = dS_i + dS_e = dt\left(\frac{\partial S_i}{\partial t}\right)_{U,V,\Sigma n_B} + \frac{\delta Q_H}{T}. \tag{5.73}$$

The rate of change of internal entropy with respect to time at constant temperature can arise from the flux $\mathbf{J}_i$ of a particular quantity caused by an appropriate driving force $\mathbf{X}_i$ that corresponds to $\mathbf{J}_i$. The contribution to the change in S_i for a number of such fluxes is given by

$$\frac{T}{V}\left(\frac{\partial S}{\partial t}\right)_{U,V,\Sigma n_B} = \sum_i \mathbf{J}_i \bullet \mathbf{X}_i. \tag{5.74}$$

To elucidate the chemical implication for $\mathbf{J}_i$ and $\mathbf{X}_i$ three examples can be mentioned: (1) for the case when $\mathbf{J}_i$ is the flux of a substance i then $\mathbf{X}_i$ is the negative of the chemical potential gradient, (2) when $\mathbf{J}_i$ is the flux for an ionic species then $\mathbf{X}_i$ is the negative gradient of the electrochemical potential, and (3) when $\mathbf{J}_i$ is the flux of energy then $\mathbf{X}_i$ is the temperature gradient. In the remainder of this section, consistently with the remainder of the chapter, we will consider only isothermal systems. We also only consider a flux and its associated force that is along one coordinate axis because this makes the treatment easier but no less illustrative.

Assuming that the gradient $\mathbf{X}_i$ is small the flux $\mathbf{J}_i$ can be considered a linear function of $\mathbf{X}_i$ and the two are interrelated by

$$\mathbf{J}_i = \sum_k L_{ik} X_k, \tag{5.75}$$

where L_{ik} is a constant. This equation illustrates how the same flux of a quantity can be the result of a number of different driving forces. The quantities L_{ik} are known as transport coefficients. The L_{ik} for all i and k are according to the Onsager reciprocal relations are related by

$$L_{ik} = L_{ki}. \tag{5.76}$$

In the particular case when there are just two driving forces then Equation 5.75 can be written as

$$J_1 = L_{11}X_1 + L_{12}X_2, \tag{5.77}$$

and

$$J_2 = L_{21}X_1 + L_{22}X_2, \tag{5.78}$$

and for L_{ik} from Equation 5.76 we can write

$$L_{12} = L_{21}. \tag{5.79}$$

In the two sections that follow we consider a number of electrochemical phenomena in this context.

5.5 WHAT ARE GALVANIC CELLS?

When metallic zinc Zn(s), which is a silver colored material, is placed in an aqueous solution of copper sulfate with a chemical formula $CuSO_4$(aq) the color of the zinc will in time change to brown. The color change is the result of Cu(s) depositing on the outer surface of the Zn(s). In this solution the reaction

$$\mathrm{Zn(s)} + \mathrm{Cu^{2+}(aq)} = \mathrm{Zn^{2+}(aq)} + \mathrm{Cu(s)}, \tag{5.80}$$

has occurred. Clearly, after a short time the Zn(s) is plated with Cu(s) and the reaction ceases. This can be prevented by separating the Zn(s) from Cu^{2+}(aq), which can be achieved with a high molality aqueous solution of copper sulfate placed in the bottom of a beaker with an aqueous solution of zinc sulfate {$ZnSO_4$(aq)} of relatively low molality carefully poured atop the $CuSO_4$(aq) to reduce mixing with the $ZnSO_4$(aq). The $ZnSO_4$(aq) floats atop the $CuSO_4$(aq) because of a difference in density of the two solutions. A Cu(s) electrode is then placed in the bottom of the jar in contact with Cu^{2+}(aq), while a Zn(s) electrode is suspended in the upper layer of $ZnSO_4$(aq) in contact with Zn^{2+}(aq). This arrangement forms a battery and was used to provide electricity for telephone systems.

A more convenient method of separating the two aqueous solutions is shown in Figure 5.1 and is known as a galvanic cell. The left hand beaker of Figure 5.1 contains $ZnSO_4$(aq) of molality about 1 $mol \cdot kg^{-1}$ in contact with metallic Zn(s), while the right hand beaker contains $CuSO_4$(aq) also of molality of about 1 $mol \cdot kg^{-1}$ in contact with metallic Cu(s). In the absence of a connection between the two beakers nothing happens. However, when the metallic electrodes are, as shown in Figure 5.1, interconnected by a cable and the solutions

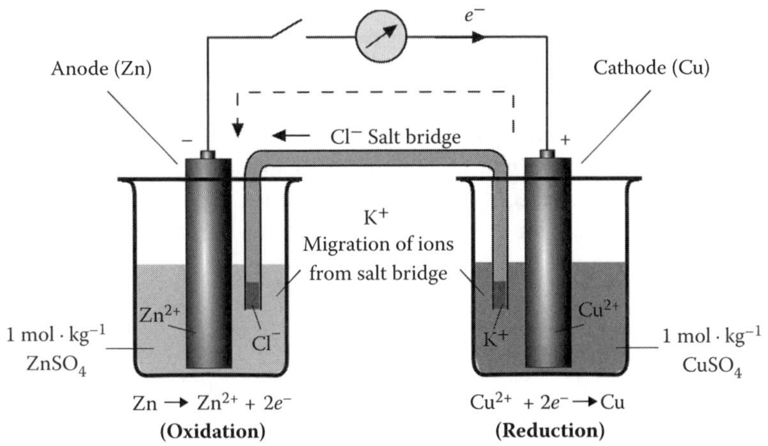

Figure 5.1 Schematic of a galvanic cell. LEFT: a beaker containing $ZnSO_4$(aq) and Zn(s). RIGHT: a beaker containing $CuSO_4$(aq) and Cu(s). The two beakers are interconnected by an electrically conducting wire with an on/off switch in this case, shown with an open circuit and a galvanometer. A KCl salt bridge also interconnects the two beakers. When the circuit is closed the reaction given by Equation 5.81 occurs in the left hand beaker and the reaction of Equation 5.82 occurs in the right hand beaker resulting in the flow of electrons.

by a salt bridge, in this case potassium chloride KCl, electrons flow from the left hand side to the right hand side according to Equation 5.80. In the left hand beaker (the anode) the reaction

$$\mathrm{Zn(s)} = \mathrm{Zn}^{2+}(\mathrm{aq}) + 2e^{-}, \tag{5.81}$$

occurs, while in the right hand beaker (the cathode) the reaction

$$\mathrm{Cu}^{2+}(\mathrm{aq}) + 2e^{-} = \mathrm{Cu(s)}, \tag{5.82}$$

takes place. The salt bridge contains an electrolyte and completes the electrical circuit so that current can flow but the solutions cannot mix and contaminate the Zn(s) with Cu(s). The electromotive force (the potential difference obtained as the current tends to zero), denoted by emf, can be used in thermodynamics to provide a method of determining the chemical potential difference and it is to this that we now turn.

In general, a discussion of galvanic cells should treat, for example, the speed with which ions move in a gradient of electric field. It thus involves transport phenomena, which are beyond the scope of this book. Furthermore, galvanic cells cannot be at equilibrium because the gradients of chemical potential that exist within the cell always ensure diffusion occurs. However, there are certain specific conditions that permit the electromotive force to be used to calculate the affinity of a chemical reaction. It is to the discussion of these conditions that we now turn.

The galvanic cell, for example, shown in Figure 5.1, is usually replaced by a simplified diagram that contains the solid metals, solutions, and bridge electrolyte. For a general galvanic cell containing Cu(s) electrodes, an unspecified solution in which reduction and oxidation occur a bridge solution is typically written as

$$\mathrm{Cu}\left|\begin{matrix}\mathrm{Re}\\\mathrm{Ox}\end{matrix}\right|\text{bridging solution}\left|\begin{matrix}\mathrm{Re}\\\mathrm{Ox}\end{matrix}\right|\mathrm{Cu}. \tag{5.83}$$

In the general case of Equation 5.83 the emf can be represented exactly by the equation

$$\begin{aligned}-FE &= \mu(\mathrm{Re}^{\mathrm{R}}) - \mu(\mathrm{Ox}^{\mathrm{R}}) - \mu(\mathrm{Re}^{\mathrm{L}}) + \mu(\mathrm{Ox}^{\mathrm{L}})\\ &\quad + \sum_{\mathrm{i}}\int_{\mu_{\mathrm{i}}^{\mathrm{L}}}^{\mu_{\mathrm{i}}^{\mathrm{R}}}\left(\frac{t_{\mathrm{i}}}{z_{\mathrm{i}}}\right)\mathrm{d}\mu_{\mathrm{i}},\end{aligned} \tag{5.84}$$

that uses an expression for the zero current and Onsager's reciprocal relations (Equation 5.76). In Equation 5.84, t_i is the transport number and z_i is the charge number of the ion i. The transport number of an ion is the fraction of the electric current arising from the flow of that ion. Equation 5.84 is often written as

$$\begin{aligned}-FE &= \mu(\mathrm{Re}^{\mathrm{R}}) - \mu(\mathrm{Ox}^{\mathrm{R}}) - \mu(\mathrm{Re}^{\mathrm{L}}) + \mu(\mathrm{Ox}^{\mathrm{L}})\\ &\quad + \frac{\mu_{\mathrm{i}}^{\mathrm{R}}}{z_{\mathrm{i}}} - \frac{\mu_{\mathrm{j}}^{\mathrm{L}}}{z_{\mathrm{j}}} + \sum_{\mathrm{i}\neq\mathrm{j}}\int_{\mu_{\mathrm{i}}^{\mathrm{L}}}^{\mu_{\mathrm{i}}^{\mathrm{R}}} t_{\mathrm{i}}\left(\frac{\mathrm{d}\mu_{\mathrm{i}}}{z_{\mathrm{i}}} - \frac{\mathrm{d}\mu_{\mathrm{j}}}{z_{\mathrm{j}}}\right),\end{aligned} \tag{5.85}$$

which is more useful when one of the ions j is present in each part of the cell.

To illustrate the use of Equation 5.85 a specific example is considered of the galvanic cell given by

$$\mathrm{Pt}|\mathrm{Ag}\left|\begin{matrix}\text{solution}\\\text{of AgNO}_3\end{matrix}\right|\begin{matrix}\text{bridging solution}\\\text{of AgNO}_3,\mathrm{Fe(NO_3)_2}\\\text{and Fe(NO}_3)_3\end{matrix}\left|\begin{matrix}\text{solution}\\\text{of Fe(NO}_3)_2\\\text{and Fe(NO}_3)_3\end{matrix}\right|\mathrm{Pt} \tag{5.86}$$

for which NO_3^- is found in each part of the system and is therefore chosen for j with $z_j = -1$. Equation 5.85 can be written for Equation 5.86 as

$$-FE = \mu\{Fe(NO_3)_2, R\} - \mu\{Fe(NO_3)_3, L\} - \mu(Ag, L) + \mu(AgNO_3, L)$$
$$+ \int_{\mu^L_{AgNO3}}^{\mu^R_{AgNO3}} t(Ag^+)\, d\mu(AgNO_3) + \int_{\mu^L_{Fe(NO3)_2}}^{\mu^R_{Fe(NO3)_2}} \frac{t(Fe^{2+})}{2}\, d\mu\{Fe(NO_3)_2\}$$
$$+ \int_{\mu^L_{Fe(NO3)_3}}^{\mu^R_{Fe(NO3)_3}} \frac{t(Fe^{3+})}{3}\, d\mu\{Fe(NO_3)_3\}. \tag{5.87}$$

Platinum is present to act as a nonreacting electrical conductor between the solution and the copper wires. In some cases, the platinum can also act as a catalyst. The transport numbers t and thus the integrals in Equation 5.87 can be made sufficiently small to be eliminated by the introduction of so-called swamping. That requires small molalities of reactants and the addition of a nonreacting electrolyte, for example, KNO_3 with relatively high molality. This approach also introduces an additional integral in Equation 5.87 which is eliminated because $d\mu(KNO_3)$ is almost constant, and Equation 5.87 can then be written as

$$-FE = \mu\{Fe(NO_3)_2\} - \mu\{Fe(NO_3)_3\} - \mu(Ag) + \mu(AgNO_3). \tag{5.88}$$

The right hand side of Equation 5.88 is given by Equation 3.32 of

$$\boldsymbol{A} = -\sum_B \nu_B \mu_B. \tag{5.89}$$

In this case the galvanic cell provides E independent of the bridging solution and a thermodynamic quantity

$$-FE = -\boldsymbol{A} = \left(\frac{\partial G}{\partial \xi}\right)_{T,p}. \tag{5.90}$$

In the specific case of Equation 5.88 the electron transfer reaction is

$$Fe(NO_3)_3 + Ag(s) = Fe(NO_3)_2 + AgNO_3, \tag{5.91}$$

and the thermodynamic quantity $\boldsymbol{A}$ is obtained from Equation 5.90.

5.5.1 What Is a Standard Electromotive Force?

For the electron transfer reaction

$$\mathrm{AgCl(s)} + \frac{1}{2}\mathrm{H_2(g)} = \mathrm{Ag(s)} + \mathrm{H^+} + \mathrm{Cl^-}, \tag{5.92}$$

the galvanic cell can be represented by

$$\mathrm{Pt}|\mathrm{H_2(g)}\left|\begin{matrix}\text{solution of H}^+\text{ and Cl}^-\\ \text{saturated with H}_2\end{matrix}\right|\begin{matrix}\text{solution of H}^+\\ \text{and Cl}^-\end{matrix}\left|\begin{matrix}\text{solution of H}^+\text{ and Cl}^-\\ \text{saturated with AgCl}\end{matrix}\right|\mathrm{AgCl(s)}|\mathrm{Ag}|\mathrm{Pt}. \tag{5.93}$$

For this cell the equation analogous to Equation 5.87 contains integrals of transport numbers and these vanish because the $t(\mathrm{Ag^+})$ is very small, and the HCl is uniform throughout the cell so that the emf is given by

$$-FE = \mu(\mathrm{Ag, s}) - \mu(\mathrm{AgCl, s}) + \mu(\mathrm{HCl, solute}) - \frac{1}{2}\mu(\mathrm{H_2, g}). \tag{5.94}$$

The expressions given for the standard chemical potential of solids, solutions, and perfect gases in Question 1.10 can be substituted in to Equation 5.94 to give

$$\begin{aligned}-FE &= \mu^\ominus(\mathrm{Ag, s}) - \mu^\ominus(\mathrm{AgCl, s}) + \mu^\ominus(\mathrm{H^+, solute}) + \mu^\ominus(\mathrm{Cl^-, solute})\\ &\quad -\frac{1}{2}\mu^\ominus(\mathrm{H_2, g}) + 2RT\ln\left(\frac{m\gamma_\pm}{m^\ominus}\right) - \frac{1}{2}RT\ln\left\{\frac{x(\mathrm{H_2, g})p}{p^\ominus}\right\}\\ &= -FE^\ominus + 2RT\ln\left(\frac{m\gamma_\pm}{m^\ominus}\right) - \frac{1}{2}RT\ln\left\{\frac{x(\mathrm{H_2, g})p}{p^\ominus}\right\}.\end{aligned} \tag{5.95}$$

In Equation 5.95 $m^\ominus = 1\ \mathrm{mol\cdot kg^{-1}}$, $p^\ominus = 0.1$ MPa, and $\gamma_\pm$ is the activity coefficient of the HCl electrolyte (see Question 5.6). For substances that are solids and liquids the differences in pressure between p and $p^\ominus$ can be ignored in Equation 5.95. Equation 5.95 contains

$$\begin{aligned}E^\ominus &\overset{\text{def}}{=} \left(\frac{RT}{F}\right)\ln K^\ominus\\ &= -\frac{\left\{\mu^\ominus(\mathrm{Ag, s}) - \mu^\ominus(\mathrm{AgCl, s}) + \mu^\ominus(\mathrm{H^+, solute}) + \mu^\ominus(\mathrm{Cl^-, solute}) - \frac{1}{2}\mu^\ominus(\mathrm{H_2, g})\right\}}{F},\end{aligned} \tag{5.96}$$

and $E^{\ominus}$ is called the *standard electromotive force* that is tabulated for electron transfer reactions. For the electron transfer reaction Equation 5.92 the standard electrode potential at $T = 298.15$ K is given by

$$E^{\ominus}\left\{\mathrm{AgCl(s)} + \frac{1}{2}\mathrm{H_2(g)} = \mathrm{Ag(s)} + \mathrm{H^+} + \mathrm{Cl^-}\right\} \approx 0.22\ \mathrm{V}. \quad (5.97)$$

When E of Equation 5.93 is measured at two molalities of HCl of m_1 and m_2 the measured electromotive force is E_1 and E_2, respectively, and subtraction of the two measurements gives the difference

$$\mu(\mathrm{HCl}, m_2) - \mu(\mathrm{HCl}, m_1) = \frac{E_1 - E_2}{F}, \quad (5.98)$$

and provides another route to determining the chemical potential difference.

5.6 WHAT IS SPECIAL ABOUT ELECTROLYTE SOLUTIONS?

Ions in solution can be considered as separate components of the system subject to the requirement for electrical neutrality given by

$$\sum_{\mathrm{i}} m_{\mathrm{i}} z_{\mathrm{i}} = 0, \quad (5.99)$$

where m_{i} is the molality and z_{i} the charge of the ion i. The Gibbs–Duhem equation (Equation 3.23) also applies to solutions of electrolytes subject to compliance with Equation 5.99 and, at constant temperature and pressure, is given by

$$\mathrm{d}\left\{(1-\phi)\sum_{\mathrm{i}} m_{\mathrm{i}}\right\} + \sum_{\mathrm{i}} m_{\mathrm{i}}\, \mathrm{d}\ln\gamma_{\mathrm{i}} = 0. \quad (5.100)$$

For a electrolyte $\mathrm{A}_{\nu+}\mathrm{B}_{\nu-}$ the molality of each ion is given by

$$m_+ = \nu_+ m \quad (5.101)$$

and

$$m_- = \nu_- m. \quad (5.102)$$

In view of Equations 5.101 and 5.102 Equation 5.100 can be written as

$$(\nu_+ + \nu_-)\mathrm{d}\{(1-\phi)m\} + \nu_+ m\, \mathrm{d}\ln\gamma_+ + \nu_- m\, \mathrm{d}\ln\gamma_- = 0, \quad (5.103)$$

and by definition of the activity coefficient $\gamma_\pm$ of the electrolyte of

$$(\nu_+ + \nu_-)\ln\gamma_\pm = \nu_+\ln\gamma_+ + \nu_-\ln\gamma_-, \quad (5.104)$$

Equation 5.103 becomes

$$\mathrm{d}\{(1-\phi)m\} + m\,\mathrm{d}\ln\gamma_\pm = 0. \tag{5.105}$$

Equation 5.105 is important because only the activity coefficient of the ion pair that complies with Equation 5.99 can be measured.

For Equation 4.235 it was possible to state $\phi = 1$ and $\gamma_\mathrm{B} = 1$ for a solution that was dilute, that is for which $\Sigma_\mathrm{B} m_\mathrm{B} < 1\ \mathrm{mol\cdot kg^{-1}}$. This was because the pair interaction energy of nonelectrolytes in solution decreased approximately as the (molality)^2. In electrolyte solutions the pair interaction energy decreases only as the cube root of concentration and so it is not possible to make the same assumptions. However, the Debye–Hückel theory (Robinson and Stokes 2002) provides the form of $\gamma_\pm$ in both the limit $\lim_{m_\mathrm{i}\to 0} \Sigma_\mathrm{i} m_\mathrm{i}$ and at finite m.

In the limit $\lim_{m_\mathrm{i}\to 0} \Sigma_\mathrm{i} m_\mathrm{i}$ the Debye–Hückel law states

$$\ln\gamma_\pm = (2\pi L\rho_\mathrm{A}^*)^{1/2}\left(\frac{e^2}{4\pi\varepsilon_\mathrm{A}^* kT}\right)^{3/2}|z_+z_-|\left(\frac{1}{2}\sum_\mathrm{i} m_\mathrm{i} z_\mathrm{i}^2\right)^{1/2}. \tag{5.106}$$

In Equation 5.106 the term $2^{-1}\Sigma_\mathrm{i} m_\mathrm{i} z_\mathrm{i}^2$ is called the ionic strength and is often given the symbol I_i, ρ_A^* is the density of the pure solvent, e is the charge on a proton, and ε_A^* is the electric permittivity of the solvent; $\varepsilon = \varepsilon_\mathrm{r}\varepsilon_0$, where ε_r is the relative electric permittivity and $\varepsilon_0\{=(\mu_0 c^2)^{-1}\}$ is the electric constant given using the magnetic constant and the speed of light in vacuum as 8.854 187 817 $\ldots \times 10^{-7}\ \mathrm{m^{-3}\cdot kg^{-1}\cdot s^4\cdot A^2}$ exactly (Mohr et al. 2008).

For solutions of electrolytes of finite molalities the Debye–Hückel approximation is given by

$$\ln\gamma_\pm = (2\pi L\rho_\mathrm{A}^*)^{1/2}\left(\frac{e^2}{4\pi\varepsilon_\mathrm{A}^* kT}\right)^{3/2}|z_+z_-| \times \frac{\left(\frac{1}{2}\sum_\mathrm{i} m_\mathrm{i} z_\mathrm{i}^2\right)^{1/2}}{1+d\left(\frac{1}{2}\sum_\mathrm{i} m_\mathrm{i} z_\mathrm{i}^2\right)^{1/2} 2(2\pi L\rho_\mathrm{A}^*)^{1/2}\left(\frac{e^2}{4\pi\varepsilon_\mathrm{A}^* kT}\right)^{1/2}}. \tag{5.107}$$

In Equation 5.107 d is an adjustable parameter called the *mean diameter* of the ions. The term $2(2\pi L\rho_\mathrm{A}^*)^{1/2}\{e^2/(4\pi\varepsilon_\mathrm{A}^* kT)\}^{1/2}$ is approximately $3.3\cdot 10^9\,\mathrm{m^{-1}\cdot kg^{1/2}\cdot mol^{-1/2}}$ and d is about $0.3\cdot 10^{-9}\,\mathrm{m}$ so that the product of these two quantities is about unity, and Equation 5.107 can be written as

$$\ln \gamma_{\pm} \approx (2\pi L \rho_{\mathrm{A}}^{*})^{1/2} \left(\frac{e^2}{4\pi\varepsilon_{\mathrm{A}}^{*} kT} \right)^{3/2} |z_+ z_-| \times \frac{\left(\frac{1}{2} \sum_{\mathrm{i}} m_{\mathrm{i}} z_{\mathrm{i}}^2 \right)^{1/2}}{1 + \left(\frac{1}{2} \sum_{\mathrm{i}} m_{\mathrm{i}} z_{\mathrm{i}}^2 \right)^{1/2}}. \tag{5.108}$$

Equation 5.108 can be extended empirically to even higher m by adopting the form

$$\ln \gamma_{\pm} \approx (2\pi L \rho_{\mathrm{A}}^{*})^{1/2} \left(\frac{e^2}{4\pi\varepsilon_{\mathrm{A}}^{*} kT} \right)^{3/2} |z_+ z_-| \times \frac{\left(\frac{1}{2} \sum_{\mathrm{i}} m_{\mathrm{i}} z_{\mathrm{i}}^2 \right)^{1/2}}{1 + \left(\frac{1}{2} \sum_{\mathrm{i}} m_{\mathrm{i}} z_{\mathrm{i}}^2 \right)^{1/2} + \frac{1}{2} \sum_{\mathrm{i}} m_{\mathrm{i}} z_{\mathrm{i}}^2}. \tag{5.109}$$

The use of the definition of ionic strength I

$$I = \frac{1}{2} \sum_{\mathrm{i}} m_{\mathrm{i}} z_{\mathrm{i}}^2, \tag{5.110}$$

and of

$$\alpha = (2\pi L \rho_{\mathrm{A}}^{*})^{1/2} \left(\frac{e^2}{4\pi\varepsilon_{\mathrm{A}}^{*} kT} \right)^{3/2}, \tag{5.111}$$

as well as

$$\beta = 2(2\pi L \rho_{\mathrm{A}}^{*})^{1/2} \left(\frac{e^2}{4\pi\varepsilon_{\mathrm{A}}^{*} kT} \right), \tag{5.112}$$

permits Equation 5.107 to be written as

$$\ln \gamma_{\pm} = \alpha |z_+ z_-| \frac{I^{1/2}}{1 + dI^{1/2}\beta}, \tag{5.113}$$

and the osmotic coefficient can then be obtained from

$$1 - \phi = \frac{1}{3} \alpha |z_+ z_-| I^{1/2} \sigma(dI^{1/2}\beta), \tag{5.114}$$

where $\sigma(dI^{1/2}\beta)$ is given by

$$\sigma(dI^{1/2}\beta) = 3(dI^{1/2}\beta)^{-3}\left\{1+dI^{1/2}\beta-(1+dI^{1/2}\beta)^{-1}-2\ln(1+dI^{1/2}\beta)\right\}. \quad (5.115)$$

As $\lim_{m_i\to 0} \Sigma_i m_i$ then $dI^{1/2}\beta \ll 1$ and with Equation 5.108, Equation 5.114 becomes

$$1-\phi=\frac{1}{3}\alpha|z_+z_-|I^{1/2}\approx-\frac{1}{3}\ln\gamma_\pm. \quad (5.116)$$

At $(z_+z_-)^2 I \le 0.01$ mol·kg^{-1} Equation 5.106 provides estimates of $\gamma_\pm$ that are within about ±5 % of experimental determinations, while at $(z_+z_-)^2 I \le 0.1$ mol·kg^{-1} Equation 5.108 provides estimates of $\gamma_\pm$ that differ from measurements also by about ±5 %. The reader interested in electrolyte solutions should consult the work of Robinson and Stokes (2002).

5.7 WHAT CAN BE UNDERSTOOD AND PREDICTED FOR SYSTEMS NOT AT EQUILIBRIUM?

The equilibrium of chemical reactions was discussed in Question 5.2. If a chemical reaction has not reached equilibrium there is a continuous change of the amount of substance of both reactants and products with respect to time. Thermodynamics makes no attempt to describe the stages through which the reactants pass on their way to reach the final products, nor does it calculate the rate at which equilibrium is attained. This is the subject of chemical kinetics that provides information about the rate of approach to equilibrium and the mechanism for the conversion of reactants to products.

To discuss chemical kinetics we will consider the reaction

$$a\mathrm{A}+b\mathrm{B}+\cdots+l\mathrm{L}=m\mathrm{M}+n\mathrm{N}+\cdots+w\mathrm{W}. \quad (5.117)$$

The rate of consumption of reactant A is given by

$$r_\mathrm{A}=-\frac{\mathrm{d}c_\mathrm{A}}{\mathrm{d}t}, \quad (5.118)$$

where c_A is the amount-of-substance concentration or simply concentration of A and is given as molarity (mol·m^{-3} or more usually mol·dm^{-3}). The rate at which product M is produced is given by

$$r_\mathrm{M}=\frac{\mathrm{d}c_\mathrm{M}}{\mathrm{d}t}. \quad (5.119)$$

The rate of consumption of a reactant A can be expressed empirically by an equation of the form

$$r_A = k_A c_A^{\alpha} c_B^{\beta} \cdots c_L^{\lambda}. \tag{5.120}$$

Similarly, the rate of production of a product M may be expressed as

$$r_M = k_M c_A^{\alpha} c_B^{\beta} \cdots c_L^{\lambda}, \tag{5.121}$$

where the quantities k_A, k_M, α, β, and λ are independent of amount-of-substance concentration and time. In Equations 5.120 and 5.121, k_A and k_M are known as the rate constants or rate coefficients and α, β, and λ are called the *orders of reaction* with respect to A, B, and L, respectively. From the stoichiometry of the reaction given by Equation 5.117 it is evident that $k_A = k_M a/M$. Rate equations are of practical importance because they are required to predict the course of the reaction and to determine the time required for reaction, yields, and for obtaining the optimum economic conditions for the reaction.

The differential rates of reactions are usually integrated before use to describe experimental data. In this context, examples of a first-order and a second-order reaction will now be discussed.

For a first-order reaction given by

$$A \rightarrow B + C, \tag{5.122}$$

where the initial amount-of-substance concentration of A is c_A. After a time t the remaining concentration of A is c_{A-X} and the concentrations of B and C are both c_X. Thus, near the beginning of the reaction when A is present in very large excess compared with the amounts of products B and C (assuming no appreciable reverse reaction), application of Equations 5.118 and 5.120 yields

$$\frac{dc_X}{dt} = k_A (c_A - c_X). \tag{5.123}$$

By separating the variables and integrating, the variation of c_X with time is obtained as

$$\ln\left(\frac{c_A}{c_A - c_X}\right) = k_A t, \tag{5.124}$$

or

$$c_X = c_A \{1 - \exp(-k_A t)\}. \tag{5.125}$$

As an example, consider the decomposition of N_2O_5 in CCl_4 according to the reaction

$$2N_2O_5 \xrightarrow{CCl_4} 4NO_2 + O_2 \tag{5.126}$$

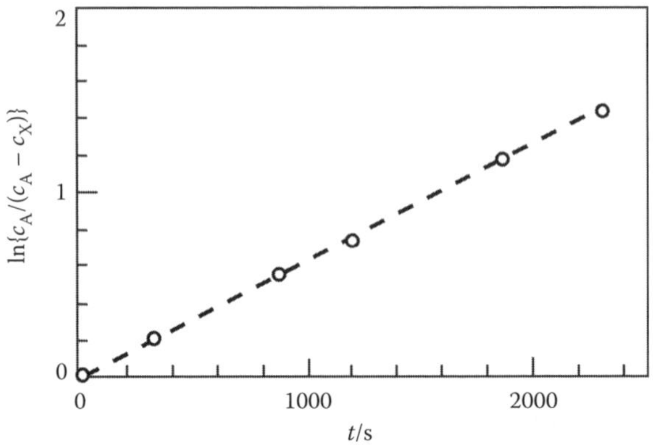

Figure 5.2 Variation of the amount-of-substance concentration of N_2O_5 as a function of time showing the decomposition of N_2O_5 in CCl_4 of Equation 5.126 is a first-order reaction.

for which results reported by Maskill (2006) were, as Figure 5.2 shows, well represented by Equation 5.126.

The results shown in Figure 5.2 confirm that the reaction of Equation 5.126 is first order with respect to N_2O_5 with a rate constant $k_A = 6.22 \cdot 10^{-4}\ s^{-1}$. This rate constant when used with Equation 5.125 gives the amount of N_2O_5 decomposed as a function of time. For example, 99 % decomposition of N_2O_5 is obtained after a time of about 7,403 s. The rate of reaction is used to determine the size of the reactor necessary in a chemical engineering process for an appropriate product specification as well as the necessary heat transfer rate through the reactor.

Similarly, we can consider the second-order reaction given by

$$A + B \rightarrow C + D \tag{5.127}$$

for which the initial concentrations of A and B are c_A and c_B, respectively. After a time t, a concentration c_X of A and B have reacted, forming C and D with concentrations of c_C and c_D, respectively. For Equation 5.127 we assumed as we did for Equation 5.126 that there is no appreciable reverse reaction. If the reaction is second order with respect to concentration the rate of reaction is given by

$$\frac{dc_X}{dt} = k_{AB}(c_A - c_X)(c_B - c_X). \tag{5.128}$$

Separating the variables and integrating the partial fractions the following expression is obtained:

$$(c_A - c_B)^{-1} \ln\left\{\frac{c_B(c_A - c_X)}{c_A(c_B - c_X)}\right\} = k_{AB}t. \tag{5.129}$$

In reality, once the amount-of-substance concentration of the products becomes appreciable the rate of the reverse reaction also becomes significant and must also be taken into account. For example, consider the first-order reaction

$$\mathrm{A} = \mathrm{B}\cdot \tag{5.130}$$

The rate of the forward and reverse reactions are given by

$$\frac{\mathrm{d}c_\mathrm{X}}{\mathrm{d}t} = k_\mathrm{A}(c_\mathrm{A} - c_\mathrm{X}), \tag{5.131}$$

and

$$-\frac{\mathrm{d}c_\mathrm{X}}{\mathrm{d}t} = k_\mathrm{B}(c_\mathrm{B} - c_\mathrm{X}), \tag{5.132}$$

respectively. The rate of reaction for 5.130 is therefore given by the sum of Equations 5.131 and 5.132 as

$$\frac{\mathrm{d}c_\mathrm{X}}{\mathrm{d}t} = k_\mathrm{A}(c_\mathrm{A} - c_\mathrm{X}) - k_\mathrm{B}(c_\mathrm{B} - c_\mathrm{X}). \tag{5.133}$$

Integration of Equation 5.133 gives

$$\ln\left\{\frac{c_\mathrm{B}(k_\mathrm{A}c_\mathrm{A} - k_\mathrm{B}c_\mathrm{B})/(k_\mathrm{A} - k_\mathrm{B})}{(k_\mathrm{A}c_\mathrm{A} - k_\mathrm{B}c_\mathrm{B})/(k_\mathrm{A} - k_\mathrm{B}) - c_\mathrm{X}}\right\} = (k_\mathrm{A} - k_\mathrm{B})t\cdot \tag{5.134}$$

5.8 WHY DOES A POLISHED CAR IN THE RAIN HAVE WATER BEADS? (INTERFACIAL TENSION)

The height h a fluid α of density ρ^α rises in a capillary tube of internal radius r, shown in Figure 5.3, above the bulk fluid and into the surrounding fluid β of density ρ^β is determined by the interfacial tension γ through

$$\gamma = \frac{gh(\rho^\alpha - \rho^\beta)}{2}\frac{r}{\cos\theta}, \tag{5.135}$$

where θ is the angle of contact between fluid α and the wall of the tube. If the fluid wets the tube, as it does for most normal fluids, so that the surface is concave then $\theta = 0$ and Equation 5.135 becomes

$$\gamma = \frac{rgh(\rho^\alpha - \rho^\beta)}{2}. \tag{5.136}$$

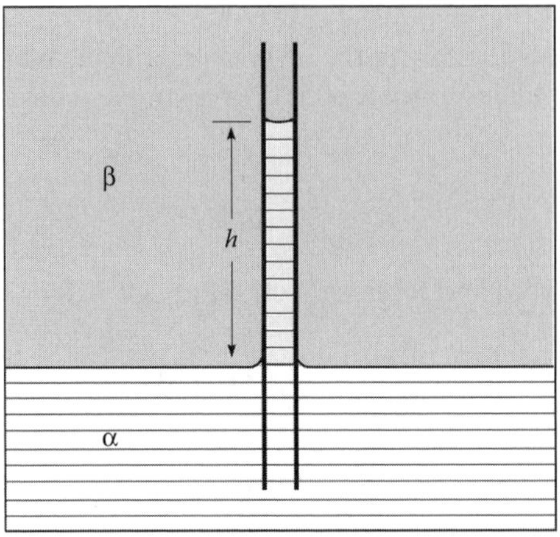

Figure 5.3 Capillary-rise method to measure interfacial tension.

If the phase β has a density $\rho^\beta \ll \rho^\alpha$ as is the case for air then Equation 5.136 can be approximated by

$$\gamma \approx \frac{\rho^\alpha rgh}{2}, \tag{5.137}$$

and in this special case γ is called the *surface tension*. In the unusual but still plausible case that the surface of the fluid in the tube is convex then $\theta = \pi/2$ and $h < 0$ and the surface in the capillary will be below the surface of the bulk phase; if phase α was mercury this scenario would be observed. The application of particular coatings on a surface can be used to alter the chemical characteristics and change the contact angle θ. An example of this is the effect of car polish on the painted surface that causes water to bead because $\theta = \pi/2$; on an unpolished painted metallic surface the water sheds.

The plane inhomogeneous surface phase σ lies between the homogeneous bulk phases α and β. If we assume the interface phase has an area of A^σ and has a thickness of d then its volume is given by $V^\sigma = A^\sigma d$. In the homogeneous bulk phases α and β the force acting on the phases is equal to the pressure p applied. The same force p is present in the plane inhomogeneous surface phase parallel to the interface. In the surface phase σ the force parallel to the interface acting over a length x is given by

$$F = pdx - \gamma x. \tag{5.138}$$

If the volume of the surface phase is increased, the work done is give by

$$W = -pA^{\sigma}\mathrm{d}d - pd\,\mathrm{d}A^{\sigma} + \gamma\,\mathrm{d}A^{\sigma}$$
$$= -p\,\mathrm{d}V^{\sigma} + \gamma\,\mathrm{d}A^{\sigma}. \quad (5.139)$$

Equation 3.23 can be recast with the additional work given by Equation 5.139 to give the Gibbs–Duhem equation of a plane surface phase as

$$0 = S^{\sigma}\,\mathrm{d}T - V^{\sigma}\,\mathrm{d}p + A^{\sigma}\,\mathrm{d}\gamma + \sum_{\mathrm{B}} n_{\mathrm{B}}^{\sigma}\,\mathrm{d}\mu_{\mathrm{B}}. \quad (5.140)$$

If the surface is curved, as it would be for a droplet of oil immersed in water, then the pressure inside the droplet of radius r formed of phase α will be greater than that outside in the phase β by $2\gamma/r$, that is,

$$p^{\alpha} - p^{\beta} = \frac{2\gamma}{r}. \quad (5.141)$$

When phase α is a gas and phase β a liquid the pressure difference is given by

$$p^{\alpha} - p^{\beta} = \frac{4\gamma}{r}, \quad (5.142)$$

because there are now two gas-to-fluid surfaces of virtually the same r.

5.9 REFERENCES

Atkins P.W., and de Paula P., 2006, *Physical Chemistry*, Oxford University Press, Oxford, pp. 49–56.

Clausius R., 1865, *The Mechanical Theory of Heat—with Its Applications to the Steam Engine and to Physical Properties of Bodies*, John van Voorst, London. de Groot S.R., and Mazur P., 1984, *Non-Equilibrium Thermodynamics*, Dover, London.

Ewing M.B., Lilley T.H., Olofsson G.M., Rätzsch M.T., and Somsen G., 1994, "Standard quantities in chemical thermodynamics. Fugacities, activities, and equilibrium constants for pure and mixed phases (IUPAC recommendations 1994)," *Pure Appl. Chem.* **66**:533–552.

Guggenheim E.A., 1967, *Thermodynamics*, 5th ed., North-Holland, Amsterdam.

Kjelstrup S., and Bedeaux D., 2010, *Applied Non-Equilibrium Thermodynamics*, Chapter 14, in *Applied Thermodynamics of Fluids*, eds. Goodwin A.R.H., Sengers J.V., and Peters C.J., for IUPAC, RSC, Cambridge.

Maskill H., Ed., 2006, *The Investigation of Organic Reactions and Their Mechanisms*, Blackwell Publishing Ltd., Oxford, UK.

Mohr P.J., Taylor B.N., and Newell D.B., 2008, "CODATA recommended values of the fundamental physical constants: 2006," *J. Phys. Chem. Ref. Data* **37**:1187–1284.

Quack M., Stohner J., Strauss H.L., Takami M., Thor A.J., Cohen E.R., Cvitas T., Frey J.G, Holström B., Kuchitsu K., Marquardt R., Mills I., and Pavese F., 2007, *Quantities, Units and Symbols in Physical Chemistry,* 3rd ed., RSC Publishing, Cambridge.

Robinson R.A., and Stokes R.H., 2002, *Electrolyte Solutions*, 2nd ed., Dover Publications, New York.

Thermodynamic Research Center (TRC), (1942–2007), *Thermodynamic Tables Hydrocarbons,* ed. Frenkel M., National Institute of Standards and Technology Boulder, CO, Standard Reference Data Program Publication Series NSRDS-NIST-75, Gaithersburg, MD.

Thermodynamic Research Center (TRC), (1955–2007), *Thermodynamic Tables Non-Hydrocarbons,* ed. Frenkel M., National Institute of Standards and Technology Boulder, CO, Standard Reference Data Program Publication Series NSRDS-NIST-74, Gaithersburg, MD.

Chapter 6

Power Generation, Refrigeration, and Liquefaction

6.1 INTRODUCTION

In this chapter we explore a number of examples of the application of thermodynamics to the design of thermal machines intended to accomplish particular tasks. In this endeavor we seek to explain how thermodynamics and the properties of fluids guide the selection of operating conditions. We are less concerned about the detailed mechanical design of the machines than we are with an exposition of how simple thermodynamic principles, that we have covered in earlier chapters, guide the overall strategy with respect to optimal performance and design. We also explore some machines with which the reader is familiar from everyday life to indicate arguments in favor of some designs over others.

Thus, we first consider various kinds of heat engine that generate mechanical work from heat, usually generated through combustion of fossil fuels. We contrast the diesel and petrol engine with which the reader will be familiar as well as a power plant with turbines driven by steam. We then consider refrigerators and heat pumps together since they are essentially two sides of the same coin. Finally, we consider the process whereby it is possible to liquefy substances that are normally gaseous under ambient conditions in order to be able to exploit them to practical ends.

All of these various machines have at their heart cyclic thermodynamic processes and it is with the definition of a cycle that we begin.

6.2 WHAT IS A CYCLIC PROCESS AND ITS USE?

As the name implies a cyclic process (see Question 1.3.5 for a definition of a thermodynamic process) consist of a series of steps that result in a closed cycle.

Thus in all characteristic thermodynamic diagrams, such as p as a function of v denoted (p, v), temperature as a function of specific entropy (T, s), and specific enthalpy as a function of specific entropy (h, s), the lines describing the individual process steps form a closed loop. The term "cyclic process" may refer either to a closed system or to a series of open systems.

For a closed system the fluid within the system undergoes a series of processes so that at the end of the cycle the fluid and the system are returned to the initial state. In an open system a fluid flows through a series of mechanical components and at the end of the cycle the fluid is returned to its initial thermodynamic state, for example, the fluid in a steam-driven power plant. In some processes, the surroundings are also regarded as one of the system components. This extension applies, for example, to an open-cycle gas-turbine engine, where air initially at ambient temperature flows through the engine components and the air temperature increases, the air is then discharged from the engine to the surroundings, where it mixes with ambient air and the temperature returns to the original ambient value. In this example, the heat exchanger for heat rejection is the air surrounding the engine.

Two types of cyclic processes may be distinguished and these are illustrated in Figure 6.1. In the first process, shown in Figure 6.1a, heat is provided to the system to obtain a net output of work or power. These power cycles constitute the origin of engineering thermodynamics and are the subject of Question 6.3. The lines describing the process steps constitute a closed clockwise loop. The second type of process arises when net work input is required to realize

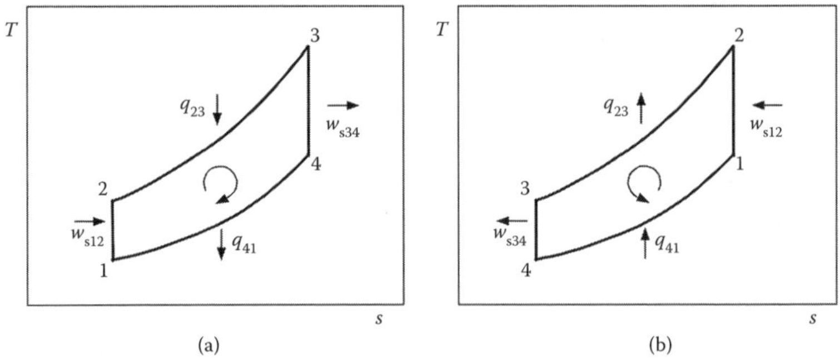

Figure 6.1 Schematic of the (T, s) diagram for the following: (a) for a power cycle the process steps are performed clockwise and heat q_{23} is required to withdraw work w_{s34} from the process; (b) for either refrigeration or heat pump cycle the process steps are performed counterclockwise and work w_{s12} is required to transport heat from a lower temperature to a higher temperature, where it is rejected from the process. For a refrigeration cycle the purpose of the process is to cool a space by transferring heat $q_{41,}$ while for a heat pump the purpose is to heat a space by transferring heat q_{23}.

the uptake of heat at a temperature and its rejection at a higher temperature. A refrigeration cycle, where heat is removed from a space that is to be cooled and the heat is finally rejected to the surroundings, is one example that is addressed in Question 6.4. Another example is a heat pump in which heat from the surroundings is brought into the space and is used for heating. In the characteristic diagram, shown in Figure 6.1b, the individual steps in the process are expressed by lines that form a closed counterclockwise loop.

6.3 WHAT ARE THE CHARACTERISTICS OF POWER CYCLES?

Because it is a defining characteristic of any state variable X that its value is independent of the history of the process by which a specific state is reached, it is then obvious for a cyclic process that $X_{\text{final}} = X_{\text{initial}}$, that is, the variable has the same value at the beginning and at the end of the process, as given by

$$\oint \mathrm{d}X = 0. \tag{6.1}$$

Considering the first law for a series of open systems we may write

$$\dot{Q}_{12} + P_{12} = \dot{m} \cdot \left\{ (h_2 - h_1) + \frac{1}{2} \cdot (c_2^2 - c_1^2) + g \cdot (z_2 - z_1) \right\}, \tag{6.2}$$

and

$$\dot{Q}_{23} + P_{23} = \dot{m} \cdot \left\{ (h_3 - h_2) + \frac{1}{2} \cdot (c_3^2 - c_2^2) + g \cdot (z_3 - z_2) \right\}. \tag{6.3}$$

In Equations 6.1 and 6.2 the subscript numerals refer to the process step (see Equation 1.40). Equations 6.1 and 6.2 are for the two steps in the process that have additional steps with similar equations, which vary only by the subscript numerals defining the process step. These equations are not given here. The final step in the process returns the fluid from state n to state 1 and is given by

$$\dot{Q}_{n1} + P_{n1} = \dot{m} \cdot \left[(h_1 - h_n) + \frac{1}{2} \cdot (c_1^2 - c_n^2) + g \cdot (z_1 - z_n) \right]. \tag{6.4}$$

Here we use the form of Equation 1.49 in which we use the time derivatives of the quantities, rather than the batch steps expressed by Equation 1.49, and $\dot{m}$, which is the mass flow rate that is evidently conserved. The reader will also discern that we are using specific quantities in this analysis consistent with the discipline that makes the most use of this material. Summing all forms of Equations 6.2 through 6.4 we obtain

$$\sum_{i=1}^{n-1} \dot{Q}_{i(i+1)} + \sum_{i=1}^{n-1} P_{i(i+1)} + \dot{Q}_{n1} + P_{n1} = 0, \tag{6.5}$$

where the sum of all heat and work fluxes (remembering that a flux has a positive sign when entering into the system, a negative sign when leaving it) is zero. The *net power* produced is equal in magnitude to the sum of all heat fluxes crossing the system boundaries and is given by $P = \Sigma_{i=1}^{n-1} P_{i(i+1)} + P_{n1}$. Usually, the heat fluxes are grouped into the following: (1) $\dot{Q}$, the sum of all heat fluxes entering into the system (heat provided) and (2) $\dot{Q}_0$, the sum of all heat fluxes leaving the system (heat rejected). In view of this definition, we can now rewrite Equation 6.5 as

$$P = -\dot{Q} - \dot{Q}_0. \tag{6.6}$$

Accordingly, the net (shaft) work produced is

$$w_s = \frac{P}{\dot{m}}, \tag{6.7}$$

and may be written as

$$w_s = -q - q_0. \tag{6.8}$$

Power cycles may make use of closed systems (such as in an internal combustion engine) or a series of open systems (such as in a gas turbine), the working fluid may be a gas (e.g., in a Stirling engine) or a fluid undergoing phase transitions (e.g., in a steam cycle). However, at least the basic and idealized variants of many practical processes may be described by four process steps that are characterized by an alternating sequence of steps involving the transfer of solely (or sometimes mainly) work or heat, respectively, as shown in Figure 6.2. The power cycle illustrated in Figure 6.2 includes the following steps:

Step 1–2: work is input as w_{s12} (increasing system pressure)
Step 2–3: heat is input as q_{23} (heat provided)
Step 3–4: work is output as w_{s34} (that is to be maximized)
Step 4–1: heat is rejected as q_{41} (waste heat)

The net work produced is these steps is given by

$$w_s = w_{s12} + w_{s34}, \tag{6.9}$$

where $w_{s12} > 0$, $w_{s34} < 0$ and $w_s < 0$ and $|w_s|$ is to be maximized.

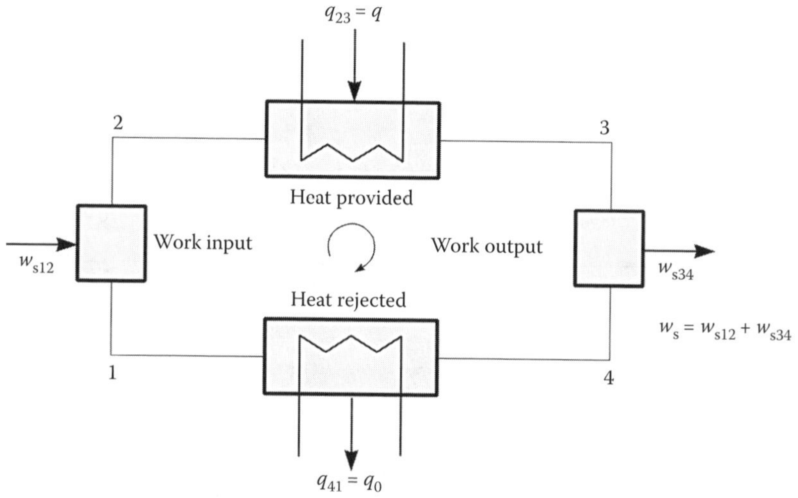

Figure 6.2 Typical scheme of a basic, idealized power cycle: the cycle consists of a series of steps, where only work or heat are transferred, respectively.

Before turning to some specific processes in the questions following this, we examine some generic and characteristic features of processes by reference to the four-step process illustrated in Figure 6.2. However, there is no requirement for exactly these four steps to occur, and in an actual process there are often many more than four.

We begin by examining two characteristic diagrams: (p, v) and (T, s). For simplicity we assume that all process steps are reversible, a restriction that if lifted does not alter the conclusions reached. Figure 6.3 shows both a (p, v) and a (T, s) diagram. Figure 6.3a illustrates the shaft work, which is the central quantity of a power cycle given by

$$(w_{s,\mathrm{ij}})_{\mathrm{rev}} = \int_{\mathrm{i}}^{\mathrm{j}} v\,\mathrm{d}p. \tag{6.10}$$

The total work output given by

$$|w_s| = q - |q_0|, \tag{6.11}$$

may therefore also be written, omitting the index for the reversible process in Equation 6.10, as follows

$$w_s = \oint v\,\mathrm{d}p. \tag{6.12}$$

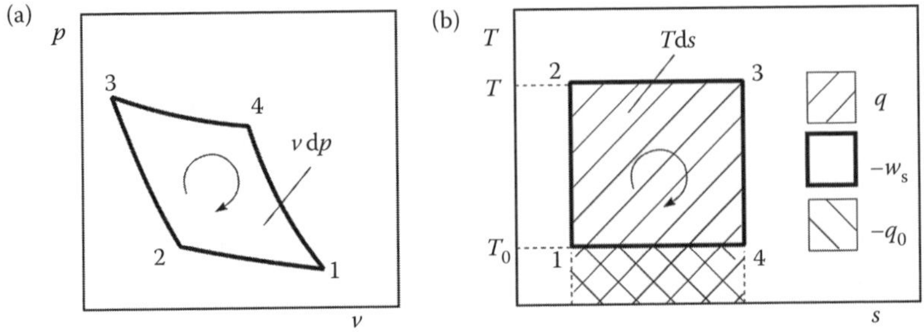

Figure 6.3 A power cycle: (a) (p, v) diagram and (b) a (T, s) diagram. For a reversible process the area enclosed by the cycle represents the net work output. The example depicts a Carnot cycle, where heat is provided at constant temperature T and is rejected at constant temperature T_0.

Equation 6.12 implies that a change of volume during the process is a prerequisite for net work output. The work output increases with the area enclosed by the process as shown in Figure 6.3. In the case when there are irreversible steps the work output is diminished by dissipation, but the result does not change in principle.

Considering the (T, s) diagram shown in Figure 6.3b we have for the total differential of the specific enthalpy

$$dh = T\,ds + v\,dp, \tag{6.13}$$

and because enthalpy is a state variable it must assume the initial value at the end of the cycle so that

$$\oint dh = 0. \tag{6.14}$$

As a consequence of Equation 6.14, Equation 6.13 becomes

$$0 = \oint dh = \oint T\,ds + \oint v\,dp, \tag{6.15}$$

so that

$$\oint T\,ds = -\oint v\,dp = -w_s. \tag{6.16}$$

It follows from Equation 6.16 that a temperature change is required if a cycle process is to deliver a net work output. These two results, rather vividly, illustrate how very general and far-reaching results can be derived from thermodynamic analysis.

The particular (T, s) diagram shown in Figure 6.3b depicts a special process that consists of two (reversible) isothermal and two (reversible) adiabatic, that is, isentropic steps. This is a *Carnot* cycle (Carnot 1872), where heat is provided at constant temperature T, and rejected at constant temperature T_0. The elegant rectangular shape shown in Figure 6.3b is a result of the specific process steps adopted. The two isentropic steps are also adiabatic, that is, no heat is transferred, and because the whole process is assumed to be reversible the change of entropy within the two isothermal steps is solely connected with heat transfer and not due to any irreversibilities. The heat provided to the process is thus

$$q = \int_2^3 T\,\mathrm{d}s = T(s_3 - s_2), \tag{6.17}$$

and the heat rejected is given by

$$q_0 = \int_4^1 T\,\mathrm{d}s = T_0(s_1 - s_4). \tag{6.18}$$

We now consider the efficiency of a power cycle in which the central goal is to maximize the output (net work w_s) for a given input (heat q provided).

The *thermal efficiency*, η_{th} characterizing the overall quality of the process is defined by

$$\eta_{\text{th}} = \frac{-w_s}{q} = \frac{-P}{\dot{Q}}. \tag{6.19}$$

The minus sign in the numerator of Equation 6.19 arises solely from the desire for the thermal efficiency to be positive; the work delivered is negative. The η_{th} ranges between zero and one. The definition of w_s is the net work output summed over all work steps in the process. All steps in a process that require work input diminish the total work output. The net work argument is particularly useful because of the direct connection between the respective components of the process. For example, in a gas turbine shown in Figure 6.4 the compressor stage of the engine is driven by the same shaft as the gas turbine. Thus, the work generated by the gases expanding through the turbine is partially offset by the work done in compressing the gases.

In contrast q only refers to the sum of all heat provided; q_0, the sum over all heat rejected, does not reduce the expended effort. Heat is rejected at a lower temperature. If a process is poorly designed so that too much heat is rejected, this does not reduce the heat provided to the process, for example, from combustion of coal or gas.

As the *Carnot* cycle is an ideal and reversible process it constitutes a reference process in engineering thermodynamics. It provides an upper limit for the thermal efficiency that can be obtained in a power cycle. Equation 6.19 can be

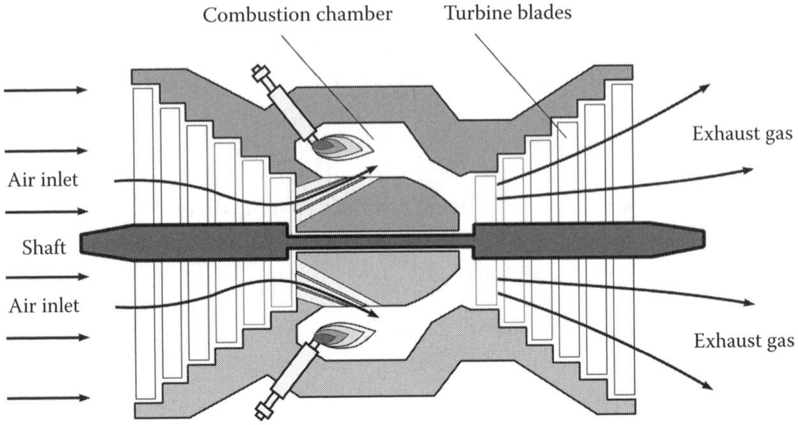

Figure 6.4 A schematic diagram of a gas turbine. A common shaft is used for both compressor and turbine blades.

written in the general form as

$$\eta_{\mathrm{th}} = \frac{q - |q_0|}{q} = 1 - \frac{|q_0|}{q}. \tag{6.20}$$

It is a consequence of the second law (as discussed in Question 3.8) that the heat provided cannot be transformed completely into useful work and that, necessarily, part of the heat must be discarded at a lower temperature.

The heat and the net power output in the Carnot cycle, as illustrated in Figure 6.3, exhibit the property that heat is provided and rejected at constant temperature, which makes the evaluation of Equation 6.20 particularly easy. The fundamental characteristic of this process, however, is that all steps are performed in a reversible manner. The thermal efficiency of Carnot cycle, which is also the maximum thermal efficiency of a power cycle, which is only obtained in a reversible process, is given by

$$\eta_{\mathrm{th,C}} = \eta_{\mathrm{th,rev}} = 1 - \frac{|q_0|}{q} = 1 - \frac{T_0(s_1 - s_4)}{T(s_3 - s_2)} = 1 - \frac{T_0}{T}, \tag{6.21}$$

where the subscript C denotes the Carnot cycle and the subscript rev denotes that the steps in the process are reversible. The fundamental consequence of this formula is that all power cycles should be designed so that heat is provided at a temperature as high as possible and heat is rejected at a temperature as close to the ambient temperature as possible to yield the highest efficiency.

As a specific example, if heat was provided at a temperature of $T = 773$ K ($t = 500$ °C) and rejected at ambient temperature of $T_0 = 298$ K ($t_0 = 25$ °C), a maximum thermal efficiency of $\eta_{th,rev} = 0.61$ would result. This value of efficiency is the base line to which an engineer has to compare his design, recognizing that the real values for the thermal efficiency will be considerably lower (often by almost a factor of two) because of the inevitable losses and irreversibility within a process. However, if one could increase the base temperature T, by choice of materials that withstand higher temperatures, the efficiency would increase. A temperature of $T = 873$ K ($t = 600$ °C) would result in a "reference" thermal efficiency of $\eta_{th,rev} = 0.69$; it is also probable that the real efficiency is higher.

6.3.1 Why Does a Diesel Car Have a Better Fuel Efficiency Than a Gasoline Car?

The balance between the use of diesel engines or gasoline engines to power freight vehicles or passenger cars has varied considerably over the lifetime of fossil-fueled vehicles. The incentives have been fashion, climate change arguments, and performance. However, in the context of this book we will concern ourselves solely with an examination of the relative fuel efficiency of the two hydrocarbon sources of energy to power the car of diesel and gasoline (also commonly known as petrol). In Europe, car manufacturer specifications cite average fuel consumption for a car required to travel a distance of 100 km, which has been determined under well-defined test conditions. As an example, consider a diesel engine of power of about 100 kW consumes 5.5 dm^3 (or 5.5 liters) of fuel to travel a distance of 100 km (i.e., equivalent to about 43 m.p.g.). For a car with a gasoline engine, also with a power of about 100 kW, the fuel consumption is about 7.0 dm^3 of fuel to travel a distance of 100 km (i.e., equivalent to about 34 m.p.g.) and about 30 % lower than for a car powered with diesel. The question to pose is then as follows: What is the thermodynamic reason for this considerable difference?

The specific energy content of gasoline and diesel fuels is about 43 $MJ \cdot kg^{-1,}$ while the mass densities are 0.74 $kg \cdot dm^{-3}$ for gasoline and 0.82 $kg \cdot dm^{-3}$ for diesel that result in volumetric energy content of 32 $MJ \cdot dm^{-3}$ for gasoline and 35 $MJ \cdot dm^{-3}$ for diesel. Thus, the energy content by volume is about 10 % of the observed difference in fuel economy between a car powered by diesel compared with a petrol version. The additional 20 % difference arises from the thermal efficiencies of the two engine types that we will now consider.

In a car, both types of engines operate in a four-stroke manner involving the following process steps (shown in Figure 6.5)

Step 0–1: intake of the mixture of air and fuel (1st stroke)
Step 1–2: compression of the gas mixture (2nd stroke)
Step 2–3: ignition through either a spark plug in a gasoline engine or autoignition in a diesel engine and then combustion

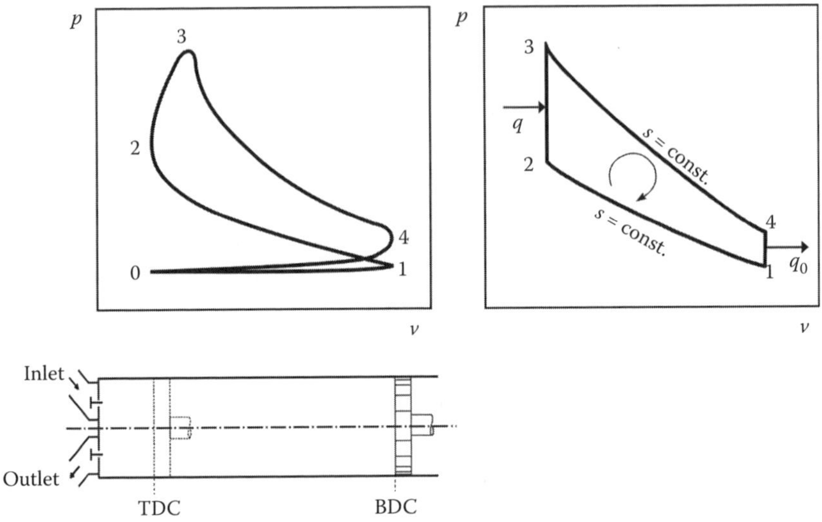

Figure 6.5 Schematic of a gasoline engine (Otto cycle) in a (p, v) diagram. The exhaust stroke $(4 \rightarrow 0)$ and the intake stroke $(0 \rightarrow 1)$ of the real cycle at left are replaced by an isochoric heat discharge in the idealized cycle shown at right. The acronyms TDC and BDC stand for top dead center and bottom dead center, respectively.

Step 3–4: expansion of the gas mixture (3rd stroke)
Step 4–0: expulsion of the burnt gas (4th stroke)

A general modification of this four-step process is that in modern cars no air-fuel mixture is sucked into the combustion chamber because fuel is injected directly. Direct injection provides a tremendous improvement in the performance of diesel engines when the fuel is injected at pressures up to about 200 MPa to ensure proper mixture formation. Direct injection has become increasingly popular for gasoline engines also.

To model these processes thermodynamically there are a number of assumptions that will be introduced to permit the simplified treatment given here. However, these assumptions do not change the overall outcome of the arguments for either engine.

First, to avoid the changes in the chemical composition of the working fluid we assume that the fluid is air (which in a first step is justified because of the relatively large mass fraction of nitrogen). Second, despite pressures of up to an order of 10 MPa we assume that the air behaves as an ideal gas. Third, the heat released by combustion and the energy removal by the discharge of burnt gases are replaced by heat transfer across the system boundaries, so that steps 0–1 and 4–0 are omitted, and the system is now considered as a piston-cylinder closed system. With this transition to a closed system the primary quantity describing the work in the system is now the boundary work and not the shaft work as for an open system. This formal problem, however, is resolved when we take into account the total flow work for the process is zero, $\oint \mathrm{d}(pv) = 0$, rendering the boundary work and the shaft work identical. Finally, to find a mathematical description for the process the actual steps, with rounded shapes between them, are replaced by idealized, well-defined steps.

With these definitions we may now provide the thermodynamic analyses of both Diesel and gasoline cycles and answer the question posed regarding their thermal efficiencies. In this context, we define the characteristic property that is called the *compression ratio* given by

$$\varepsilon = \frac{v_2}{v_1} = \frac{V_{\mathrm{TDC}}}{V_{\mathrm{BDC}}}. \tag{6.22}$$

Equation 6.22 is the ratio between the volumes when the piston is at top dead center (TDC), V_{TDC}, where the volume enclosed in the cylinder is a minimum, and often termed the *clearance volume*, and when it is at bottom dead center (BDC), V_{BDC}, where the volume is a maximum. The difference between V_{TDC} and V_{BDC} is the displacement volume.

We begin with a closer look at the idealized cycle for a gasoline engine, which is often called an *Otto cycle* after Nikolaus Otto, who in 1876 built the first engine of this type. Parenthetically, it is worth remarking that those early versions had little in common with modern *Otto* engines apart from the basic working principle. The idealized Otto cycle consists of the following processes:

Step 1–2: reversible adiabatic (i.e., isentropic) compression
Step 2–3: isochoric addition of heat
Step 3–4: reversible adiabatic (i.e., isentropic) expansion
Step 4–1: isochoric rejection of heat

The thermal efficiency of this process is given by

$$\eta_{\mathrm{th,O}} = 1 - \frac{|q_0|}{q} = 1 - \frac{|q_{41}|}{q_{23}}. \tag{6.23}$$

Because

$$q_{23} = c_v(T_3 - T_2), \tag{6.24}$$

and

$$q_{41} = c_v(T_1 - T_4), \tag{6.25}$$

Equation 6.23 becomes, when we assume a constant heat capacity c_v,

$$\eta_{\text{th,O}} = 1 - \frac{T_4 - T_1}{T_3 - T_2} = 1 - \frac{T_1}{T_2} \cdot \frac{(T_4/T_1) - 1}{(T_3/T_2) - 1}. \tag{6.26}$$

From Chapter 1, Question 1.7.6, the expression for a reversible adiabatic process is given by Equation 1.64 for an ideal gas and when applied to the Otto engine it gives

$$\frac{T_3}{T_4} = \left(\frac{v_4}{v_3}\right)^{\gamma-1} = \left(\frac{v_1}{v_2}\right)^{\gamma-1} = \frac{T_2}{T_1}. \tag{6.27}$$

In Equation 6.27 γ is the ratio of specific heat capacities at constant pressure to that at constant volume and, as in Chapter 1, it is given by $\gamma = c_p/c_v$. Because $v_1 = v_4$ and $v_2 = v_3$ Equation 6.27 becomes

$$\frac{T_4}{T_1} = \frac{T_3}{T_2}. \tag{6.28}$$

Thus, the thermal efficiency of an Otto engine of Equation 6.26 is then

$$\eta_{\text{th,O}} = 1 - \frac{T_1}{T_2}, \tag{6.29}$$

or when the compression ratio defined by Equation 6.22 is used Equation 6.29 becomes

$$\eta_{\text{th,O}} = 1 - \frac{1}{\varepsilon^{\gamma-1}}. \tag{6.30}$$

Examination of Figure 6.6 reveals the efficiency of the ideal Otto cycle increases steeply at first with increasing compression ratio ε and then flattens off. Nevertheless, from a thermodynamic view point alone it would be desirable to increase the compression ratio as far as possible.

However, increasing the compression ratio leads to a marked increase in the temperature at the end of the compression stroke. This results in auto ignition of the fuel and uncontrolled combustion (engine knock) that can damage the engine. Use of higher octane gasoline and fuel injection permits Otto engines to reach compression ratios between 10 and 12. The desire to increase the compression ratio while also avoiding uncontrolled combustion and the resultant engine knock will be used in our discussion of the diesel engine in which fuel vapor auto ignites.

One of our assumptions was that the fluid contained in the process was air for which as an ideal gas $\gamma = 1.4$. We can now determine how variations of chemical composition alter the thermal efficiency η_{th}. The gases resulting from combustion are mainly water vapor and carbon dioxide with nitrogen as an "almost" inert gas; chemists would denote this as $\{H_2O(g) + CO_2(g) + N_2(g)\}$. At room temperature ($T = 298$ K) and at low pressure ($p = 0.1$ MPa) water vapor and carbon dioxide have a heat capacity ratio of $\gamma \approx 1.3$. As Figure 6.6 shows, η_{th} varies with chemical composition, but because the mole fraction of nitrogen is the largest in the whole process the effect will be small.

The Diesel cycle, named after Rudolf Diesel, who presented the first prototype of his engine in 1897, permits the use of higher compression ratios (and thus pressures). The (p, v) diagram for the Diesel cycle, shown in Figure 6.7, is similar

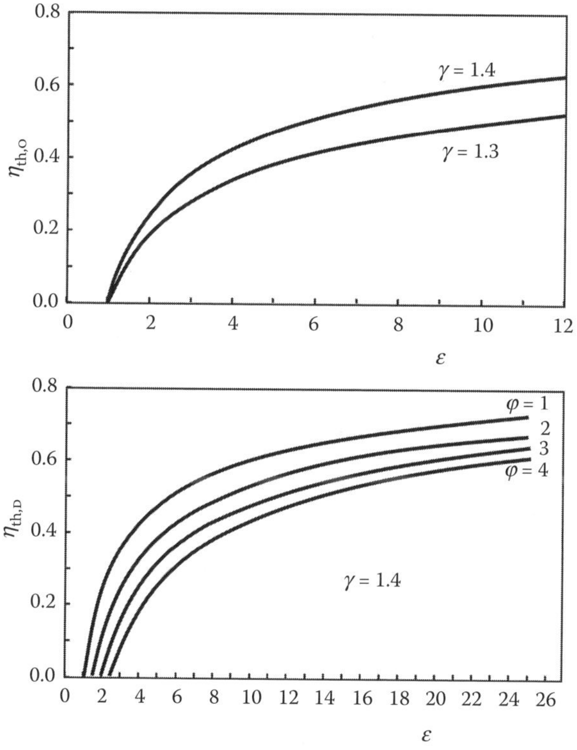

Figure 6.6 Thermal efficiencies η_{th} of internal combustion engines. (a): The thermal efficiency $\eta_{th,O}$ of an Otto cycle for $\gamma = 1.3$ and $\gamma = 1.4$. (b): The thermal efficiency $\eta_{th,D}$ of a Diesel cycle under the assumption of a constant isentropic exponent $\gamma = 1.4$ but for a range of cut-off ratio $\varphi = V_3/V_2$ from 1 to 4; for a Diesel cycle φ indicates the duration of the heat release (at constant pressure). The compression ratio ε is defined by Equation 6.22. The higher efficiency of diesel engines arises from the possibility of realizing a compression ratio of about 20 compared with about 12 for a gasoline engine.

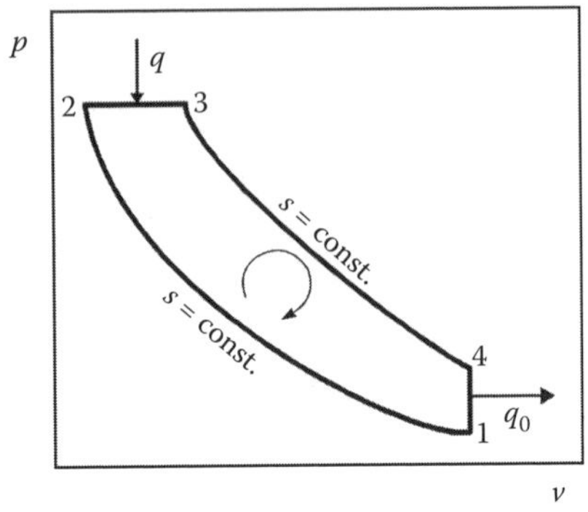

Figure 6.7 Schematic of a diesel engine as a (p, v) diagram. In contrast to the Otto cycle the heat released by combustion, modeled as heat input (2–3), is now realized by an isobaric process instead of an isochoric one.

to the (p, v) diagram of the Otto cycle shown in Figure 6.5. The major difference is that for Figure 6.7 the process of heat addition is now modeled as one at constant pressure. A more refined model of either the Otto or the Diesel process splits the combustion phase into two processes, namely a constant-volume and a constant-pressure process, where the precise segmentation depends on the cycle, and is called a *dual* cycle or *Seiliger* cycle. The basic model of the Diesel cycle, however, follows the process depicted in Figure 6.7, where there are two isentropic processes, one isochoric process and one isobaric process that makes the treatment more complicated than that for the Otto cycle.

Because the heat addition in the Diesel cycle is at constant pressure the thermal efficiency of this cycle $\eta_{\text{th,D}}$ is given by

$$\eta_{\text{th,D}} = 1 - \frac{|q_0|}{q} = 1 - \frac{|q_{41}|}{q_{23}} = 1 - \frac{c_v (T_4 - T_1)}{c_p (T_3 - T_2)} = 1 - \frac{1}{\gamma} \frac{(T_4 - T_1)}{(T_3 - T_2)} = 1 - \frac{1}{\gamma} \frac{T_1}{T_2} \frac{(T_4 / T_1) - 1}{(T_3 / T_2) - 1}. \tag{6.31}$$

Using the (p, T) relations for reversible adiabatic and isochoric processes with an ideal gas and the fact that $p_2 = p_3$, we obtain from Equation 6.31 after manipulation

$$\frac{T_4}{T_1} = \frac{T_3}{T_2} \left(\frac{p_4 p_2}{p_3 p_1} \right)^{\frac{\gamma-1}{\gamma}} = \frac{T_3}{T_2} \left(\frac{p_4}{p_1} \right)^{\frac{\gamma-1}{\gamma}}, \tag{6.32}$$

or

$$\frac{T_4}{T_1} = \left(\frac{T_3}{T_2} \right)^{\gamma}. \tag{6.33}$$

The ratio between the volumes after and before combustion is defined as the *cut-off ratio* φ and is given as

$$\varphi = \frac{V_3}{V_2} = \frac{v_3}{v_2} = \frac{T_3}{T_2}. \tag{6.34}$$

Using the definition of Equation 6.34 in Equation 6.31 and also of the compression ratio given by Equation 6.22 we obtain

$$\eta_{\text{th,D}} = 1 - \frac{1}{\gamma \cdot \varepsilon^{\gamma-1}} \cdot \frac{\varphi^{\gamma} - 1}{\varphi - 1}. \tag{6.35}$$

The efficiency of the diesel engine depends on the cut-off ratio φ which itself depends on the volume change during combustion (where the volume is expanded) and therefore depends upon the quantity of fuel burnt; in turn this depends on the quantity of fuel injected (accelerator depression).

Figure 6.6 shows the thermal efficiency of the diesel engine as function of the compression ratio with the cut-off ratio as a parameter and reveals that $\eta_{\text{th, D}}$ decreases with increasing φ. Using L'Hospital's rule to examine the limit as $\varphi \to 1$ we find the efficiency of the diesel engine approaches as a limiting case that of the Otto engine. That is, the thermal efficiency of the diesel engine (with $\varphi > 1$) is inferior to that of the Otto engine. However, the diesel engine permits compression ratios up to about 20, and it is these high compression ratios that enable the overall efficiency of a diesel engine to be higher than that of an Otto engine. This is the case for the idealized cycle considered but also for the real one.

6.3.2 Why Do Power Plants Have Several Steam Turbines?

We begin our discussion with an idealized scheme for a simple steam power plant that, as Figure 6.8 shows, consists of a series of process steps alternately involving transfer of work and heat, respectively. The working fluid water undergoes the following processes:

Step 1–2: adiabatic compression (pumping) of liquid water to the boiler pressure

Step 2–3: constant-pressure addition of heat in the boiler through the heating of subcooled water to its vaporization temperature, complete vaporization and then superheating of the water vapor

Step 3–4: adiabatic expansion of the vapor in a steam turbine usually into the two-phase region close to the saturated vapor line

Step 4–1: heat rejection and complete condensation at constant pressure

These four processes are characteristic of a basic steam power plant, which is also called a *Rankine* or sometimes *Clausius-Rankine cycle*. Figure 6.8 shows a Rankine cycle and includes typical values for T, p, P, and $\dot{Q}$ in each process.

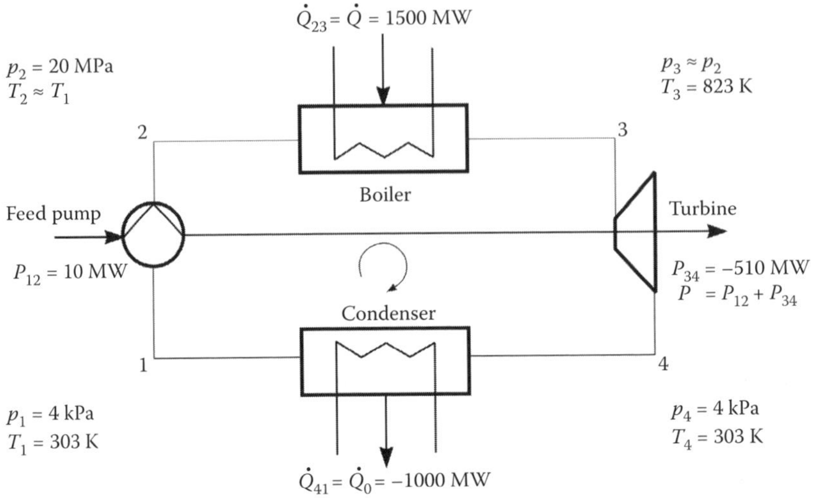

Figure 6.8 Schematic of a basic Rankine cycle.

A general requirement for the thermal efficiency discussed in Question 6.3 is that heat should be provided at the highest possible temperature and rejected at the lowest possible temperature. On the basis of the upper temperature limits imposed by materials used to construct the machinery the highest practical temperature is about 550 °C. The lower temperature where heat is rejected is determined by the temperature of the surroundings where the power plant is located. For the purpose of this example we assume a condensation temperature of 30 °C, which corresponds to a water vapor pressure of about 4 kPa. The heat and power fluxes listed in Figure 6.8 are for a power plant with a net output power of 500 MW, where a part fraction (albeit small) of the power available at the turbine (shown in Figure 6.4) is consumed by the feed pump. As a rule of thumb we may assume an overall thermal efficiency of 1/3 (state-of-the-art power plants achieve a thermal efficiency of >0.4). The efficiency of 1/3 means that a heat flux of 1500 MW must be provided, of which a fraction of two-thirds is discharged at low temperature, mainly as a consequence of the second law, but also because of inevitable irreversibilities and losses within the process.

One of the major losses within a Rankine cycle is the necessarily nonideal operation of the steam turbine. In a perfect turbine, process step 3–4 would be reversible and, thus, isentropic, resulting in an ideal state denoted by 4^s in Figure 6.9. In a real process the entropy of the fluid is increased, yielding fluid at a higher temperature and enthalpy as shown in Figure 6.9. Therefore, not all of the available energy (equal to the exergy discussed in Question 3.9) of the fluid at state 3 is exploited in the real process. Similar considerations hold for the feed pump that operates in the step 1–2 of the process. To illustrate the salient points

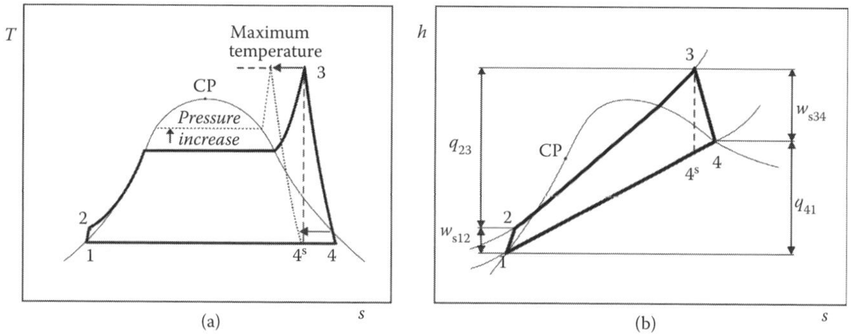

Figure 6.9 (a): (T, s) diagram for a Rankine cycle (thick solid line). (b): (h, s) diagram for a Rankine cycle (thick solid line). Point 4^s denotes the state after expansion in an idealized (isentropic) turbine (dashed lines). Increasing the boiler pressure $p_2 = p_3$ at a given maximum temperature T_3 (dotted line) results in a moisture content that is too high for the turbine.

Figure 6.9 is not drawn to scale in this region of the diagram. The rise in both temperature and enthalpy are relatively small; the process is operated close to the saturation line and the differences would be practically indistinguishable on the overall scale of Figure 6.9. For an ideal process the step involving work would be reversible (and thus isentropic) giving a vertical line; in reality the entropy of the fluid increases and the line is not vertical.

The thermal efficiency of the Rankine cycle can be obtained from the (h, s) diagram of Figure 6.9 as

$$\eta_{\text{th,R}} = \frac{-w_s}{q} = \frac{-(w_{s34} - w_{s12})}{q_{23}} = \frac{(h_3 - h_4) - (h_2 - h_1)}{h_3 - h_2} \approx \frac{(h_3 - h_4)}{h_3 - h_2}, \quad (6.36)$$

where the last step follows because of the relatively small enthalpy change that accompanies the liquid compression. From a fundamental thermodynamic point of view the obvious measure to improve the efficiency is to increase the spread of temperatures between the levels where heat is provided and where heat is discharged. Because the upper temperature is determined by the materials used for construction of the power plant and the lower temperature by the ambient value the margin for efficiency improvement from this source is small. There are of course always efforts to develop materials that could enhance the upper temperature.

In the remaining discussions we provide reasons why particular design features are incorporated in power plants. It is important to recognize it is the average temperature of heat provision that is of paramount importance. On the basis of this fact it is therefore desirable to obtain a high temperature in the two-phase region for water, and this can be realized by increasing the boiler pressure. Recent developments in steam turbine plants also use pressures >22 MPa that are supercritical and result in step 2–3 of the process extending outside the two-phase region. However, increasing the $p_2 = p_3$ at a maximum temperature T_3 will shift point 3 to the left of Figure 6.9 in both the (T, s) and (h, s) diagrams. As a consequence, after expansion of the vapor, point 4 lies further into the two-phase region with a higher fraction of liquid water present and this leads to the formation of larger water droplets which lead to increased erosion of the turbine blades. Indeed, it is because of erosion that the steam quality x is maintained >0.9 at the end of expansion.

Combining the requirements for a high boiler pressure and temperature with the need for a state after expansion near to that of the saturated vapor leads to what is termed the *reheat power plant design* for which the schematic is shown in Figure 6.10 and the corresponding (T, s) in Figure 6.11. After the steam is expanded to a medium pressure in a high-pressure turbine it is reheated to about the original maximum temperature. In a second step, the steam is expanded again, this time to the condenser pressure in a low-pressure turbine. However, additional turbines increase the complexity of a plant and a large number of turbines are neither beneficial nor economical. Consequently, a second reheat step and thus a third turbine operating at an additional intermediate pressure level are normally introduced only in the case of boiler pressures close to or above the critical pressure of water of 22 MPa. For all steam turbines, increasing the boiler pressure and temperature in a reheat process is one of the most important thermodynamic methods used to increase the efficiency of a steam power plant.

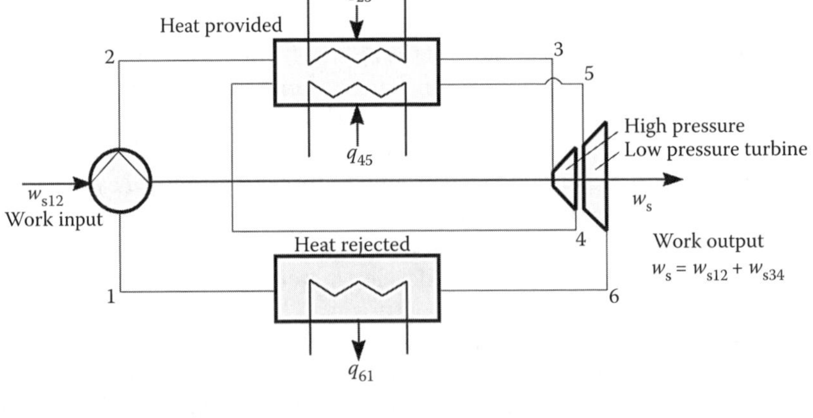

Figure 6.10 Schematic of a reheat Rankine cycle. After expansion in a high-pressure turbine the steam is reheated and expanded again in a second turbine.

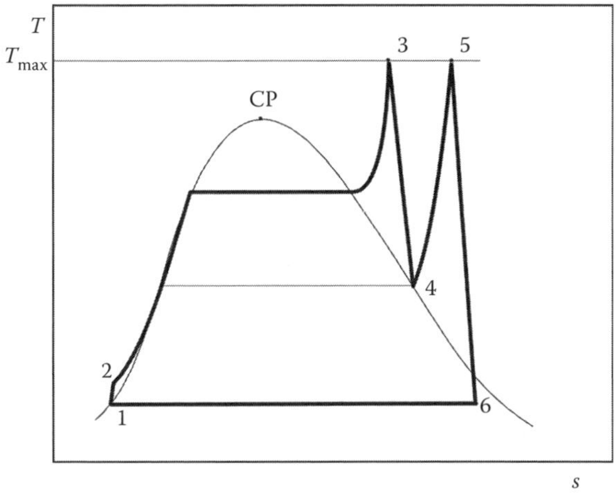

Figure 6.11 (*T, s*) diagram of a reheat Rankine cycle. The maximum temperature is determined by the materials of construction. The average temperature of heat provision can be increased, while the quality *x* after the final expansion is large.

For the sake of completeness we mention another important variation that also increases the average temperature of heat provision. In the Rankine cycle discussed so far liquid water at low temperatures is fed into the boiler after compression. However, it is advantageous to heat the water in a regenerative scheme. In this case, steam from a turbine is extracted and is either directly mixed with the feedwater or used for preheating via a heat exchanger. Steam power plants use a series of feedwater heaters each at a different temperature that use steam bleed at appropriate points of the turbine stages.

6.3.3 What Is a Combined Cycle?

The term "combined cycle" commonly refers to a combination of a gas-turbine cycle and a steam power cycle; the introduction of the combination is intended to increase the overall efficiency. The key feature of this approach is the use of the waste heat from the gas turbine as a partial replacement for the heat that must be provided to a steam cycle, normally from the combustion of fossil fuel.

We start with the operation scheme of the gas-turbine cycle, which consists of the following three steps as shown in Figure 6.12:

Step 1–2: adiabatic compression of ambient air to a pressure of up to 2 MPa (through a common shaft the compressor is directly driven by the turbine as shown in Figure 6.4)

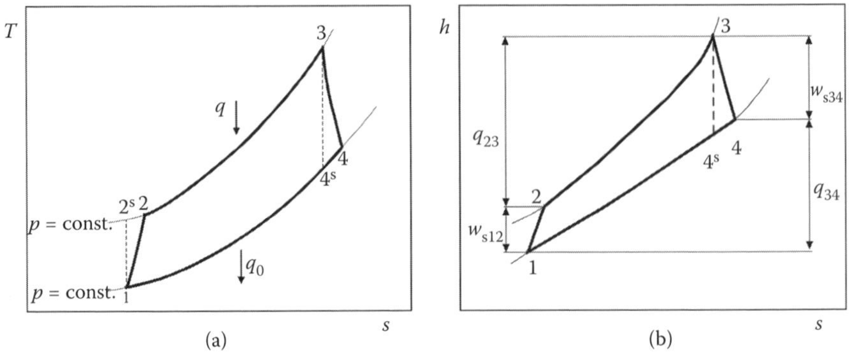

Figure 6.12 The Brayton cycle. (a): (T, s) diagram. (b): (h, s) diagram. The superscript s denotes the idealized (isentropic) processes.

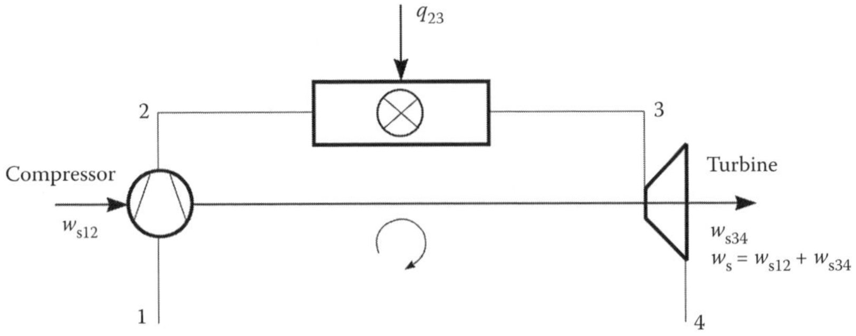

Figure 6.13 Scheme of a basic Brayton cycle (open cycle). Closure of the cycle is realized through the cooling down of exhaust gases in ambient air.

Step 2–3: combustion of gas in the chamber modeled as a constant-pressure heat addition

Step 3–4: adiabatic expansion of hot compressed gas in a turbine

The gas turbine cycle is commonly referred to as either a *Brayton* or a *Joule* cycle. The term *Joule* cycle is normally used only for the particular case when both compression and expansion are performed reversibly.

This "open-cycle" arrangement is normally utilized within a gas turbine for the generation of electricity and, as illustrated in Figure 6.13, at first sight overlooks the closure of the cycle. There are closed cycles where a fourth process is used to reject heat with a heat exchanger (and the combustion chamber is replaced by an additional heat exchanger). This closed-loop system often uses helium as the working fluid, and only finds limited application, because it is impractical and uneconomic for large-scale power generation from the combustion of gas. In an open-cycle gas turbine the final step of heat rejection at constant pressure is omitted, resulting in the elimination of an additional mechanical component. Practically, closure is obtained by heat rejection to ambient air, and the prerequisite for a thermodynamic cycle (the thermodynamic properties before and after the cycle must be identical) is accomplished by the surroundings: air at ambient conditions occurs at the beginning and end of the cycle.

From the (T, s) and (h, s) diagrams for this cycle (shown in Figure 6.12) the thermal efficiency of the system can be determined. In both diagrams we have already accounted for the irreversibilities in the operation of both compressor and turbine. In both cases the pressure change is connected with an increase in entropy. In an ideal Brayton (or Joule) cycle both compression and expansion are isentropic and are represented by vertical lines in the diagrams.

By analogy to the Rankine cycle the thermal efficiency of the Brayton cycle is given by

$$\eta_{\text{th,B}} = \frac{-w_s}{q} = \frac{-(w_{s34} - w_{s12})}{q_{23}} = \frac{(h_3 - h_4) - (h_2 - h_1)}{h_3 - h_2} = 1 - \left(\frac{h_4 - h_1}{h_3 - h_2}\right). \tag{6.37}$$

For simplicity, we assume the working fluid is an ideal gas for which the heat capacity at constant pressure c_p is constant and the enthalpy differences are given by

$$h_y - h_x = c_p(T_y - T_x), \tag{6.38}$$

so that Equation 6.37 becomes

$$\eta_{\text{th,B}} = 1 - \left(\frac{T_4 - T_1}{T_3 - T_2}\right). \tag{6.39}$$

Again, the assumption of a perfect gas does not affect the conclusions obtained from the analysis.

For an ideal Brayton cycle with isentropic compression and expansion (and a constant heat capacity ratio γ) Equation 6.39 can be simplified utilizing the pressure ratio $\Pi = p_2/p_1 = p_3/p_4$ and

$$\frac{T_2}{T_1} = \left(\frac{p_2}{p_1}\right)^{\frac{\gamma-1}{\gamma}} = \Pi^{\frac{\gamma-1}{\gamma}} = \left(\frac{p_3}{p_4}\right)^{\frac{\gamma-1}{\gamma}} = \frac{T_3}{T_4} \tag{6.40}$$

to obtain

$$\eta_{\text{th,B,id}} = 1 - \frac{T_1}{T_2} \cdot \frac{(T_4/T_1) - 1}{(T_3/T_2) - 1} = 1 - \frac{T_1}{T_2} = 1 - \frac{1}{\Pi^{(\gamma-1)/\gamma}}. \tag{6.41}$$

For the idealized process Equation 6.41 implies that the efficiency increases with increasing pressure ratio Π. Other parameters that influence the performance of a gas-turbine cycle are the temperature T_3, which is the maximum temperature of the process, where the gas enters the turbine and the temperature ratio $\tau = T_3/T_1$. In view of the materials used to construct the turbine the inlet temperature is limited to about 1,500 °C (1,800 K), and operation at these high temperatures is only possible with the use of air-cooled turbine blades. It is uneconomic to raise the pressure ratio >20 for a given maximum temperature T_3 because the net work output $-w_s = -(w_{s34} - w_{s12})$ has a maximum, and a further increase of the pressure ratio Π results in a decrease in the net work output. This observation can be rationalized by considering that for fixed temperatures T_1 and T_3 (and thus a fixed enthalpy difference $h_3 - h_1$) an increase in Π results in an increase in the compressor work w_{s12} and a reduction of the heat q_{23}. Closer inspections show that the compressor work w_{s12} increases at a rate greater than the turbine work output $-w_{s34}$, resulting in a maximum for the net work output $-w_s$. From the condition $\mathrm{d}w_s/\mathrm{d}\Pi = 0$ the optimum pressure ratio is given by

$$\Pi_{\text{opt}} = \tau^{\gamma/\{2(\gamma-1)\}}, \qquad (6.42)$$

which is equivalent to the condition $T_2 = T_4$. Table 6.1 lists the variation of w_s and q as a function of Π. The derivation of Equation 6.42 is discussed in detail in the literature, for example, by Burghardt and Harbach (1993). The results listed in Table 6.1 reveal that the thermal efficiency gradually increases with the increasing pressure ratio with a maximum (albeit shallow) for the net work output at a pressure ratio $\Pi_{\text{opt}} = 19.75$.

The thermal efficiency of the gas-turbine cycle can be increased by several methods and two significant ones are as follows: (1) multistage compression with repeated intercooling and reheating between stages and (2) utilization of the exhaust gas to preheat the air before entering the combustion chamber in a counterflow heat exchanger. Item 1 reduces the overall work required

TABLE 6.1 THE WORK w_s AND HEAT q FOR EACH STEP OF A BRAYTON CYCLE AS A FUNCTION OF THE PRESSURE RATIO Π. THE VALUES ARE BASED ON FIXED INTAKE TEMPERATURE $T_1 = 290$ K AND MAXIMUM TEMPERATURE $T_3 = 1{,}595$ K ($\tau = T_3/T_1 = 5.5$) AND AIR, THE WORKING FLUID, ASSUMED TO BE A PERFECT GAS

Π	w_{s12}/kJ·kg^{-1}	q_{23}/kJ·kg^{-1}	w_{s34}/kJ·kg^{-1}	w_s/kJ·kg^{-1}	η_{th}
14.00	328	983	−848	−520	0.53
16.00	352	958	−876	−524	0.55
18.00	374	936	−900	−526	0.56
19.75	392	919	−919	−527	0.57
20.00	394	916	−921	−527	0.58
22.00	413	897	−939	−526	0.59
24.00	431	879	−956	−525	0.60

for compression because the process approaches isothermal compression and requires less work than adiabatic compression (compare Question 1.7.6). Item 2 is used when the compression ratio Π and, thus, the compressor exit temperature T_2 are not too high, and the exhaust gas at temperature T_4 preheats the air before entering the combustion chamber.

It is the high outlet temperatures of about 800 K for a modern gas turbine that leads to the combined cycle. The exhaust gases of the gas turbine may be used either to preheat the water for the steam cycle or to act as the sole heat source for the steam turbine through a heat exchanger (boiler). One example for such a combination is depicted in Figures 6.14 and 6.15 for which the heat rejected from the gas turbine and given by

$$q_{45} = \int_4^5 T\,\mathrm{d}s, \tag{6.43}$$

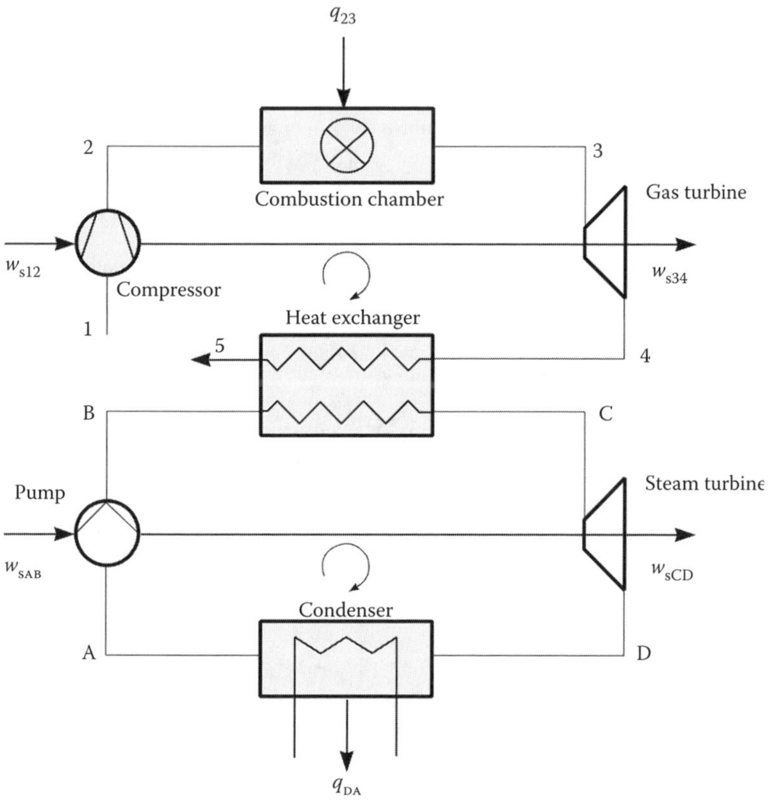

Figure 6.14 Scheme of a combined cycle. In this configuration the hot exhaust of the gas turbine is used as the sole heat source for the steam process.

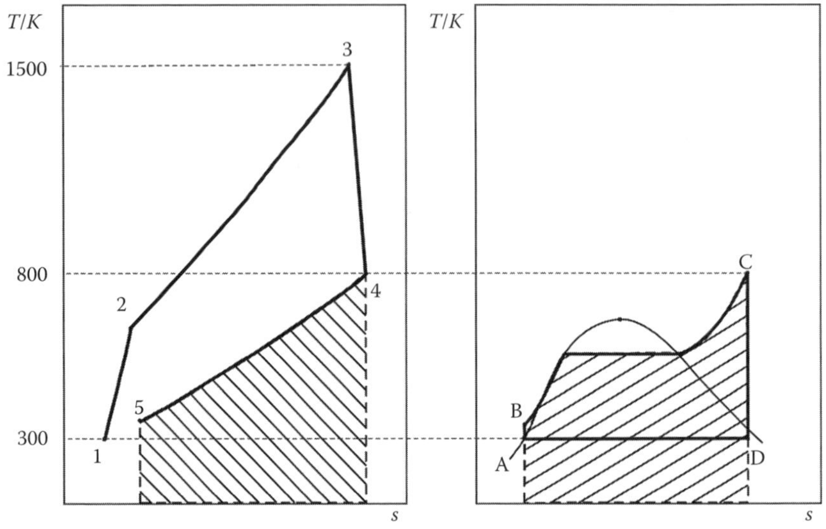

Figure 6.15 (T, s) diagrams of a combined cycle that incorporates a Brayton cycle (a) and a Rankine cycle (b). The heat q_{45} rejected from the gas turbine at a comparatively high temperature provides the heat input q_{BC} for the steam cycle.

is used completely to provide the heat for the steam cycle that is given by

$$q_{BC} = \int_{B}^{C} T\,ds. \tag{6.44}$$

Practically, of course, there are some losses owing to imperfect heat transfer; also to avoid corrosion, the exhaust gases are not completely cooled to ambient temperature. However, for our current purpose we can assume that the exhaust gases are completely utilized and that there is no additional heat for the steam cycle. In that case, the maximum thermal efficiency of the combined cycle is given by

$$\eta_{th,max} = \frac{-w_{s,B} - w_{s,R}}{q} = \eta_{th,B} + \eta_{th,R}(1 - \eta_{th,B}). \tag{6.45}$$

In practice, combined cycles may attain an overall thermal efficiency of about 60 %.

6.4 WHAT IS A REFRIGERATION CYCLE?

Refrigeration is the process of removing heat from one zone and rejecting it to another zone. The primary purpose of refrigeration is to lower the temperature of the one zone and then to maintain it at that temperature. In this case heat is transferred from a high to a low temperature that requires a machine and a thermodynamic cycle, which are called *refrigerators* and *refrigeration cycles*, respectively.

In Chapter 1, within Questions 1.7.6 and 1.8.7, we discussed the temperature drop of a working fluid after its flow through a constriction in an isenthalpic process. We now consider how that phenomenon can be exploited in a closed thermodynamic cycle to produce continuous cooling. Refrigerators and heat pumps are essentially the same devices that differ only in their specific objective. We discuss two types of refrigeration cycles in the remainder of this question: the vapor-compression cycle and the (ammonia) absorption cycle.

6.4.1 What Is a Vapor-Compression Cycle?

The refrigeration cycle is a closed loop of four processes using a working fluid known as the refrigerant. Typical refrigerants are fluorocarbons, hydrofluorocarbons, and hydrocarbons with the specific choice of working fluid dependent on the application. The refrigerant in the vapor refrigeration cycle undergoes four stages. A schematic diagram of this refrigeration cycle is given in Figure 6.16. Figure 6.16a shows a cycle where the refrigerant expands through a turbine, which is owing to cost unusual and occurs solely in large installations, and Figure 6.16b shows a cycle where the refrigerant expands through a valve is the most widely used for refrigerators, air conditioning, and heat pumps.

The four stages of a simplified process illustrated in Figure 6.16 are as follows:

Step 1–2: the refrigerant is adiabatically compressed raising the pressure so that the corresponding saturation temperature is above ambient temperature

Step 2–3: the refrigerant rejects heat to the environment through a heat exchanger

Step 3–4: the refrigerant is expanded through either a turbine at constant entropy (as shown in Figure 6.16a) or through a throttling valve at constant enthalpy (as shown in Figure 6.16b) and condenses

Step 4–1: the fluid evaporates as it absorbs heat from the space to be cooled

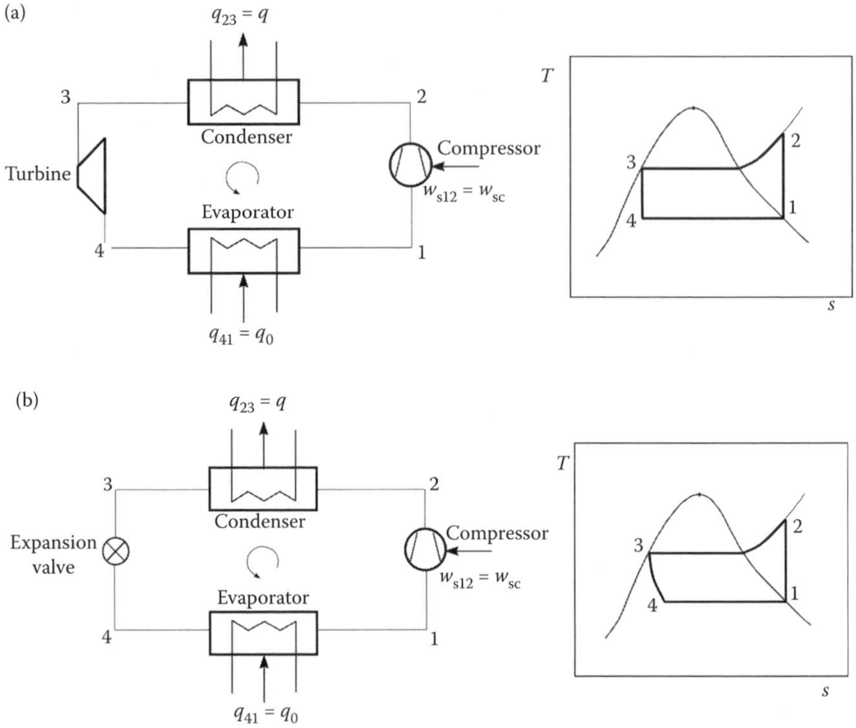

Figure 6.16 Schematic (a) and (T, s) diagram (b) of two idealized vapor refrigeration cycles: (a) a turbine is used to expand the working fluid and (b) expansion occurs through a valve.

In the idealized refrigerator the compression in the first stage is isentropic and the work required is given by

$$w_{s12} = (h_2 - h_1), \tag{6.46}$$

and the fluid temperature is raised from temperature T_1 to temperature T_2 so that the refrigerant enters the condenser at a temperature higher than the surroundings. The refrigerant enters the condenser where heat is rejected to the surroundings and then the refrigerant condenses completely leaving as a saturated liquid at temperature T_3. The heat transferred is given by

$$q = h_3 - h_2. \tag{6.47}$$

The refrigerant then enters either a turbine, as shown in Figure 6.16a, where an isentropic expansion occurs for which

$$s_4 = s_3, \tag{6.48}$$

and produces work according to

$$w_{s34} = h_4 - h_3. \tag{6.49}$$

When an expansion valve is used, as is the case in Figure 6.16b, an isenthalpic expansion occurs as defined by

$$h_4 = h_3. \tag{6.50}$$

In both cases, the refrigerant temperature is reduced to temperature T_4 below the temperature of the object to be cooled. Typically, the refrigerant leaves either the turbine or the expansion valve at a temperature and pressure within the two-phase region with a low quality factor x so that more heat can be absorbed in the next step.

The refrigerant then passes to an evaporator where heat is absorbed from the object to be cooled, and in this process the enthalpy returns to h_1 so that

$$q_0 = h_1 - h_4. \tag{6.51}$$

In the ideal case, the refrigerant leaves the evaporator as a saturated vapor, however, in the actual cycle, the vapor is superheated to prevent liquid droplets entering the compressor and causing damage to it and it leaves the condenser subcooled so as to provide greater cooling capacity. The (T, s) diagram for a real refrigeration cycle is shown in Figure 6.17 and should be compared with Figure 6.16.

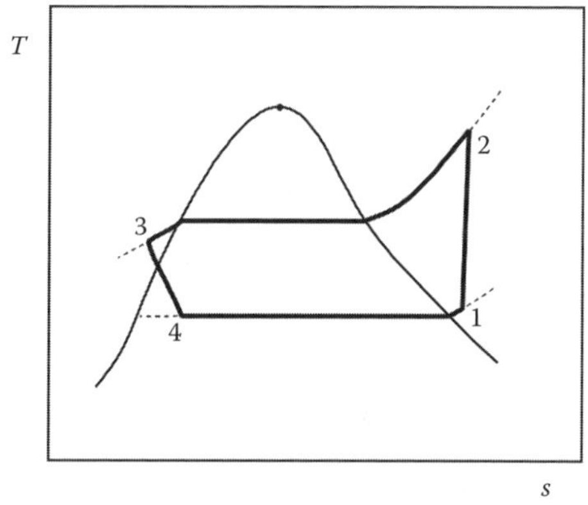

Figure 6.17 (T, s) diagram of a real vapor refrigeration cycle.

We now consider the coefficient of performance (COP) for these cycles. If the purpose of using these cycles is to cool a space then the COP is defined as the ratio of cooling effect to the required net work. The net work required in any cycle is found by the application of the first law:

$$\left|w_{\mathrm{s,net}}\right| = \left|q\right| - \left|q_0\right|. \tag{6.52}$$

Thus, for the cycle shown in Figure 6.16a the COP is given by

$$\mathrm{COP_c} = \frac{\left|q_0\right|}{\left|w_{\mathrm{s,net}}\right|} = \frac{h_1 - h_4}{(h_2 - h_3) - (h_1 - h_4)}, \tag{6.53}$$

while for the idealized cycle shown in Figure 6.16b and the real cycle shown in Figure 6.17 (no work production) the COP is given by

$$\mathrm{COP_c} = \frac{\left|q_0\right|}{\left|w_{\mathrm{s,net}}\right|} = \frac{h_1 - h_4}{h_2 - h_1}. \tag{6.54}$$

If these cycles are used to heat a space (with a heat pump) then the COP is defined as the ratio of heating effect to the required net work. Thus, for the cycle of Figure 6.16a the COP is given by

$$\mathrm{COP_h} = \frac{\left|q_0\right|}{\left|w_{\mathrm{s,net}}\right|} = \frac{h_2 - h_3}{(h_2 - h_3) - (h_1 - h_4)}, \tag{6.55}$$

while for the idealized cycle shown in Figure 6.16b and the real cycle shown in Figure 6.17 the COP is given by

$$\mathrm{COP_c} = \frac{\left|q\right|}{\left|w_{\mathrm{s,net}}\right|} = \frac{h_2 - h_3}{h_2 - h_1}. \tag{6.56}$$

For a Carnot cycle the COP is

$$\mathrm{COP_c} = \frac{\left|q_0\right|}{\left|w_{\mathrm{s,net}}\right|} = \frac{\left|q_0\right|}{\left|q\right| - \left|q_0\right|} = \frac{T_0}{T - T_0}, \tag{6.57}$$

for cooling and the COP for heating is

$$\mathrm{COP_h} = \frac{\left|q\right|}{\left|w_{\mathrm{s,net}}\right|} = \frac{\left|q\right|}{\left|q\right| - \left|q_0\right|} = \frac{T}{T - T_0}. \tag{6.58}$$

The coefficients of performance given by Equations 6.53 through 6.56 are less than those of the ideal, reversible Carnot cycle given by Equations 6.57 and 6.58.

The cooling capacity of the refrigerator cycle is given by

$$\dot{Q}_c = \dot{m}_{ref} \cdot q_0, \tag{6.59}$$

where $\dot{m}_{ref}$ is the rate at which the mass of the refrigerant working fluid is circulated around the refrigerator cycle and q_0 is the heat absorbed in the evaporator per mass of refrigerant.

The power required to move and compress the refrigerant is given by

$$P = \dot{m}_{ref} w_{s12}, \tag{6.60}$$

where w_{s12} is the work required for compression per unit mass of circulating refrigerant.

How do you choose the right refrigerant for an application? The evaporation and condenser temperatures are fixed for given refrigeration tasks by the temperatures of the space to be cooled and the temperature of the surroundings. The choice between the several refrigerants available depends on many factors and generally the most important are as follows: (1) the vapor pressure in the evaporator and condenser, (2) the specific enthalpy of vaporization should be as high as possible to obtain the greatest cooling effect per kilogram of fluid circulated, (3) the specific volume of the refrigerant should be as low as possible to minimize the work required per kilogram of refrigerant circulated, (4) chemical stability, (5) toxicity, (6) cost, and (7) environmental factors. For item 1 the vapor pressure in the evaporator should not be lower than atmospheric pressure to avoid air leaking in, while the vapor pressure at the condenser should not be much greater than atmospheric pressure to avoid refrigerant leaking out of the system. For example, refrigerant 1,1,1,2-tetrafluoroethane (commonly known in the refrigeration industry by the American Society of Heating, Refrigerating and Air-Conditioning Engineers [ASHRAE] Standard 34 nomenclature as R-134a) should be avoided in a refrigeration cycle at low temperature because, at an evaporator temperature of 233 K, the corresponding vapor pressure is 51.64 kPa and can permit air to leak into the system from the surroundings. On the other hand the use of 1,1,1-trifluoroethane (otherwise known by the by the ASHRAE Standard 34 nomenclature as R-143a) is acceptable because at a temperature of 233 K the vapor pressure is 140 kPa, which is greater than normal atmospheric pressure of about 101 kPa at sea level. It is also important from the perspective of energy (usually electrical) that the compression ratio should be as low as possible to reduce the energy required for compression.

There are special cases where extra compression is required to obtain higher temperature in the condenser or lower temperature in the evaporator. This extra compression requires the use of two or more compressor stages accompanied by intercooling between the stages to reduce the refrigerant volume and, thus, reduce the required compression power as illustrated in the (T, s) diagram of Figure 6.18.

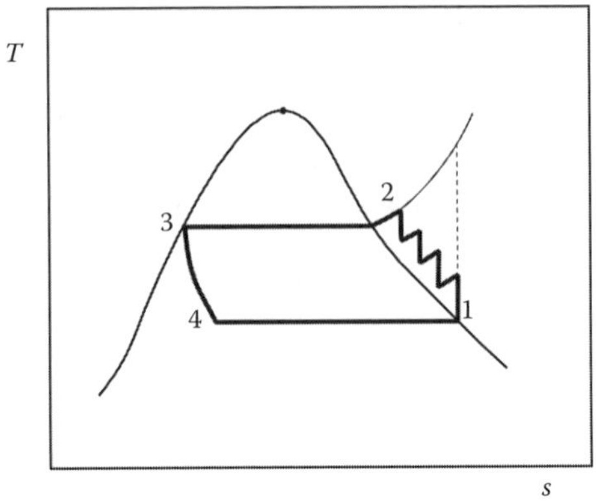

Figure 6.18 *T-s* diagram of a vapor refrigeration cycle with a multistage compressor.

When temperatures lower than 243 K are required a single refrigerator cycle results in a high pressure ratio between the evaporator and the condenser for which the compressor has a low-energy efficiency. For example, for a cycle that uses an almost azeotropic mixture of $\{CH_2F_2 + CHF_2CF_3\}$ that is difluoromethane + 1,1,2,2,2-pentafluoroethane (or using the ASHRAE Standard 34 nomenclature R32 + R125) to achieve an evaporator temperature of 233 K (where the pressure of the vapor in equilibrium with the liquid, that is, the bubble pressure is about 174 kPa) and a condenser temperature of 318 K (where the pressure of the vapor in equilibrium with the liquid, that is, the dew pressure is about 2721 kPa) the compression ratio is 15.6. This pressure ratio can be reduced by use of a cascade cycle system that consists of two cycles completely independent of each other except that the evaporator of the higher temperature cycle acts as the condenser for the low temperature cycle. For example, the low temperature cycle may have an evaporator temperature of 233 K and a condenser temperature of 278 K (where the pressure of the vapor in equilibrium with the liquid that is about 932 kPa) and the higher temperature cycle may have an evaporator temperature of 268 K (where the pressure of the vapor in equilibrium with the liquid that is about 677 kPa) and a condenser temperature of 318 K. The compression ratio of the first cycle is 5.3 and the compression ratio of the second cycle is 4.0, and both of these values are considered acceptable for common compressors.

6.4.2 What Is an Absorption Refrigeration Cycle?

The absorption refrigeration cycle, shown schematically in Figure 6.19, is similar to the vapor-compression cycle shown in Figure 6.16 with the major difference

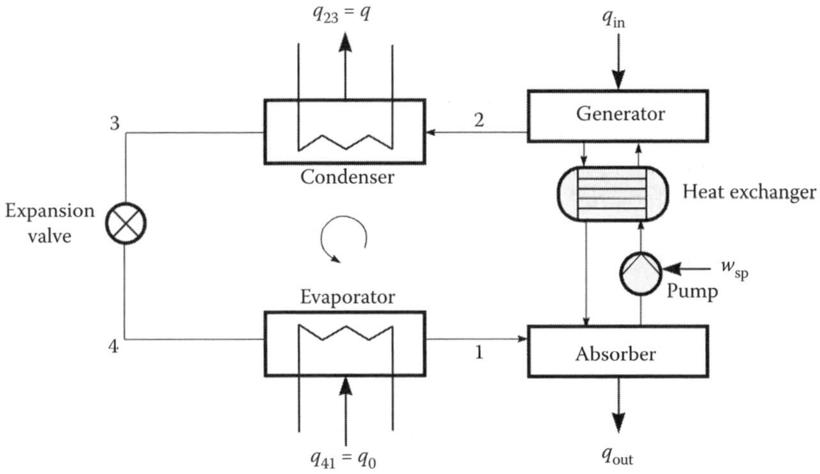

Figure 6.19 Schematic diagram of the absorption refrigeration cycle.

being the method used for refrigerant compression. In the vapor-compression refrigerator, the working fluid is compressed to a high pressure by a compressor, while in the absorption refrigerator cycle the refrigerant is first absorbed into water and then the liquid solution is compressed to a high pressure by a pump from which the absorbed gas is subsequently released by heating.

The large-scale application of absorption refrigerators has been confined to the use of ammonia as a refrigerant because it has a high solubility in water (at a temperature of 298 K and a pressure of 0.1 MPa about 320 g of $NH_3(g)$ are soluble in 1 dm^3 of $H_2O(l)$, that is, a molality of 18.8 mol · kg^{-1}) and the ammonia reacts with water in a reversible reaction of

$$NH_3(g) + H_2O(l) \xrightleftharpoons{K_b(298\,K) = 1.8 \cdot 10^{-5}\,mol \cdot kg^{-1}} NH_4^+(aq) + OH^-(aq). \quad (6.61)$$

to give a basic solution. The steps in the process shown in Figure 6.19 are as follows:

Step 1–2: ammonia vapor at a temperature T_1 is dissolved or absorbed in liquid water

Step 2–3: the refrigerant rejects heat to the environment through a heat exchanger

Step 3–4: the refrigerant is expanded in the throttling process at constant enthalpy and condenses

Step 4–1: the fluid evaporates as it absorbs heat from the environment to be cooled

In step 1–2, heat is rejected to the environment to maintain the temperature as low as possible so as to increase the amount of substance of NH_3 that can dissolve in water; the solubility increases with decreasing temperature, for example, at a temperature of 273 K 900 g of $NH_3(g)$ dissolve in 1 dm^3 of $H_2O(l)$, that is, a molality of about 53 $mol \cdot kg^{-1}$ assuming a mass density of 1 $kg \cdot dm^{-3}$. The liquid solution is then pumped to the high pressure of the generator and heat transferred to produce $NH_3(g)$ from the water solution. In the absorption refrigerator the work required to pump the liquid solution (which is essentially an incompressible fluid) is much less than for compression if the ammonia were gaseous as it would be in a vapor-compression cycle. The work for this liquid pumping is given by

$$w_{sp} = v(p_2 - p_1), \tag{6.62}$$

where v is the specific volume of the liquid solution and p_1 and p_2 are the pressures of the evaporator and condenser, respectively. The energy required to operate the pump is much less than the energy required to evolve $NH_3(g)$ from the $NH_3(aq)$.

The steps of the absorption refrigeration cycle are the same as those of a vapor-compression cycle except for step 1–2. The heat transferred in the condenser is given by Equation 6.47 and that of the evaporator is given by Equation 6.51. There is isenthalpic expansion through a throttling valve. Absorption refrigeration becomes economically attractive when there is a source of inexpensive heat energy to evolve $NH_3(g)$ from the $NH_3(aq)$.

The COP for absorption refrigeration is defined as the ratio of the cooling effect to the required energy input (heat input to the generator plus pump work); this differs from the definition of COP used for a vapor-compression cycle given by Equation 6.54. The work input to the pump is relatively small so that the COP is given by

$$\mathrm{COP_c} = \frac{|q_0|}{|q_{in}| + |w_{sp}|} \approx \frac{|q_0|}{|q_{in}|} = \frac{h_1 - h_4}{|q_{in}|}. \tag{6.63}$$

6.4.3 Can I Use Solar Power for Cooling?

The maximum radiant flux on the surface of the earth of about 1 $kW \cdot m^{-2}$ that arises from the sun can only be achieved on a clear sunny day at noon. The resultant mean radiant flux over a time of 1 d is about 250 $W \cdot m^{-2}$. This energy can be used in a number of ways, and here we consider how it may be used in refrigeration by partial substitution for the heat provided to a refrigeration cycle, for example, the ammonia absorption described in Question 6.3.2 that would otherwise be obtained from another source. Solar energy is particularly appropriate for cooling buildings because the demand for cooling during a day is essentially in phase with the energy available from the sun; unfortunately, at the time of writing this the cost of solar-powered refrigeration equipment prohibits deployment but this may be overcome by the requirement to reduce carbon dioxide emission that result from combustion of fossil fuel in the fullness of time.

For solar-powered absorption refrigeration water is used as the working fluid and a solution of an alkali metal halide, for example, LiBr, as the absorbent that relies on the solubility of LiBr in H_2O of 1.67 kg in 1 dm^3 of water at a temperature of 298 K and pressure of 0.1 MPa; that is, a molality of about 18.8 $mol \cdot kg^{-1}$ similar to ammonia in water. From the safety and environmental perspectives water is an extremely advantageous refrigerant. The heat required to separate the water from the aqueous solution of lithium bromide requires high temperature in the generator that is provided by the sun with special solar energy collectors. One form of solar collector uses evacuated tubes made of glass, where the round profile favors the near-perpendicular incidence of the sun rays on the tube during the whole day. In addition, the vacuum within the tubes reduces convection and conduction heat loses and thus achieves high thermal efficiency and temperature. Another form of solar collector relies on mirrors or lenses to focus the energy and to obtain the temperatures required. Figure 6.20 shows the schematic diagram of the basic elements of this absorption cycle.

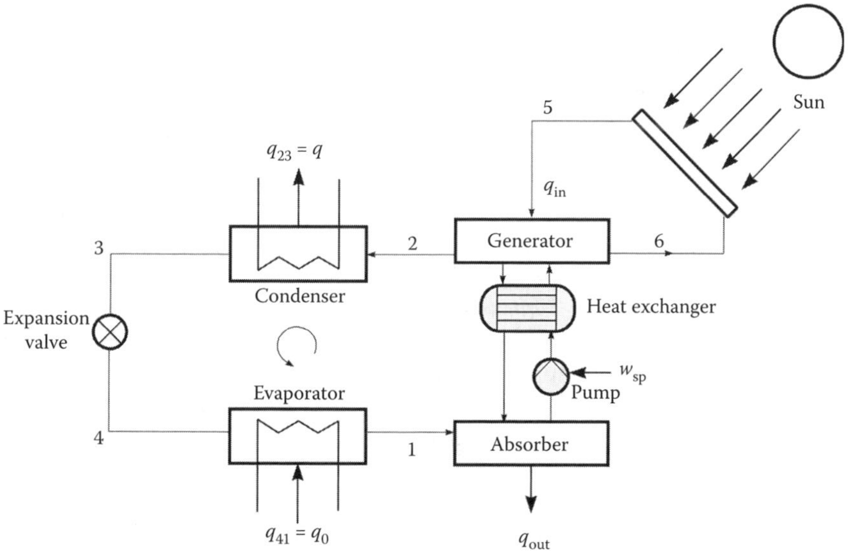

Figure 6.20 Schematic diagram of a solar cooling cycle.

An analytical description of the absorption cycle was given in Question 6.3.2. Heat is provided to the generator by another working fluid circulating between the solar collector and the generator. Thus, the energy from the sun is used to reduce the energy required from other sources and, thus, the cost of operating the cooling system.

The definition of COP for this cooling system is defined as the radio of the cooling effect to the required energy input (heat plus pump work). Since the work input to the pump is usually small the COP is given by

$$\mathrm{COP_c} = \frac{|q_0|}{|q_{\mathrm{sun}}| + |w_{\mathrm{sp}}|} \approx \frac{|q_0|}{|q_{\mathrm{sun}}|} \approx \frac{h_1 - h_4}{|q_{\mathrm{sun}}|}. \tag{6.64}$$

The COP for the solar-powered absorption refrigeration cycle is usually between 0.6 and 0.75. Manufacturers give an average value of specific collector surface area between 3 and 4.5 m^2 for each kiloWatt of cooling capacity. The electric energy for pumping the aqueous solution of the absorption refrigeration cycle is much less than the one required for compression of the gaseous refrigerant in the vapor-compression cycle.

6.5 WHAT IS A LIQUEFACTION PROCESS?

Liquefied gases are used in many practical situations: for example, liquid oxygen is shipped and stored in many hospitals in chilled tanks until required, and then allowed to boil to release oxygen gas for patients; liquid chlorine is shipped and stored for sterilization of water; and liquefied natural gas is shipped and stored to be used as fuel. The reduction in volume per unit amount of substance from the gaseous to the liquid phase is significant and further, because of the reduction in volume, the cost of transportation is also reduced substantially. As an example, let us consider liquefied natural gas for which the major chemical component is methane so that we can assume, for the purpose of this discussion, that liquefied natural gas (LNG) is entirely methane. Gaseous methane at a temperature of 298 K and pressure of 0.1 MPa has a molar volume of 24.7 $dm^3 \cdot mol^{-1}$, while at a temperature of 111 K and pressure of 0.1 MPa the molar volume of the liquid is about 0.038 $dm^3 \cdot mol^{-1}$, that is, about 650 times less than the same amount of substance in the gas phase. Liquefaction is the process whereby a material in the gas phase is converted to the liquid phase.

A gas can be liquefied only at temperatures below the critical temperature (see Question 4.2 and Figure 4.1). At temperatures above the critical temperature, a substance will remain in the gaseous state irrespective of the applied pressure. There are certain substances commonly used as liquefied gases that have a very low critical temperature and examples of these substances with

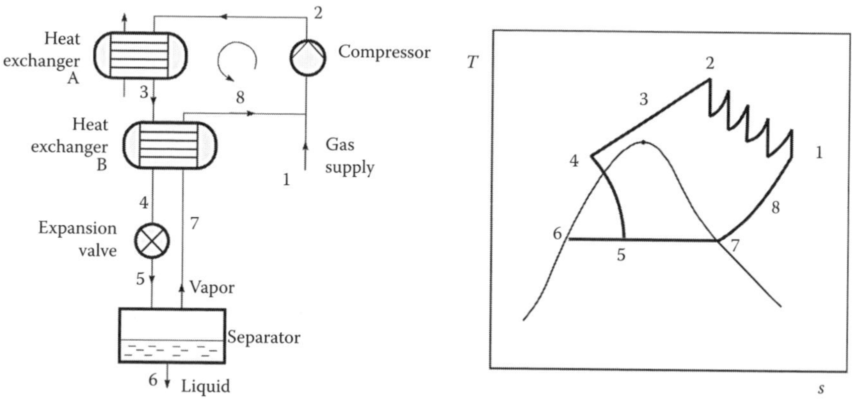

Figure 6.21 Linde liquefaction process.

their critical temperatures T_c are as follows: hydrogen (T_c = 33.145 K), oxygen (T_c = 154.58 K), helium (T_c = 5.1953 K), nitrogen (T_c = 126.19 K), and methane (T_c = 190.56 K). Liquefaction of these gases can be achieved only at temperatures below T_c and these temperatures cannot be obtained with ordinary refrigeration techniques because of the low efficiency and high power (energy) consumption. The most widely used liquefaction cycle is known as the Linde process and it is to this that we now turn.

The Linde process is shown schematically in Figure 6.21 and during this cycle the following steps occur:

Step 1–2: Gas supplied is mixed with the gas (at state 8), which was not liquefied during its pass through the Linde process, and then enters a multistage compressor with intercooling to avoid the compressed gas reaching elevated temperatures and therefore to reduce the power required for compression

Step 2–3: the high-pressure gas is cooled passing through a heat exchanger from state 2 to state 3

Step 3–4: the gas is cooled further by passage through another heat exchanger cooled with low temperature gas discharged from steps 5–7 and 7–8

Step 4–5: the high-pressure gas is throttled through an expansion valve to low pressure and low temperature as a saturated liquid + gas mixture

Step 5–6 and 7: the (liquid + gas) mixture is separated to give a saturated gas (at state 7) and saturated liquid (at state 6) that is removed from the process as the required product

Step 7–8: the low-temperature saturated gas is returned to the start of the Linde process after passing through a heat exchanger that cools the high-pressured gas stream in step 3–4

The Linde process is, for example, used to liquefy natural gas (or LNG) at a temperature of 111.65 K.

6.6 REFERENCES

Burghardt M.D., and Harbach J.A., 1993, *Engineering Thermodynamics*, Harper Collins College Publishers, New York.

Carnot S., 1872, "Reflections sur la puissance motrice du feu et sur les machines propres à développer cette puissance," *Annales scientifiques de l'École Normale Supérieure ser. 2*, **1**:393–457.

Chapter 7

Where Do I Find My Numbers?

7.1 INTRODUCTION

The practical application principles of thermodynamics to any of the fields of science and engineering ultimately depend upon the physical properties of the materials that make up the thermodynamic system discussed in Chapter 1. Some general notions such as the idea of equilibrium, the ultimate efficiency which can be achieved in a heat engine and the description of phase behavior in multi-component systems can be accommodated without recourse to particular materials, but if one wants to build a real machine or design a real separation process properties such a density, enthalpy and entropy of components and mixtures really matter. Those properties of materials that are of concern are collectively known as thermophysical properties and include those characteristic of the equilibrium state (thermodynamic properties) and of the nonuniform state (transport properties) for gases, liquids, and solids. These properties are the subject of considerable international research (e.g., *Experimental Thermodynamics* 1968, 1975, 1991, 1994, 2000, 2003, 2005, and 2010) involving both experimental effort to measure them directly and theoretical effort to provide a sound physical basis for their prediction from first principles or to at least supplement the available experimental information.

In this chapter we seek to set out some of the issues that surround the supply and use of such thermophysical properties; in particular where are the numerical values of material properties best found and how can one assess the reliability of such numbers, their pedigree and how should one proceed if there are no sources of the particular information sought. The chapter is again aimed at a general audience encompassing students engaged in projects to design engineers who are not specialists in the field of thermophysical properties.

7.2 WHAT KIND OF NUMBERS ARE WE SEARCHING FOR?

Before proceeding to the main question of where to find the value of a required property, in this section we will try to specify the type of value we are interested in. That means, what uncertainty should this value have, should it be an internationally agreed upon value, must it be an experimental value or an estimated value?

7.2.1 How Uncertain Should the Values Be?

In any calculation, design, simulation, of any sort, thermophysical properties are required to complete the computation. Before we examine where one can find such numbers, we need first to discuss the uncertainty required of each property. That is, what should be the uncertainty of the property for the calculation required. This point is quite important; even if in many cases, most students are so happy about finding the value of a particular property they do not bother about its uncertainty. We can illustrate this by two examples.

In Figure 7.1 we show a schematic diagram of a typical methanol catalytic reactor. The reaction takes place at a temperature of 610 K and a pressure of 20 MPa. The feed, hydrogen and carbon monoxide, enter at the bottom of the reactor vessel and are preheated to the required temperature by the hot product

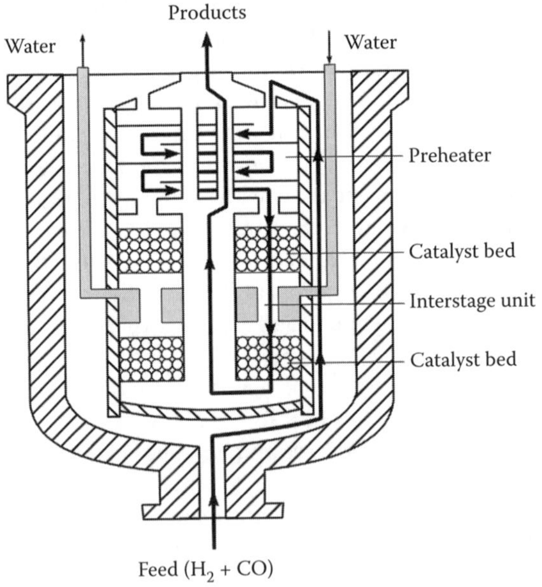

Figure 7.1 Methanol catalytic reactor.

gases. Hydrogen and carbon monoxide react in two catalyst beds according to the simplified reaction

$$2H_2(g) + CO(g) = CH_3OH(g), \tag{7.1}$$

that is, exothermic $\Delta H_m^{\ominus}(610\ K) \approx -226\ kJ \cdot mol^{-1}$; for the reaction $2H_2(g) + C(s) + \frac{1}{2}H_2(g) = CH_3OH(1)$, $\Delta H_m^{\ominus}(298.15\ K) = 238.7\ kJ \cdot mol^{-1}$.

To keep the reaction temperature low a water-cooled interstage unit is employed between the two catalyst beds. This type of interstage heat exchanger is very common when exothermic reactions take place as a means of keeping the temperature low. The area of the interstage heat exchange will be a function of the viscosity η and the thermal conductivity λ of the gases; the area is proportional to $(\eta/\lambda)^{1/3}$ (Kern 1950; Assael et al. 1978). Hence, irrespective of any optimized design procedure employed, if the gas viscosity was underestimated by 20 % and if its thermal conductivity was overestimated by an equal amount, then the area of the interstage heat exchanger will be underestimated by 13 % and the reactor will fail to operate as required. Obviously, the uncertainty of the viscosity and thermal conductivity of the feed and the product gases at $T = 610$ K and $p = 30$ MPa can be quite large.

So does this mean that on average we should aim for an uncertainty of better than ± 20 %? Unfortunately, there is no "general answer" to this question. The answer is directly related to a sensitivity analysis of the uncertainty of the value of the property to the final outcome of the calculation. This is imperative and clearly defines the level of uncertainty that can be accepted.

Furthermore, one ought to remember that it was a small discrepancy in the measurement of the mass of nitrogen that led to the discovery of argon by Lord Rayleigh (Nobel Prize 1904). Lord Rayleigh wrote in 1895 (Rayleigh 1970)

> One's instinct at first is to try to get rid of a discrepancy, but I believe that experience shows such an endeavor to be a mistake. What one ought to do is to magnify a small discrepancy with a view to finding out the explanation; and, as it appeared in the present case that the root of the discrepancy lay in the fact that part of the nitrogen prepared by the ammonia method was nitrogen out of ammonia, although the greater part remained of common origin in both cases, the application of the principle suggested a trial of the weight of nitrogen obtained wholly from ammonia.

In that case the difference in the mass obtained was under 0.5 % (atmospheric nitrogen 2.3102 g, chemical nitrogen 2.2990 g).

7.2.2 Should the Numbers Be Internationally Agreed upon Values?

In some cases the prescribed uncertainty of the required property may not be enough. In Figure 7.2, the Magnox nuclear power plant is shown; Magnox

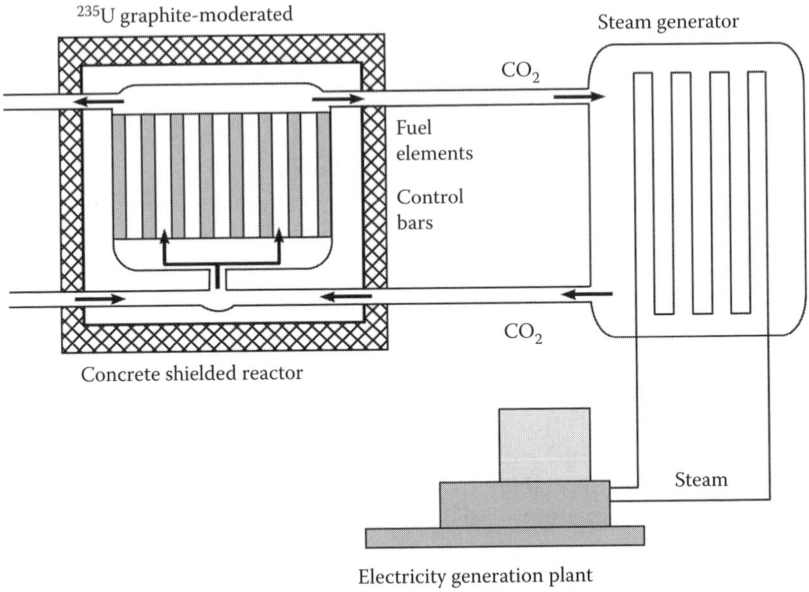

Figure 7.2 Magnox nuclear power plant reactor.

reactors are now obsolete and the name comes from that of the metal alloy* used to clad the fuel rods. The typical Magnox reactor has a diameter of 14 m, is 8 m high, and shielded in thick concrete walls. The enthalpy of reaction is removed from the system by circulating carbon dioxide gas at a temperature of 670 K and a pressure of 2 MPa. This $CO_2(g)$ is transported through the system and then used to heat the steam that drives the turbines in the electricity generation plant. As in the methanol catalytic reactor discussed previously, for the design of the plant it is imperative that the properties of carbon dioxide and steam are known with the required uncertainty. In this case, however, in addition to the design calculations, a very important roles are played by safety calculations, quality assurance, and validation of the plant. For the latter three factors, the uncertainty of the thermophysical properties is insufficient to provide the solution. Values must also be internationally accepted and validated. In the particular case of the Magnox reactor, the properties of steam to be employed are those proposed by International Association for the Properties

* Magnox, which is short for magnesium nonoxidizing, is an alloy formed mostly from magnesium with aluminum and one of its advantageous characteristics, at least for the nuclear power industry, is a relatively small neutron capture cross-section.

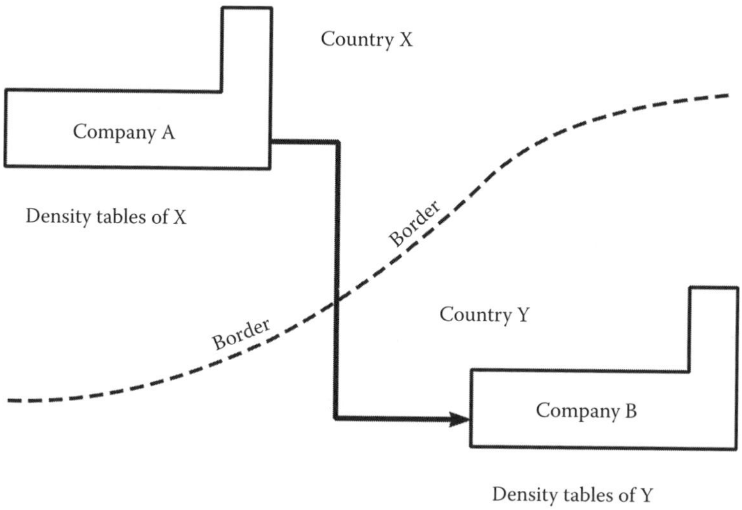

Figure 7.3 Custody transfer.

of Water and Steam (IAPWS), while for carbon dioxide the properties are those proposed by International Union of Pure and Applied Chemistry (IUPAC).

The international dimension of the value of a thermophysical property can be further easily illustrated by considering custody transfer shown in Figure 7.3. We can assume that Company A in Country X sells a fluid to Company B in Country Y. In both cases, the quantity delivered by Company A and the quantity received by Company B is measured by volume but is paid for by mass. Since the same volume crosses the border the options are as follows:

(a) If both countries employ the same density tables, then the mass calculated in both countries is the same and hence payments requested will be equal to payments to be paid.
(b) If, however, different density tables are employed, different masses will be calculated, and clearly payments requested and paid will not agree resulting in a payment dispute.

It is thus evident that in the case of custody transfer, the uncertainty of the density and any correlation used to determine it is not of primary importance but what is important is whether the values are accepted internationally.

Consider, for example, the pipeline from Burghas, Bulgaria, to Alexandroupolis, Greece, which will annually transport about 35 Gkg of crude oil that originates in Russia from the port of Novorossyk. This mass is similar to that transferred through the so-called Trans Alaska Pipeline. If the density of the crude oil originating in Russia is assumed to be $\rho = 900\ \mathrm{kg \cdot m^{-3}}$, the volume of oil transferred annually will be about 38 Mm^3 (about $244 \cdot 10^6$ barrels).* At an oil price of \$629 m^{-3} (\$100 bbl^{-1}),* this is equivalent to \$24.4 G (24 billion USD); a convenient list of unit conversion factors is provided at http://physics.nist.gov/Pubs/SP811/appenB8.html#top. Hence, a difference of 1 % in density of the crude oil used by Company A and Company B, shown in Figure 7.3, will result in a \$0.24 G difference, and presumably a dispute with potentially at least legal ramifications if not more! Based solely on this one example it is not difficult to see the importance of internationally accepted values.

A similar argument can be put forward in the case of technology transfer. A process or plant developed in one country and sold to another must meet detail specifications and design methodology that will also include the data used for the engineering calculations. An example of this fact is provided by returning to discuss the Magnox nuclear power reactor shown in Figure 7.2. In this case, internationalization of the thermophysical properties of water and steam was recognized as highly significant to the generation of electricity from steam-driven power plants; the properties of steam are an essential part of the design as well as form the basis for estimating the energy efficiency of the system that will ultimately be compared with measurements, albeit too late by that stage for major changes because the generator has been designed, constructed, and commissioned. It was usual for each country to have their own values for the thermophysical properties of steam that are often referred to as steam tables. The measurements that underpinned these steam tables were combined, in some cases complimented with new measurements, and then fit by a correlation all under the auspices of the IAPWS. This organization has spent more than 40 years developing what are now internationally accepted steam tables and correlations otherwise known as formulations.

A final point an engineer will almost certainly be called upon to consider arises from quality assurance, that is, the requirement to satisfy regulatory requirements imposed for safety and environmental reasons. These may be imposed by a National Regulatory body or an international organization. The requirements of these organizations must be satisfied; in some cases national regulatory bodies, perhaps for the purpose of trade, comply with regulations of other nations. Quality assurance of a plant or a process can often require a demonstrable pedigree for each number used in the design calculations, one example is the calculation of the energy (heat) transfer that would be required during a meltdown of a nuclear reactor.

The discussion above clearly demonstrates that in such cases the user must search for internationally accepted thermophysical data, which is data that are used by the majority of the world as a basis for trade, regulation, or

* 1 U.S. barrel of liquid contains 42 U.S. gallons that is equivalent to 0.159 m^3 as provided in http://physics.nist.gov/Pubs/SP811/appenB8.html#B

standardization. This refers to supranational bodies that propose such standards. Such bodies include the following:

- International Association for the Properties of Water and Steam (IAPWS)
- International Association for Chemical Thermodynamic (IACT)
- International Association for Transport Properties (IATP)
- International Union of Pure and Applied Chemistry (IUPAC)

International accepted values or standard or reference values can be found in reference journals or textbooks concerned with reference data, for example, the *Journal of Physical and Chemical Reference Data* to name but one.

7.2.3 Should I Prefer Experimental or Predicted (Estimated) Values?

Having discussed the uncertainty associated with property values as well as the international dimension, one obvious question that can arise is whether the reader should be looking specifically for experimental values or for predicted ones? The answer to this question is relatively easy.

Let us assume that we have a need to measure only 10 properties at just 10 temperatures and 10 pressures, for 15 pure fluids and all their mixtures, at 5 compositions in the liquid and gas phases; we will assume there are no values reported in primary tables of the standard equilibrium constant and molar enthalpy of formation that would provide a means of determining the required properties. The total number of measurements required is $3.3 \cdot 10^8$ $(10 \cdot 10 \cdot 10 \cdot 32{,}766 \cdot 5 \cdot 2)$. If one further assumes that three measurements can be obtained for each normal 8 h working day and that a person works for 48 weeks (or 240 days per year) then the number of years the task of measurements requires is about 457,000; alternatively one might employ 457,000 people working for 1 year. In view of this estimate, it is rather obvious that we cannot rely solely on measurements. In reality, some of the required values can be reliably estimated at least for most purposes from primary tables of standard thermodynamic properties perhaps when combined with data from secondary tables; these have been discussed in Chapter 1. These values can be used because they have been validated and checked before publication and relate the properties required as described in Chapter 1, Question 1.8 and in Chapter 4; these tables are maintained, for example, by the Thermodynamic Research Center now located at the National Institute of Standards, which also maintain the JANAF tables; JANAF is the acronym for Joint Army Navy and Air Force.

If one ends up searching the archival literature and is indeed fortunate to find measured values of the required property then the question arises, should we trust it? Unfortunately nothing is that simple. There are, of course, just as with every human endeavor, good and bad measurements, and the fact that a measurement exists does not imply that the value is correct. Of course, the measurement can be evaluated, for example, to be deemed consistent with other data and discussion of this point will be left for Question 7.4. Instead, let us assume (as is generally the case for most systems of engineering interest) that there are no measurements of the required property, and we must then resort to a method of estimation.

Let us consider the case of low-density transport properties. In a low-density gas, the diffusion of a group of molecules in the gas will characterize the mass, momentum, and heat transport, and consequently the diffusion, the viscosity, and the thermal conductivity coefficients, respectively. In this case, kinetic theory is well defined and although its mathematical formulation is complicated, in 1950s a team of 30 clerical assistants armed solely with mechanical calculators succeeded in determining the transport properties of monatomic gases from the Lennard-Jones potential (see Chapter 1, Question 1.4.3.3). That is, an intermolecular potential was combined with kinetic theory to calculate the transport properties. In the 1980s for low-density monatomic gases, theoretical progress permitted the inverse procedure (Maitland et al. 1981); measurements of one property were used to determine the intermolecular potential. From any intermolecular potential all thermophysical properties (thermodynamic and transport) at many temperatures and pressures can be calculated. It is unfortunate that this kind of approach is still restricted to monatomic and simple molecular gases at low density. However, as we discussed in Chapter 1, transport properties depend upon the intermolecular potential and for monatomic gases this is a function of both length and energy scaling parameters. From these so-called scaling parameters we obtain a corresponding-states procedure whereby the transport properties of monatomic gases can be obtained (Maitland et al. 1981). This indeed is an excellent example where a few precise measurements have been combined with theory based firmly on the principles of physics and have then permitted development of a procedure by which many properties at different conditions can be predicted. Values obtained from such procedures are usually found to differ insignificantly from the measured value, at least from the normal requirements, for uncertainty imposed by engineering calculations (see Chapter 2).

In summary, during the quest for the value of a specific thermophysical property, measurements can sometimes be available. If the measurement satisfies the criteria of quality laid down for the experimental technique then the measurement results are preferred. In the absence of such measurements, predicted values should be sought but that does not absolve the user from the obligation to conduct an assessment of the uncertainty of the values so obtained.

7.3 IS THE INTERNET A SOURCE TO FIND ANY NUMBER?

Having established what kind of data we require and with what uncertainty the next question is where can one find these data? The answer for all such questions today, for many people, seems to be the Internet. In the following subsections we will try to investigate this answer by examining the following plausible sources: (a) web pages, (b) archival scientific and engineering journals, and (c) encyclopedias and compilations.

7.3.1 What about Web Pages?

To search the apparently infinite number of web pages that exist today, search engines are employed. The most common and probably the most powerful one is the Google Search Engine. This is powered by the PageRank technology, which was developed at Stanford University by Larry Page (hence the name *Page*Rank http://en.wikipedia.org/wiki/Page_rank) and later by Sergey Brin as part of a research project about a new kind of search engine. The project started in 1995 and led to a functional prototype, named Google, in 1998.

It is interesting to look briefly into this technology. PageRank reflects the importance of web pages by considering more than 500 million variables and 2 billion terms. Pages that are believed to be important receive a higher PageRank and are more likely to appear at the top of the search results. PageRank also considers the importance of each page that casts a vote, as votes from some pages are considered to have greater value, thus giving the linked page greater value. The search engine also analyzes the full content of a page and factors in fonts, subdivisions, and the precise location of each word.

What all these mean in essence is that it searches for popular pages where any of your key words appear but not necessarily all of them! Hence the importance of the results of a scientific search is quite small. Let's demonstrate this by a simple example. For this, our search will be for the viscosity of decane with a preference for measurements. The following results were obtained according to the key words given:

(a)	Viscosity decane	:	310,000 results
(b)	"viscosity of decane"	:	581 results
(c)	"viscosity of *n*-decane"	:	1,830 results
(d)	"viscosity of *n*-decane" + measurements	:	1,160 results

The two words without quotations imply that we are looking for web pages where at least one of them appears. The result is quite useless. Words inside quotations force the search engine to look for exactly this combination of words, while the "+" sign in front of a word requires a search for exactly this word, excluding synonyms. Our search for measurements for the viscosity of decane was thus restricted to 1,160 results although, of course, the number obtained will vary as a function of time. These 1,160 results included

- Abstracts of scientific journals, which were useful, but not available through the search engine.
- One article from the USA Department of Energy that included a useful correlation.
- Irrelevant abstracts of scientific journals, not available through the search engine.
- Irrelevant web pages.
- A reference to a book on viscosity.

Hence, from the initial 310,000 results, no real relevant answer could be found. Of course it is not always like this. If one looks for very common properties such as, the density of water at 20 °C, even if the words are inside quotations, 8,300 results appear. Furthermore, more or less, all contain the correct answer.

Perhaps, the question really is, how often do we need the "density of water at a temperature of 20 °C?" The answer is, not very often. What we do need is properties of not such common fluids under not such common conditions. These results are not easy to find within the internet.

7.3.2 What about Encyclopedias and Compilations (Databases and Books)?

In addition to web pages, the internet hosts encyclopedias and compilations such as either databases or books. We have to distinguish between these two.

The most interesting example of a web encyclopedia today is Wikipedia (http://www.wikipedia.org/). Wikipedia is a free, multilingual, open content encyclopedia operated by the United States-based nonprofit Wikimedia Foundation. Its name is a portmanteau of the words *wiki* (a technology for creating collaborative websites) and *encyclopedia*. Launched in 2001 by Jimmy Wales and Larry Sanger, it attempts to collect and summarize all human knowledge in every major language. As of April 2008, Wikipedia had over 10 million articles in 253 languages, about a quarter of which are in English. Wikipedia's articles have been written collaboratively by volunteers around the world, and nearly all of its articles can be edited by anyone with access to the Wikipedia web site (http://en.wikipedia.org/wiki/Wikipedia - cite_note-7). Having steadily risen in popularity since its inception, it is currently the largest and most popular general reference work on the internet.

Although the growth of Wikipedia is amazing, it is the principle by which this growth is achieved that is of concern. Anybody can contribute to Wikipedia by creating an account, which means that specific knowledge is not really checked; it only reflects the opinion and the knowledge of the writer, who is not necessarily a professional or a well-known scientist. Hence, for specific data, care must be taken, and values obtained from Wikipedia: should be traced to the original source if available, or double checked with another source.

A source of data that has been compiled and reviewed by leading experts for the thermodynamic and transport properties of gases, liquids, and solids is that which is now known as Kaye and Laby. This was originally published in 1911 as a text book entitled "Tables of Physical and Chemical Constants." The last printed edition was the sixteenth, which was published in 1995. Kaye and Laby is now available online at http://www.kayelaby.npl.co.uk/. The reader interested in fluid phase equilibrium calculations for (vapor + liquid) that relate to phase behavior require critical temperature, pressure, vapor pressure, and acentric factors will find this source invaluable. Calculations of the equilibrium between vapor and liquid phases are essential in a number of areas and the acronym VLE is used routinely in the field and is shorthand that the reader will often encounter. We shall use the acronym here in what follows for the same reason.

Finally, we should also mention e-books, and these certainly have great value, providing full reference to scientific papers. An example is the virial coefficients of gases (Dymond et al. 2002) and gaseous mixtures (Dymond et al. 2003) also available in an e-book (solely for purchase).

7.3.3 What Software Packages Exist for the Calculation of Thermophysical Properties?

A number of software packages claim to calculate or predict the thermophysical properties of fluids and much of this work has been conducted by the National Institute of Standards and Technology (NIST), USA, and in the subquestions of this question we list a few examples.

7.3.3.1 What Is the NIST Thermo Data Engine?

NIST Standard Reference Database 103a available from http://www.nist.gov/srd/nist103a.htm for pure fluids and NIST Standard Reference Database 103b available from http://www.nist.gov/srd/nist103b.htm for mixtures. These data provide about 50 properties for pure fluids (Database 103a) and about 120 properties for mixtures (Database 103b), including density, vapor pressure, heat capacity, enthalpies of phase transitions, critical properties, melting and boiling points, and so on. It fills the gaps in experimental data by deployment of automated group-contribution and corresponding-states prediction schemes and most of all emphasizes the consistency between properties (including those obtained from predictions), and provides for flexibility in selection of default data models depending on the particular data scenario. The Thermo Data Engine supports several equations of state for pure compounds (original and modified volume-translated Peng-Robinson, Sanchez-Lacombe, PC-SAFT, and Span-Wagner) and allows the user to fit parameters to experimental and predicted data. Enthalpies of formation are evaluated on the basis of stored experimental enthalpies of combustion and the modified Benson group-contribution method.

7.3.3.2 What Is the NIST Standard Reference Database 23, REFPROP?

The NIST Standard Reference Database 23 is commonly known by the acronym REFPROP and provides estimates of the thermophysical properties of pure fluids and mixtures and is available from http://www.nist.gov/srd/nist23.htm. REFPROP employs correlations or models that represent experimental data. It includes 84 pure fluids, 5 pseudo-pure fluids (such as air) and mixtures with up to 20 components (natural gas, hydrocarbons, refrigerants, alternative and natural refrigerants, air, noble elements, and many predefined mixtures). The properties calculated are as follows: density, energy, enthalpy, entropy, C_V, C_p, sound speed, compressibility factor, Joule-Thomson coefficient, quality, 2nd and 3rd virial coefficients, Helmholtz function, Gibbs function, heat of vaporization, fugacity, fugacity coefficient, K value, molar mass, thermal conductivity, viscosity, kinematic viscosity, thermal diffusivity, Prandtl number, surface tension, dielectric constant, isothermal compressibility, volume expansivity, isentropic coefficient, adiabatic compressibility, specific heat input, exergy, and many others. REFPROP incorporates "high accuracy" Helmholtz function and MBWR equations of state, including many international standard equations, the Bender equation of state for several of the refrigerants, an extended corresponding-states model for fluids with limited data, an excess Helmholtz function model for mixture properties, while experimentally based values of the mixture parameters are available for hundreds of mixtures. Finally, predictions of both viscosity and thermal conductivity are provided by fluid-specific correlations (where available): a modification of the extended corresponding-states model, or the friction theory model. Because the compilation was created by NIST, which is a governmental agency, and full reference to the original scientific journals are given, this compilation should be an excellent source for data for the purposes of both science and engineering. However, no program is always correct and variations in the properties predicted can be obtained from different versions of the program. For example, albeit an extreme test, the viscosity of gaseous H_2S was calculated from two versions of REFPROP under two different sets of conditions and the results obtained are listed in Table 7.1 together with a recent measurement reported in the archival literature. The calculated values differ from experiment by between −20 % and 11 %.

TABLE 7.1 PREDICTED AND EXPERIMENTAL VALUES OF THE VISCOSITY η OF $H_2S(g)$

	η/μPa s	
	$T = 273.15$ K, $p = 50$ MPa	$T = 273.15$ K, $p = 100$ MPa
REFPROP v.7.0	253.3	318.7
REFPROP v.8.0	201.7	242.5
Exp. (Gallieto and Boned 2008)	213.7	272.8

7.3.3.3 What Is the NIST Standard Reference Database 4, SUPERTRAPP?

SUPERTRAPP (available from http://www.nist.gov/srd/nist4.htm) is an interactive computer program to predict thermodynamic and transport properties of pure fluids and fluid mixtures containing up to 20 components. The components are selected from a database of 210 substances, mostly hydrocarbons. Properties that can be calculated include the following: density, compressibility factor, enthalpy, entropy, heat capacity, sound speed, Joule-Thomson coefficient, as well as, viscosity and thermal conductivity. Features include bubble and dew-point pressure or temperature calculations, flash calculations $\{(T, p), (T, S),$ and $(p, H)\}$, saturation properties for pure components and mixtures.

7.3.3.4 What Is the NIST Chemistry Web Book?

The NIST Chemistry web book (available from http://webbook.nist.gov/) is free and includes the following: thermochemical data for 7,000 organic and inorganic compounds (enthalpy of formation, enthalpy of combustion, heat capacity, entropy, phase transition enthalpies and temperatures, vapor pressure); reaction thermochemistry data for more than 8,000 reactions; infrared spectra for more than 16,000 compounds; mass spectra for more than 15,000 compounds; ultraviolet and visible spectra for more than 1,600 compounds; gas chromatography data for more than 27,000 compounds; electronic and vibrational spectra for more than 5,000 compounds; constants of diatomic molecules (spectroscopic data) for more than 600 compounds; ion energetics data for more than 16,000 compounds; and, thermophysical property data for 74 fluids at the time of writing this.

7.3.3.5 What Is the DIPPR Database 801?

The Design Institute for Physical Property Data (DIPPR) provides a database (available from http://dippr.byu.edu/) that contains evaluated thermodynamic and physical property data for process engineering. It is supported by the American Institute of Chemical Engineers (AIChE) and is run by Brigham

Young University, USA. DIPPR contains 49 thermophysical properties for 2,013 industrially relevant compounds. It also includes 15 temperature-dependent properties; contains raw data from the literature; contains critically evaluated, recommended thermophysical values; and predicts appropriate values when experimental chemical data are not available.

7.3.3.6 What Is the Landolt-Börnstein?

The Landolt-Börnstein database for pure substances incorporates the 400 Landolt-Börnstein volumes that include 250,000 substances and 1,200,000 citations available with a single keystroke. Marketed as "the world's largest resource for physical and chemical data," SpringerMaterials—The Landolt-Börnstein Database (http://www.springer.com/librarians/e-content/springer materials?SGWID=0-171102-0-0-0/) brings the print collection's content into one easy-to-access online platform (with 91,000 online documents and 3,000 properties). The core of the database is two-fold; first, it employs a user interface with a search engine, and, second, it makes the content findable. Users can search in several ways: with a Google-like search box, an advanced search tab that creates a Boolean search term automatically as the user sets up the parameters, or a color-coded periodic table.

7.3.3.7 What Is NIST STEAM?

STEAM (Harvey et al. 2008) is a computer package for the calculation of the properties of water and steam. The STEAM package employs the latest correlations developed by IAPWS (http://www.iapws.org/) for water and steam. As such they are standard values and their uncertainty is the one quoted by IAPWS.

7.3.4 How about Searching in Scientific and Engineering Journals?

The most serious source for property values is the scientific journals where those values are first published. Today the retrieval of information from scientific journals is very easy. The two most commonly used such search engines are as follows: (1) SciFinder, obtained from the Chemical Abstract Service of the American Chemical Society; (2) Scopus, an abstract and citation database of peer-reviewed literature; and (3) the Web of Science, a Thomson Reuters citation database.

These can be easily used provided the users institution is registered, the paper can be made to appear directly on the screen. Just to demonstrate their use, in the search for the viscosity of decane, where 12 web pages were found, each engine produced about 120 different papers, from which at least half of them had measurements. Hence, this is certainly the easiest method of locating values for thermophysical properties.

7.4 HOW CAN I EVALUATE REPORTED EXPERIMENTAL VALUES?

To evaluate the experimental data that one finds in literature, it is imperative to recognize that not all experimental values are of equal worth. The field of thermophysical properties, and particularly transport properties, is littered with examples of quite erroneous measurements made, in good faith, with instruments whose theory was not completely understood. It is therefore always necessary to separate all of the experimental data collected during a literature search into primary and secondary data by means of a thorough study of each paper.

Data with the lowest attainable uncertainty (e.g., density with a fractional uncertainty of ±0.001 % discussed in Question 7.4.1.1) can be used in developing correlations. These data must satisfy the following conditions:

(1) The measurements will have been carried out in an instrument for which a complete working equation is available together with a complete set of corrections;
(2) The instrument will have had a high sensitivity to the property to be measured; and
(3) The primary, measured variables will have been determined with high precision.

Occasionally, experimental data that fail to satisfy these conditions may be included in the primary data set if they are unique in their coverage of a particular region of state and cannot be shown to be inconsistent within theoretical constraints. Their inclusion is encouraged if other measurements made in the same instrument are consistent with independent, nominally lower uncertainty data. Secondary data, excluded by the above conditions, are used for comparison only.

In the following sections an attempt to critically evaluate the different measuring techniques will be presented.

7.4.1 What Are the Preferred Methods for the Measurement of Thermodynamic Properties?

Thermodynamics interrelates measurable physical quantities (see Questions 3.4 and 3.5). More generally, the physical properties of interest are called *thermophysical properties*, of which a subset are thermodynamic properties, which pertain to the equilibrium states and another subset are transport properties that refer to dynamic processes in nonequilibrium states. In the remainder of Question 7.4 information is provided regarding the methods that are used to measure both thermodynamic and transport properties. Although this book is mostly concerned with thermodynamics we have included a discussion of methods used to determine transport properties because these are required in the complete analyses of real systems that are not at equilibrium and are illustrated by the example in Question 7.2.1.

Here we continue the Question posed in 1.8, which included methods used to determine temperature, pressure, enthalpy, heat capacity, and energy, and extend our discussion to density, vapor pressure, critical properties, sound speed, viscosity, thermal conductivity, and diffusion. The methods included in our discussion are those for which complete working equations are available and have been discussed elsewhere in the series *Experimental Thermodynamics* (Vol. I 1968, Vol. II 1975, Vol. III 1991, Vol. IV 1994, Vol. V 2000, Vol. VI 2003, Vol. VII 2005, Vol. VIII 2010). In Question 1.8 we introduced the concept of uncertainty, and in this section we emphasize that measurements must have a quantifiable uncertainty so that properties deduced from them can be used in effective engineering design. For example, the design of an effective and efficient air conditioning system that performs within a set of specifications (boundary conditions). It is with these criteria in mind that we provide the methods that are preferred for the measurements of thermophysical properties.

7.4.1.1 How Do I Measure Density and Volume?

Density (and volume) has appeared repeatedly in the questions of Chapters 3 and 4, and the density $\rho_{\mathrm{B}}(p,T)$ of substance B is defined by

$$\rho_{\mathrm{B}}(p,T)=\frac{m_{\mathrm{B}}}{V(p,T)}, \tag{7.2}$$

where m_{B} is the mass of substance B contained within a volume $V(p,T)$. From Equation 7.2 it would at first sight seem that the density should be rather simple to measure, particularly given the ease with which mass can be determined with a relative uncertainty of $<\pm 0.001$ %. However, the measurement of volume required with Equation 7.2 is rather more taxing except when volumes with particular geometry are used. The volume of a densimeter can be obtained as a function of temperature from dialatometry with, for example, mercury and dimensional microwave measurements or more usually by calibration with a fluid (for which p, V, T) is known, for example, water. If the variation $V(p)_T$ is required this can be estimated with auxiliary methods such as combining the zero-pressure characteristic dimension at a pressure with the compliance of the wall evaluated using reliable values of the elastic constants of the material used to construct the cell. To avoid unnecessary expenditure of both effort and cost, it is imperative that the uncertainty with which the density is required be determined in advance. This will permit a method to be selected, which yields the appropriate uncertainty.

Measurements of the density of gas, liquid, and solid have been discussed in both *Experimental Thermodynamics Volume II* and *Volume VI*. Here we briefly outline the method of determining density from vibrating objects, piezometers, isochoric methods, Archimedes principle, and the use of silicon spheres and absolute density standards by combining both optical and mass metrology, and providing a precision in density measurement that was hitherto unforeseen.

7.4.1.1.1 Vibrating Bodies

Vibrating devices have become the instrument of choice for both routine as well as precision density measurements over a relatively wide range of temperature and pressure. These methods have also been adapted for monitoring commercial processes and fluid custody transfer. In this category of densimeter there are two important laboratory instruments and these are the vibrating tube and the vibrating wire. In the vibrating-tube technique, the density is deduced from the resonant frequency of a U-shaped tube containing the sample fluid where the fluid sample is a part of the vibrating system affecting directly its mass and, thus, also its resonant frequency. A typical vibrating-tube densimeter consists of a hollow metallic or glass thin-walled tube bent in a "U" or "V" shape and firmly clamped in a block which is, itself, fixed to a large mass to reduce the effect of recoil and isolating the tube from external mechanical perturbations. The geometrical complexity of the practical U-tube means that the densimeter requires calibration with one or more fluids for which $\rho(p,T)$ are known.

The vibrating-wire densimeter is essentially a hydrostatic weighing densimeter in which a vibrating wire has a mass suspended from it and the apparent weight of a sinker immersed in the sample fluid is determined from variations of the resonant frequency. These arise from changes in wire tension owing to the buoyancy of the mass within the fluid; the buoyancy force exerted on the sinker by the surrounding fluid reduces the tension of the wire and, thus, lowers its resonant frequency from that observed under vacuum. The geometrical simplicity of the vibrating wire has permitted the development of a working equation, that is, firmly based in fluid mechanics that relates the measured complex resonance frequency to density and viscosity. In principle, the density can be calculated directly from the theory. The vibrating wire continues to perform as a viscometer because of the damping effect, and, logically, the introduction of the buoyancy device has no effect on the results of the fluid-mechanical theory. Since the working equations require both density and viscosity this approach is rather attractive because the instrument is capable of measuring density and viscosity with an uncertainty on the order of ±0.1 % and ±1 %, respectively.

The vibrating-wire densimeter can operate at pressures >100 MPa and at temperatures up to 473 K, while the vibrating tube has been used at temperatures up to 723 K and without pressure compensation to pressures of about

50 MPa because of tube deformation; with pressure compensation, pressures on the order of 1 GPa have been attained.

For a vibrating-tube densimeter a working equation for a straight tube is clamped at both ends and filled with fluid and surrounded by either another fluid or vacuum; this analysis assumes the fluid within the tube does not flow, and, thus, the viscosity of the fluid is neglected. If negligible internal damping is assumed within the metallic U-tube and if it is surrounded by vacuum then the density of the fluid contained within is obtained from the measured frequency f with the expression

$$\rho(p,T) = \frac{K(p,T)}{f^2(p,T)} + L(p,T). \tag{7.3}$$

In Equation 7.3, K and L are parameters determined through calibrations with two reference liquids of known density, such as water and nitrogen, or with one liquid of known density, for example, water and with vacuum. Density with an uncertainty of $<\pm 0.1$ %, and in some case ± 0.001 %, can be obtained from the resonant frequency of a vibrating U-tube densimeter when combined with Equation 7.3. Typically, the calibration is performed with fluids that have viscosity <1 mPa · s, and the error arising from neglecting viscosity in the working equations when the tube is used as a densimeter at another viscosity must be determined empirically. For an Anton-Parr model 512P U-tube densimeter the correction to density for fluid viscosity is given by

$$\frac{\Delta\rho}{\rho} = \left[-0.5 + 0.45(\eta/\mathrm{mPa \cdot s})^{1/2}\right] \cdot 10^{-4} \tag{7.4}$$

and is subtracted from Equation 7.3.

7.4.1.1.2 Piezometers

There are three categories of piezometer and these are as follows: (1) devices that measure the mass or amount of substance contained within a volume and conform to Equation 7.2; (2) measurements of the change in pressure effected by a change in volume; and (3) devices that utilize one or more expansions from one volume to another. Item 2 will be discussed in the section concerned with bellows volumometers. For item 3, the sample is expanded from volume V_1 into a second volume V_2 (i.e., usually evacuated before the expansion), and the ratio of the original volume to the final volume establishes the ratio of densities before ρ_1 to that after the expansion ρ_2 through

$$\frac{\rho_1}{\rho_2} = \frac{V_1 + V_2}{V_1} = r, \tag{7.5}$$

where r is the so-called cell constant.

7.4.1.1.3 Bellows Volumometer

The bellows, which separates the substance to be compressed from the hydraulic fluid used for pressurization, prevents contamination of the fluid and transmits the applied pressure to it with only a minimal pressure loss. The compression of the fluid is determined from the linear motion of the end of the bellows with applied pressure. This approach has been used for fluid mixtures. The volume $V(p)_T$ of a bellows is determined from the volume $V(p_r)_T$ of the bellows at the same temperature T and a reference pressure p_r, typically about 0.1 MPa, and the variation of the bellows area $A(p)_T$, and a length $l(p_r)_T$ with the expression

$$V(p)_T = V(p_r)_T + \int_{p_r}^{p} A(p)_T \mathrm{d}l. \tag{7.6}$$

The area $A(p)_T$ and a length $l(p_r)_T$ can usually be determined from measurements with a fluid for which $V(p)_T$ is known.

7.4.1.1.4 Isochoric

In an isochoric densimeter a previously evacuated vessel is filled with a known mass of substance to a desired pressure then sealed and placed within a thermostat usually at the highest temperature of the proposed measurements. The temperature and pressure are measured once equilibrium is attained. The temperature is then changed (usually reduced) and the pressure and temperature determined. This process is repeated until a preset pressure or temperature is reached. The density is then obtained from Equation 7.2 at each (p,T) from m and $V(p,T)$. If $V(p,T)$ remained constant then the measurement would be isochoric. In practice, the finite elastic constants of the wall material mean that the method is almost isochoric and designated a pseudo-isochore. Isochoric methods are advantageous because the substance is contained at all times and the mass reduced only to permit measurements at a different, usually lower, isochore. Consequently, isochoric measurements are used for potentially hazardous substances. The $V(p,T)$ is measured and if necessary calculated at another temperature and pressure from knowledge of the thermal expansion and the mechanical deformation under the pressure of the material.

7.4.1.1.5 Buoyancy Densimeters

The density of fluids can be determined using the buoyancy method, which is based on Archimedes' principle. This principle states that the upward buoyant force exerted on a body (called *buoy, float,* or *sinker*) immersed in a fluid is exactly equal to the weight of the displaced fluid. For density measurements over wide ranges of temperatures and pressures four main types of buoyancy densimeters have been used: hydrostatic balance densimeters, magnetic float and magnetic suspension densimeters, hydrostatic balance densimeters in combination with the magnetic float or magnetic suspension method, and hydrostatic balance densimeters in combination with magnetic suspension couplings.

The basic principle of the hydrostatic balance densimeter is an object called a *sinker*, which is typically a sphere or cylinder, fabricated from glass or metal, is suspended by a platinum wire on the weighing hook of a commercial analytical balance. The sinker is immersed in the fluid that is contained within a thermostated container and the apparent loss in the true weight of the sinker is equal to the weight of the displaced liquid. Thus, the density of the sample liquid can be calculated by the simple relation

$$\rho = \frac{m_s - m_s^*}{V_s}, \tag{7.7}$$

where m_s is the "true" mass of the sinker, m_s^* is the "apparent" mass of the sinker immersed in the sample liquid, and V_s is the volume of the sinker. The volume V_S of the sinker can be determined by measuring its dimensions or by hydrostatic weighing in water; and the mass of the sinker can be determined by weighing in air. Hydrostatic balances with magnetic suspension couplings can be used to determine density with a fractional uncertainty on the order of 10^{-6}.

7.4.1.1.6 *Absolute Measurements of Density*

Absolute measurement of density require traceability to standards of mass and length and, in practice, includes measurements of volume of a solid object that can be related to the length standard with a small uncertainty. A cube has been used for the solid object; however, a sphere is preferred because it can be fabricated with sufficient sphericity that the volume can be calculated from the mean of optical measurements of the radius over all directions and the density of a single-crystal silicon obtained from the measurement of mass. The density of another silicon sphere can then be determined with a relative uncertainty of $<\pm 10^{-6}$ by the floatation method. In a magnetic suspension densimeter single-crystal silicon can be used as the sinker with either a spherical or cylindrical geometry to determine the density of liquids with a relative uncertainty of about $\pm 4 \cdot 10^{-6}$; that is about a factor of 10 greater because the thermal expansion coefficient of an organic liquid is much greater than that of silicon.

7.4.1.2 How Do I Measure Saturation or Vapor Pressure?

There are several methods that can be used to measure vapor pressure (defined in Question 4.2 and Figure 4.1) and the exact choice depends on the pressure range of interest relative to the critical pressure, thermal stability, and volume of substance available. Some, but by no means all, of these methods will be mentioned here. A rather more extensive list of methods along with details of their practical implementation can be found in *Experimental Thermodynamics Volume VII*

(Weir and de Loos 2004). Measurements of the vapor pressure about the normal boiling temperature can be represented by Equation 4.19 or over a slightly wider temperature range about the normal boiling temperature by Equation 4.20, while the whole vapor pressure curve can be fit by Equation 4.21.

The so-called static methods, which consist of a pressure vessel housed within a thermostat with a means of measuring the pressure of the gas phase, is routinely used to determine the saturation (or vapor) pressure. However, this approach is not optimum at ≤0.1 MPa because of the effect of impurities on the measured pressure. Air, which can be regarded as a typical impurity with a low normal boiling temperature, increases the measured vapor pressure over that of the pure compound, while impurities with a normal boiling temperature higher than the substance of interest, decreases the measured vapor pressure from the true value. These systematic differences become increasingly important as the vapor pressure decreases, and the static method is then replaced by a method where the fluid is boiled, and is known as ebulliometry, that continuously degasses the sample. One method of practical advantage particularly at low pressure (<1 MPa) is comparative ebulliometry, where the condensing temperature of the substance and a reference fluid (typically water) of known vapor pressure are boiled in separate containers maintained at the same pressure, which removes the requirement to measure pressure directly; instead the pressure is determined from the condensing temperature combined with $p^{\text{sat}}(T)$ of the reference fluid. Irregular condensing temperatures may reveal inadequacies of fluid purity except for azeotropes. The apparatus symmetry with two thermometers means that errors tend to be self-cancelling. The two boilers are interconnected through cold traps to avoid cross-contamination and are maintained at the same pressure with a buffer gas when the pressure is maintained by either a ballast volume or a pressure controller. Comparative ebulliometry constructed from appropriate materials has been used at pressures close to the critical. The lower pressure limit of ebulliometry is on the order of 1 kPa and is determined by the requirement for a steady boiling process that places significant demands upon both heat and mass transfer.

At lower pressures (<1 kPa) the Knudsen effusion method permits determination of vapor pressure p^{sat} by means of the measurement of the mass loss through an orifice of area A into a vacuum system with the relationship

$$p^{\text{sat}} = \frac{\Delta m}{kAt}\left(\frac{2\pi RT}{M}\right)^{1/2}, \tag{7.8}$$

where Δm is the change in mass of the sample in time t, k is the Clausing probability factor, R is the gas constant, T is the temperature, and M is the molar mass. The mass loss can be determined from the change in resonance frequency of a quartz crystal microbalance.

An alternative to Knudsen effusion is the method of transpiration in which an inert gas flows, at a rate sufficiently low to attain equilibrium, through a thermostatically controlled saturator packed with the pure substance for which the vapor pressure is to be determined. The substance is then trapped with sorbents or cryogenic traps and the amount of substance moved is determined. Assuming Dalton's law of partial pressure applies to the carrier gas saturated with the substance B the vapor pressure p_B^{sat} is, assuming an ideal gas, given by

$$p_B^{sat} = \frac{m_B R T_t}{V_t M_B}. \tag{7.9}$$

The transpiration method is similar to the process used in analytical chemistry of gas-liquid chromatography (GLC), and indeed GLC is used to determine vapor pressure by so-called headspace analysis. Indirect chromatographic methods of determining vapor pressure, particularly at relatively low vapor pressure, include the use of retention times that requires the use of calibrants for which $p^{sat}(T)$ is known. Assuming $pV_m = RT$ the vapor pressure is obtained from

$$p^{sat} = \frac{RT}{V_B M_B f_B^{\infty}}, \tag{7.10}$$

where f_B^{∞} is the activity coefficient in a dilute gaseous mixture in the limit as $x_B \rightarrow 0$ often referred to as the "infinite dilution activity coefficient"; for a binary it is defined by Equation 4.138.

Calorimetric determinations of the enthalpy of vaporization $\Delta_l^g H_m$ (or in IUPAC terminology $\Delta_{vap} H_m$) can be used with Clapeyron's equation (Equation 4.15)

$$\frac{dp^{sat}}{dT} = \frac{\Delta_\alpha^\beta H_m}{T\Delta_\alpha^\beta V_m}, \tag{7.11}$$

or with simplifications discussed in Question 4.2 to determine $p^{sat}(T)$. Thus, adiabatic, drop, differential, and scanning calorimetry can all be used to determine $p^{sat}(T)$ from $\Delta_\alpha^\beta H_m$; these methods are discussed in *Experimental Thermodynamics Volume IV* (Marsh and O'Hare 1994).

7.4.1.3 How Do I Measure Critical Properties?

If Equation 4.21 is used to represent vapor pressure measurements it requires a measure of the critical temperature and an estimate of the critical pressure, if known. In addition, phase borders can be estimated (as described in Questions 4.4 through 4.7) using the vapor pressure and an equation of state (discussed in Question 4.7.2) that contains substance-specific parameters, which are expressed in terms of critical temperature and critical pressure with equations such as those provided in, for example, *Experimental Thermodynamics Volume VIII* (Goodwin, Sengers, and Peters 2010). Critical parameters are, thus, of considerable use in engineering. For a pure substance, the critical properties are obtained from numerous experiments and these include those known as sealed ampoule, flow methods, spontaneous boiling, open tube, as well as from analysis of (p, V, T), sound speed measurements, and methods that rely upon the phenomena of critical opalescence that are also used for the determination of critical loci for mixtures that can also be obtained from dew- and bubble-point curves. The reader interested in these measurements should consult, for example, *Experimental Thermodynamics Volume VII* (Weir and de Loos 2004).

7.4.1.4 How Do I Measure Sound Speed?

The speed of sound in a fluid medium depends, as discussed in Question 3.5.3, primarily on the thermodynamic properties of the medium, as given by Equations 3.81 and 3.82, and can be used to obtain heat capacity, density, compression factor, while sound speed measurements in solids are a source of elastic constants. Sound attenuation measurements can be used to determine transport coefficients. The speed of sound can usually be determined with very small random uncertainty and the systematic errors to which such measurements are exposed differ markedly from those encountered in conventional calorimetry and gas imperfections.

The most common techniques for measuring the speed of sound can be categorized as variable-frequency fixed-cavity resonators, variable path-length fixed-frequency interferometers, and time-of-flight methods. To determine the speed of sound from standing-wave measurements in either a cavity or interferometer requires efficient reflection of sound at the interface between the medium and the wall of the container and this is necessarily the case when the acoustic impedances (the product of density and sound speed) of the medium differs greatly from that of the wall as, for example, is the case for a gas inside a metallic container where the ratio is on the order of 10^{-5}. However, for liquids and dense gases this impedance mismatch is not so easily achieved because the ratio of the acoustic impedances of the materials is on the order of 0.1 and, therefore, measurements of the time required for a sound wave to travel a known distance are preferred for dense gases, liquids, and solids.

The methods used to measure sound speed can be selected given a knowledge of the phase of the material and the geometry that can be employed, which may be determined, for example, by constraints imposed by overall dimensions of the device. In principle, the measurements of a single resonance frequency of a known mode of oscillation within a cavity of known dimension, or of a single time-of-flight over a known distance, is sufficient to determine the speed of sound. In practice, most techniques provide redundancy in the form of measurements over different and resolved modes of oscillation or over different frequencies and path lengths. This redundancy provides a means of identifying and reducing sources of error in the measurements.

For gases at pressures below a few MPa, three principal sources of systematic errors have been identified in an experimental measurement of the speed of sound; these arise from precondensation, viscothermal boundary layers, and molecular thermal relaxation. Fortunately, these can be rendered either negligible or small enough to model with an appropriate experimental technique. For variable-frequency fixed-geometry resonators the frequency measured for a particular mode yields the ratio of the sound speed to a characteristic dimension of the cavity. These suffice to determine gas imperfections but a knowledge of the characteristic dimension at zero pressure allows the determination of R, T, or $C_{p,\mathrm{m}}^{\mathrm{pg}}(T)$. Indeed, measurement of the speed of sound is a convenient and precise route for determining the heat capacity of polyatomic gases with sources of error that differ markedly from those encountered in conventional calorimetry (Questions 1.8.5 through 1.8.7 and *Experimental Thermodynamics Volume V* (Marsh and O'Hare 1994).

To obtain absolute values of the sound of speed requires measurements of the characteristic dimension as a function of temperature and pressure. Values of the characteristic dimension can be obtained as a function of temperature from dialatometry, dimensional microwave measurements or by the most usual means of calibration with a gas of known molar mass and perfect-gas heat capacity. For time-of-flight measurements, the distance traveled is obtained from measurements with a fluid of known sound speed. In variable path-length fixed-frequency measurements the speed of sound can be determined directly from measurements of temperature and the characteristic dimension of length. However, the exacting measurement of length can be avoided by calibrating the resonator with a gas for which the perfect-gas heat capacity is known.

For gases at pressures on the order of 1 MPa a spherical resonator is the preferred technique for which the principle advantage is the presence of radial modes, because of both the absence of viscous damping at the surface and insensitivity of the frequency to imperfections in the geometry of the cavity. Spheres constructed without recourse to special machining methods are sufficient. For radial modes only the internal volume of the cavity is required to determine the sound speed with a relative uncertainty of 10^{-6}. The absence of viscous damping and the favorable volume-to-surface ratio in the sphere leads to resonance quality factors in gases that are greater than attainable with any other geometry of similar volume and operating frequency. For gases at higher pressures, greater than on the order of 10 MPa, time-of-flight methods, with an uncertainty of about ± 0.1 %, are preferred.

There are essentially two methods that are used to determine the speed of sound in liquids and these are variable path-length fixed-frequency interferometry and time-of-flight measurements. The time-of-flight methods can be divided into single and multiple path-length devices. A single path pulse echo apparatus that was modified by a fractional uncertainty of $<\pm 0.5$ % typically operate at frequencies on the order of 10 MHz and can be operated at temperatures up to 2100 K and pressures up to 200 MPa, although more typically at temperatures of less than 500 K. The path length is determined by calibration measurement with water for which the sound speed is known with sufficient precision. Time-of-flight measurements are often used for solids albeit with methods, which differ from those adopted for liquids.

The techniques chosen to determine sound speed rely on the measurements of frequency f, one of the most accurate ($\Delta f/f < 1 \cdot 10^{-8}$) and easily reproduced physical quantities. This is also the case for measurements of the relative electric permittivity that we will consider in Question 7.4.1.5.

7.4.1.5 How Do I Measure Relative Electric Permittivity?

Measurements of the relative electric permittivity of fluids (which is one of the electrical properties of a fluid) as a function of the pressure and the temperature have found diverse applications from determining the onset of phase separation to the heating value of natural gas and studies of evaporation and condensation of ^{3}He near its liquid-vapor critical point. Provided the molar polarizability $\wp(\rho, T)$ has simple dependencies on density, temperature and frequency the Clausius-Mossotti relation can be used to obtain the amount of substance density ρ_n of the fluid from

$$\rho_n \equiv \frac{\varepsilon - 1}{\varepsilon + 2} \wp(\rho, T)^{-1}, \tag{7.12}$$

where ε is the dielectric constant. The applications listed previously rely in part on Equation 7.12. For substances that are electrically insulating $\wp(\rho, T)$ is essentially independent of frequency and for small nonpolar molecules, for example, methane. Equation 7.12 is also almost independent of temperature and density. However, for polar molecules $\wp(\rho, T)$ can have significant density and temperature dependencies.

At frequencies $f < 100$ MHz the relative electric permitivity $\varepsilon(p, T)$ can be determined from the ratio of electrical impedances of a capacitor filled with the fluid divided by the impedance of the same capacitor when it is evacuated. The impedance ratio is a complex number $\boldsymbol{\varepsilon}_\mathrm{r}$ and given by $\boldsymbol{\varepsilon}_\mathrm{r}(f) = \boldsymbol{\varepsilon}'(f) - \mathrm{i}\boldsymbol{\varepsilon}''(f)$, where $\mathrm{Re}\{\boldsymbol{\varepsilon}_\mathrm{r}(f)\}$, the real part of the impedance ratio, is the dielectric constant $\boldsymbol{\varepsilon}_\mathrm{r}$ and $\mathrm{Im}\{\boldsymbol{\varepsilon}_\mathrm{r}(f)\}$ is the imaginary part of the impedance ratio given by $\boldsymbol{\varepsilon}''(f) = \sigma/\{2\pi f \varepsilon(p = 0)\}$ that accounts for electrical dissipation within the dielectric fluid, and where σ is the electrical conductivity; $\varepsilon(p=0) \approx 8.854187 \cdot 10^{-12}$ F $\cdot$ m^{-1} is the electric constant.

The selection of the method and frequency of operation required to measure ε can be estimated from either the ratio $\varepsilon'\varepsilon(p=0)/\sigma \equiv \tau_\mathrm{d}$, which is the time required for charges within a dielectric to reach the surface of the sample, or from the quality factor given by the ratio $\varepsilon'/\varepsilon'' = Q_d$. For $Q_d \gg 1$ or $2\pi f \tau_d \gg 1$, the dielectric loss is small and the equivalent circuit of the fluid-filled capacitor is that of a capacitor in parallel with a large resistor. However, for $2\pi f \tau_d \ll 1$, the electrical equivalent circuit is that of a capacitor with a very small resistor in parallel with it; the capacitor is nearly short circuited. Resonance methods are useful for determining ε' only when $Q_d \gg 1$.

For the practical measurement of $\varepsilon(p, T)$ there are four factors that determine the design of the capacitor to be used and these are as follows: (1) the fluid electrical conductivity, (2) the measurement frequency (which is determined by item 1); the instrumentation used for the measurement, (3) the mechanical stability of the capacitor, and (4) the capacitor geometry. At frequencies in the range 1–10^6 Hz impedance measurements are used, while over the frequency range ($\approx 10^6$–10^9) reflection coefficients are determined with network analyzers.

Ratio transformer bridges are often used to measure the complex impedance of capacitors at audio frequencies for which the preferred geometry is either a coaxial cylinder or a toroid; the latter is operated as a cross-capacitor. The capacitance of both the coaxial cylinders and toroid are insensitive to small displacements of their electrodes, while the cross-capacitor is also insensitive to the presence of dielectric films, such as permanent oxide layers or condensed fluids or adsorbed gas layers, on their electrodes. Resonance methods are used to measure capacitance at frequencies over the range 10^6–10^9 Hz and use a capacitor connected to an inductor where both parts contribute equally to the resonance frequency and so the inductor must be as stable as the capacitor.

7.4.2 What Are the Preferred Methods for the Measurement of Transport Properties?

The transport of mass, momentum, and energy through a fluid are the consequences of molecular motion and molecular interaction. At the macroscopic level, associated with the transport of each dynamic variable, is a transport coefficient or property, denoted by X, such that the flux J of each variable is proportional to the gradient of a thermodynamic state variable such as concentration or temperature. This notion leads to the simple phenomenological laws such as those of Fick, Newton, and Fourier for mass, momentum, or energy transport, respectively of

$$J = -X\nabla Y. \tag{7.13}$$

In Equation 7.13, Y is the appropriate state variable conjugate to the flux J and X depends upon the thermodynamic state of the system. These linear, phenomenological laws are fundamental to all processes involving the transfer of mass, momentum, or energy but, in many practical circumstances encountered in industry, the fundamental transport mechanisms arise in parallel with other means of transport such as advection or natural convection. In those circumstances the overall transport process is far from simple and linear. However, the description of such complex processes is often rendered tractable by the use of transfer equations, which are expressed in the form of linear laws such as

$$J = -C\nabla Y. \tag{7.14}$$

In Equation 7.14 the transport coefficient C is not simply a function of the thermodynamic state of the system but may depend upon the geometric configuration of the system and the properties of the surfaces, for example. We are concerned here with the transport properties X of materials, which depend only upon thermodynamic state of the material only. In practical situations the transport coefficients C will often have been expressed as correlations with respect to dimensionless groups that characterize the problem; the dependence on the property X is then parametric (Bird et al. 1960).

The vast majority of precise transport-property measurements have been performed on molecularly simple pure fluids at conditions close to ambient pressure and temperature. As one moves away from this set of circumstances, the amount of available information decays rather rapidly and its uncertainty increases dramatically. The three transport properties of the greatest concern are the viscosity, thermal conductivity, and mass-diffusion coefficients. In each case, although measurements have been conducted over a period of at least 150 years, it was not until around 1970 that techniques of an acceptable uncertainty were developed for the relatively routine measurement of any of these properties. There is ample evidence in the literature (Millat et al. 1995; Jensen 2001) of very large discrepancies among measurements made before that date. One reason for these discrepancies lies in the conflicting requirements that, to make a transport-property measurement, one must perturb an equilibrium state but, at the same time, make the perturbation as small as possible so that the property determined does refer to a well-defined thermodynamic state. The latter requirement implies that the signals to be measured in any such experiment are always small and up against the limits of resolution. There are methods based on light scattering that, rather than rely on perturbations from equilibrium, make use of statistical microscopic fluctuations present in a macroscopic thermodynamic equilibrium (Will and Leipertz 2001). Another reason for the discrepancies arises from the failure of some experimenters to develop rigorous working equations for their instruments using the full conservation equations of continuum mechanics. The last 25 years of the twentieth century saw a very considerable refinement in measurement resolution and the theory of instruments has now obviated many of these difficulties. In the following paragraphs, the most important techniques employed at present for the measurement of the viscosity, thermal conductivity, and diffusion coefficients will be briefly presented and the reader is referred to both *Experimental Thermodynamics Volume III* (Wakeham, Nagashima, and Sengers 1991), which is devoted to the measurement of transport properties, and, because of the time elapsed between the publication of this volume and today, the recent archival literature for further details.

7.4.2.1 How Do I Measure Viscosity?

Since 1970 two generic types of viscometer have received the greatest attention, the first makes use of the torsional oscillations of bodies of revolution and the second is based upon the rather simpler concept of laminar flow through capillaries. Both reduce the measurement of viscosity to measurements of mass, length, and time.

In the case of torsional oscillating-body viscometers, an essentially exact description of the motion allows measurements of low uncertainty. In such viscometers, the characteristics of the oscillator are affected by the presence of the fluid and its properties in a way that is readily measured. Thus, oscillating-disk viscometers have found the greatest application to both gases and liquids under relatively mild conditions of temperature and pressure. They have been especially important in the determination of gases at low density and high temperature in work pioneered by Kestin and his collaborators (*Experimental Thermodynamics Volume III,* Wakeham et al. 1991) and continued by Vogel and his group (Vogel 1972). For work on molten metals at elevated temperatures (up to 1,500 °C), instruments using oscillating cylinders, with the fluid inside or outside, have been favored for operational reasons. The work of Oye and his group (Oye and Torklep 1979) is an excellent example of what can be achieved. Among the great advantages of oscillating viscometers the fact that no bulk motion of the fluid is required and that the measurements can be made absolute are paramount. Although a knowledge of the density of a fluid is necessary to evaluate the viscosity from the measurements made, it need not be known with the same accuracy required of the viscosity.

Capillary viscometers measure the time of efflux of a known volume of fluid through a circular section tube. They intrinsically determine the kinematic viscosity, (dynamic viscosity/density), so that the evaluation of the dynamic viscosity requires knowledge of the density with a comparably low uncertainty. Such viscometers are most often employed for liquids at ambient conditions when a hydrostatic head provides the driving force or for gases allowing the decay of a generated pressure difference. At elevated temperatures, the use of a fluid pump complicates the experimental installation. Although the theory of such viscometers is superficially simple, the details of some of the applicable corrections have only recently been resolved (Millat et al. 1995). Furthermore, owing to the difficulty of knowing the dimensions of the capillary tube and the uniformity of bore with a very high precision, measurements are mostly performed on a relative basis. We note the excellent work of Smith and his collaborators (Clarke and Smith 1968) in the dilute gas phase from (90 to 1,500) K, as well as the work of Nagashima and his coworkers on water (Kobayashi and Nagashima 1985).

In the case of high pressures, different types of viscometer have been employed owing to the need to reduce the volume of fluid required. The most popular have been falling-body viscometers and torsional-crystal viscometers (Wakeham et al. 1991). However, in neither cases are there completely developed physically based working equations so that their uncertainty is intrinsically limited (Wakeham et al. 1991). On the other hand, the vibrating-wire viscometer (Assael et al. 1991; Assael and Wakeham 1992) that makes use of the damping of a transverse oscillation of a thin wire enjoys a complete working equation based on the Navier–Stokes equations, that is, essentially exact (see Question 7.3.1.1.1).

The application of optical techniques such as that involving the study of the frequency and decay of surface waves on fluids (known as ripplons) (Nagasaka 2002) has seen tremendous development and routinely permits the simultaneous determination of liquid viscosity and surface tension (Fröba and Leipertz 2003).

It is now possible to achieve an uncertainty in the measurement of the viscosity of a fluid of a few parts in a thousand under near ambient conditions, which deteriorates to a few percent at extremely low and high temperatures for low densities. In very high-pressure gases and liquids the uncertainty achieved is at best a few percent and frequently very much worse.

7.4.2.2 How Do I Measure Thermal Conductivity?

In the case of the thermal conductivity there are three main techniques: those based upon Equation 7.13 and those based upon a transient application of it. Before about 1975 two forms of steady-state technique dominated the field: parallel-plate devices, in which the temperature difference between two parallel disks on either side of a fluid is measured when heat is generated in one plate, and concentric cylinder devices, which apply the same technique in an obviously different geometry. In both cases, early work ignored the effects of convection. In more recent work, exemplified by the careful work in Amsterdam with parallel plates (Mostert et al. 1989) and in Paris with concentric cylinders (Tufeu 1971), the effects of convection have been investigated. Indeed, the parallel-plate cells employed in Amsterdam by van den Berg and his coworkers (Mostert et al. 1989) have the unique feature that, because the temperature difference imposed can be very small and the horizontal fluid layer very thin, it is possible to approach the critical point in a fluid or fluid mixture very closely (to within a few mK).

Only one transient technique has enjoyed success and that is the transient hot-wire method, in a form pioneered by Haarman (Wakeham et al. 1991) and subsequently developed for a wide range of applications in gases and liquids for temperatures from (70 to 500) K and pressures up to 700 MPa. The essential features of this technique are that it measures the transient temperature increase of a thin metallic wire (a few μm) immersed in the test fluid, following the initiation of electrical heating within it. Its principal advantages are that the temperature increases need last for a time of no more than 1 s so that the inertia of the fluid inhibits the development of significant convective heat transfer. In addition, the small magnitude of the temperature difference applied (about 2 K), but the large temperature gradient (about 10^6 K · m^{-1}), means that radiative heat transfer is also generally insignificant (Wakeham et al. 1991). Although, the technique is unsuitable for work near the critical point it has been successfully employed over a very wide range of conditions for different types of fluids. In general, an uncertainty of a few parts in a thousand in the thermal conductivity around ambient conditions is possible for simple fluids, but this is degraded to several percent at extremes of temperature although relatively unaltered by pressure.

The transient hot-wire method has recently been applied to polar or electrically conducting fluids with considerable success, while a transducer made of a platinum wire embedded between two thin layers of alumina has recently been employed to measure the thermal conductivity of molten metals (Dix et al. 1998). Finally, a variety of new techniques for measuring the thermal diffusivity of fluids have been introduced, relying largely on light scattering (Will et al. 1998; Nagasaka 2002). These methods have distinct advantages in special fluids and regions of thermodynamic state, particularly near the critical state. This approach is of particular interest because the primary quantity determined is the thermal diffusivity, rather than the thermal conductivity. A combination of methods with potentially different sources of systematic error can be used to determine the consistency of measurements involving the determination of both density and heat capacity.

7.4.2.3 How Do I Measure Diffusion Coefficients?

The measurement of diffusion coefficients in either gases or liquids is a very slow process and many techniques require days to attain a single result. This is largely because of the intrinsic slowness of the process and the fact that most methods use equipment of large scale. For that reason alone, there are relatively few experimental results and the measurements do not extend over wide ranges of temperatures. On the other hand there are a great number of different techniques for the measurement of diffusion coefficients from optical interferometric methods to nuclear magnetic resonance (NMR) measurements of spin relaxation and chromatographic flow broadening (Wakeham et al. 1991). The range is too wide to treat here but it is worthwhile noting that the most precise interferometric techniques yield diffusion coefficients with an uncertainty of about ±0.1 % but only near ambient conditions. Away from these conditions, other techniques are usually employed and the uncertainty then is typically a few percent (Wakeham et al. 1991).

7.5 How Do I Calculate Thermodynamic Properties?

For the calculation of equilibrium properties required within the relationships provided in Chapters 1 through 4 for both pure fluids and mixtures two approaches are extensively employed today:

(a) The fastest and easiest way is to employ a generalized equation of state. In the case of pure fluids, equilibrium properties are calculated as a function of the critical parameters and the acentric factor, while in the case of mixtures, appropriate mixing rules of these parameters are usually incorporated. In addition, for mixtures, a binary interaction parameter k_{ij} (see Question 4.7) needs to be deduced usually by fits to measured VLE. The most commonly employed equations of state (Assael et al. 1996) were discussed in Question 4.7.2 and are as follows:
  - Peng and Robinson (PR) (Peng and Robinson 1976)
  - Benedict, Webb, and Rubin (BWR) in the Han and Starling form (Starling and Han 1972)

This approach is recommended for nonpolar fluids.

(b) When a lower uncertainty in the estimated value of the property is required the corresponding-states approach is preferred, which was discussed previously in Question 2.8. One of the most widely employed corresponding-states schemes is the three-parameter scheme proposed by Lee and Kesler (LK) (1975) and its four-parameter modification for polar molecules proposed by Wu and Stiel (1985). This approach also requires critical parameters and acentric factors, while in the case of mixtures, usually the Plöcker (Plöcker et al. 1978) mixing rules for the critical parameters are employed. The corresponding-states approach constitutes a more accurate scheme, for systems containing polar molecules; these methods involve complex and long calculations.

The aforementioned two approaches will be demonstrated with examples in Questions 7.4.1 and 7.4.2.

7.5.1 How Do I Calculate the Enthalpy and Density of a Nonpolar Mixture?

Let us consider, as an example, the calculation of the density and enthalpy of a mixture (0.33 octane + 0.67 benzene) at (a) $T = 470$ K and $p = 1.4$ MPa and (b) $T = 590$ K and $p = 9.7$ MPa.

For this example, the calculations have been performed with the PR, BWR in the Han and Starling form, and the LK scheme. The SUPERTRAPP software package (Huber 1998) supplied by NIST and based on the principle of corresponding states is also employed. The results for the density ρ and the specific

TABLE 7.2 DENSITY ρ AND SPECIFIC ENTHALPY AND DIFFERENCE IN ENTHALPY Δh BETWEEN A TEMPERATURE OF 470 K AND PRESSURE OF 1.4 MPa AND A TEMPERATURE OF 590 K AND PRESSURE OF 9.7 MPa FOR (0.33 OCTANE + 0.67 BENZENE)

	ρ (470 K, 1.4 MPa)/ kg · m^{-3}	h (470 K, 1.4 MPa)/ kJ · kg^{-1}	ρ (590 K, 9.7 MPa)/ kg · m^{-3}	h (590 K, 9.7 MPa)/ kJ · kg^{-1}	Δh/kJ · kg^{-1}
PR	579	21.3	418	363.0	340
BWR	601	17.3	433	360.6	343
LK	597	15.2	462	347.3	332
SUPERTRAPP	608		463		349
Exp. (Lenoir et al. 1971)					315

enthalpy h are listed in Table 7.2. In the same table the enthalpy difference Δh between the two states is also given and compared with the experimental value. The interaction parameter k_{ij} was equal to 0.001 for both equations of state. For the calculations the computer programs given in Assael et al. (1996) were employed. Enthalpy was arbitrarily set equal to zero at $T = 273.15$ K and $p = 0.101325$ MPa to provide h but this cancels for Δh as can be seen from the definition of standard enthalpy in Chapter 1 and how to measure the enthalpy variations with temperature and pressure in Chapters 1 and 3.

In Table 7.2, it can be seen that the density predicted by the BWR equation of state, the LK corresponding-states scheme, and SUPERTRAPP lie within 1.2 %, at a pressure of 1.4 MPa. However, at a pressure of 9.7 MPa the two corresponding-states schemes provide estimates of density that lie within 0.2 % of each other, while both of the equations of state underestimate the density between 10 % and 6 %, respectively. In the case of the enthalpy difference, comparison with the experimental value (Lenoir et al. 1971) indicates that all schemes seem to overestimate the experimental value up to 6 %.

In general, one can state the corresponding-states schemes represent the behavior of nonpolar mixtures better than the equations of state. If speed of calculation is essential, calculations using the BWR equation of state will certainly provider results faster than the two corresponding-states schemes.

7.5.2 How Do I Calculate the Enthalpy and Density of a Polar Substance?

The enthalpy and density of polar substances are usually more difficult to calculate. Generalized equations of state do not apply, as they were derived

TABLE 7.3 DENSITY ρ (250 K, 2 MPa) AND ρ (450 K, 10 MPa) ALONG WITH THE SPECIFIC ENTHALPY CHANGE Δh BETWEEN THESE TWO STATES FOR 1,1,1,2-TETRAFLUOROETHANE, A REFRIGERANT GIVEN THE ACRONYM R134a

	ρ (250 K, 2 MPa)/kg · m^{-3}	ρ (450 K, 10 MPa)/kg · m^{-3}	Δh/kJ · kg^{-1}
Lee–Kesler	1322	464	319
Wu–Steil	1371	473	318
TransP	1371	475	
Exp. (Sato et al. 1994)	1371	475	324

for nonpolar substances, while extra corrections have to be employed for corresponding-states schemes. We will demonstrate this by calculating the density of 1,1,1,2-tetrafluorethane (commonly known in the refrigeration industry by the ASHRAE Standard 34 nomenclature as R-134a) at the following temperatures and pressures: (a) $T = 250$ K and $p = 2$ MPa and (b) $T = 450$ K and $p = 10$ MPa. We will also estimate the difference in enthalpy between the temperatures and pressure of (a) and (b).

Since R-134a is a polar molecule, the corresponding-states scheme of Lee and Kesler (1975) with the Wu and Stiel (1985) modification was employed; see Question 2.8.1. The results are shown in Table 7.3 together with the experimental value (Sato et al. 1994). As expected, the value predicted by the Wu and Stiel (1985) modification to LK for polar fluids produces values for the density and enthalpy, which are in very good agreement with those obtained from experiment. Hence, in polar fluids, a corresponding-states scheme corrected for polar interactions is recommended.

In the same table, values calculated by the software TransP (Assael and Dymond 1999), which is based on hard-spheres, are also included. These values also show an excellent agreement as the scheme, although restricted in its application, has a sound theoretical basis.

7.5.3 How Do I Calculate the Boiling Point of a Nonpolar Mixture?

For VLE calculations with mixtures of nonpolar substances fugacity coefficients obtained from an equation of state can be used (see Question 4.4.1). The mole fractions in the vapor and liquid phases y_i and x_i expressed as a ratio of the fugacity coefficients for the liquid $\phi_{B,l}(T, p, x_C)$ to that of the gas $\phi_{B,g}(T, p, y_C)$ are given by Equations 4.83 and 4.69 (where the fugacity of the liquid $\tilde{p}_{B,l}(T,p,x_C)$ and gas $\tilde{p}_{B,g}(T,p,y_C)$ are given by Equations 4.82 and 4.68) by

$$K_p = \prod_B \frac{\phi_{B,l}(T,p,x_C)}{\phi_{B,g}(T,p,y_C)} = \prod_B \frac{\tilde{p}_{B,l}(T,p,x_C)y_B}{\tilde{p}_{B,g}(T,p,y_C)x_B} = \prod_B \frac{y_B}{x_B}. \quad (7.15)$$

This calculation can be performed with a cubic equation of state discussed in Chapter 4, Question 4.7.2 (Goodwin et al. 2010), and requires for the components of the mixture knowledge of the critical temperature, critical pressure, acentric factors, and binary interaction parameters.

We can demonstrate the use of Equation 7.15 by calculating the boiling point of a mixture $\{xCO_2 + (1-x)C_2H_6\}$ at $p = 3.025$ MPa with $x = 0.31$. In this example, the PR equation of state was used (See Question 4.7.2 and Goodwin et al. 2010). The mixture is of particular interest, as it will be shown in next section to exhibits azeotropic behavior (discussed in Question 4.11.4).

Equation 7.15 is of course to be applied with the requirement that the sum of mole fractions in the liquid phase must be equal to unity. As already mentioned, for mixtures, a binary interaction parameter k_{ij} is required and, when unknown, is typically set equal to zero. To demonstrate the effect of this parameter, the calculations were conducted with $k_{ij} = 0$ and $k_{ij} = 0.124$.

For $k_{ij} = 0.124$ the normal boiling temperature T_b (at a pressure of 0.1 MPa) was estimated to be 263.15 K and the vapor mole fraction y found equal to 0.31; the measured azeotrope composition at $x \approx 0.7$ was predicted. If, however, $k_{ij} = 0$ then at a pressure of 0.1 MPa the estimated T_b was found to be 271.31 K and $y = 0.73$.

For $(CO_2 + C_2H_6)$ these results show that setting $k_{ij} = 0$ results in a predicted T_b some 8 K greater than the measured value and that the azeotrope is not predicted. Of course, this effect was pronounced because of the azeotropic behavior (Question 4.11.4) of the mixture. In general, Equation 7.15 is an excellent method of estimating the normal boiling temperature at a pressure of 0.1 MPa for a mixture of nonpolar components.

7.5.4 How Do I Calculate the VLE Diagram of a Nonpolar Mixture?

The procedure adopted in this case is essentially the same as that developed in Question 7.5.3, and as an example, we can construct the (vapor + liquid) equilibrium diagram of $(CO_2 + C_2H_6)$ at $T = 263.15$ K that is the $p(x)$ section at constant T. At a given temperature the pressure is calculated as a function of liquid mole fraction x. The $p(x)_T$ section has been estimated with $k_{ij} = 0$ and $k_{ij} = 0.124$ and along with the measured values shown in Figure 7.4 (which is identical to

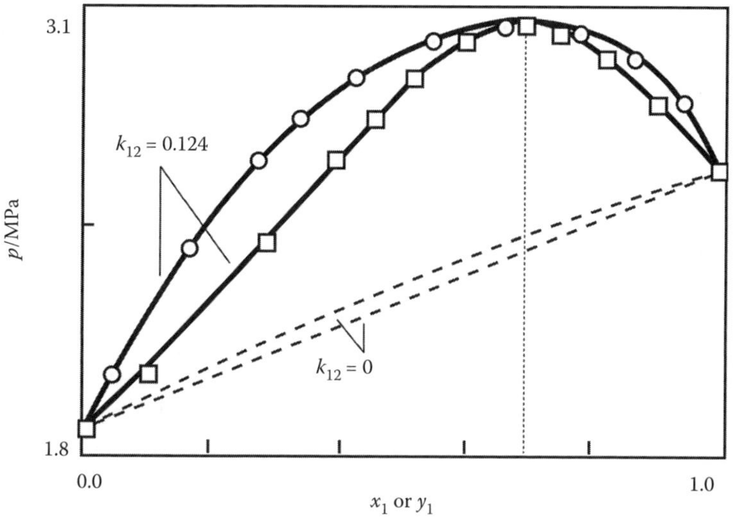

Figure 7.4 $P(x)_T$ section for the vapor + liquid equilibrium of $\{CO_2(1) + C_2H_6(2)\}$ as a function of mole fraction x of the liquid and y of the gas phases. ○, liquid phase measured bubble pressure (Fredenslund and Mollerup 1974); □, gas phase measured dew pressure (Fredenslund and Mollerup 1974) ——, estimated from the Peng-Robinson equation of state with $k_{12} = 0.124$; - - - - -, estimated from the Peng-Robinson equation of state with $k_{12} = 0$; vertical ··········, indicates the azeotropic mixture at $x = 0.7$.

Figure 4.7) illustrating the variation of prediction with the value of k_{ij}, and in this case for $k_{ij} = 0$ the azeotrope was not estimated.

7.5.5 How Do I Calculate the VLE of a Polar Mixture?

In the case of polar components, activity coefficients are introduced to describe the liquid phase (see Question 4.6). In this case, the mole fractions in the vapor and liquid phases y_i and x_i are expressed with an activity coefficient so that the VLE is determined with Equation 4.151

$$y_B \phi_{B,g}(T, p^{sat}, y_B)\, p = x_B f_{B,l}(T, p, x_B)\, \tilde{p}_{B,l}(T, p, x_B)\, p_B^{sat} F_B, \qquad (7.16)$$

that is, Equation 4.155

$$K_B = \frac{y_B}{x_B} = \frac{f_{B,l}(T, p, x_B)\, \tilde{p}_{B,l}(T, p, x_B)\, p_B^{sat} F_B}{\phi_{B,g}(T, p^{sat}, y_B)\, p}. \qquad (7.17)$$

In Equations 7.16 and 7.17 $f_{B,l}(T, p, x_B)$ is the activity coefficient (Equation 4.111), $\phi_{B,g}(T, p^{sat}, y_B)$ is the fugacity coefficient (Equations 4.68 and 4.69), $\tilde{p}_{B,l}(T, p, x_B)$

is the liquid fugacity (Equation 4.81), p_B^{sat} is the vapor pressure (obtained, e.g., from Equation 4.21 or at temperatures about T_b by Equation 4.20), the F_B is the Poynting factor (Equation 4.86), and p is the system pressure. The $\phi_{B,g}(T, p^{sat}, y_B)$ and $\tilde{p}_{B,1}(T, p, x_B)$ are usually obtained from an equation of state, while the $f_{B,1}(T, p, x_B)$ from an activity-coefficient model (such as those known as Wilson, Non-Random Two Liquid [NRTL], or Universal Functional Activity Coefficient [UNIFAC] discussed in Question 4.6.5). Since the parameters of NRTL and UNIFAC are obtained from measured VLE it is not surprising that the predictions obtained from this approach differ from experiment less than the results obtained solely from fugacity coefficients. This approach is more complex but is preferred when $f_{B,1}(T, p, x_B)$ can be determined for both nonpolar and polar fluids.

We will now use Equation 7.17 to construct the VLE diagram for the liquid and vapor phases of a mixture of (water + ammonia) at $T = 293.15$ K. The algorithm required is identical with that described in Questions 7.5.3 and 7.5.4, and the calculations were performed with both the PR and the BWR equations of state with the Wilson activity coefficient model. The results obtained are shown in Figure 7.5

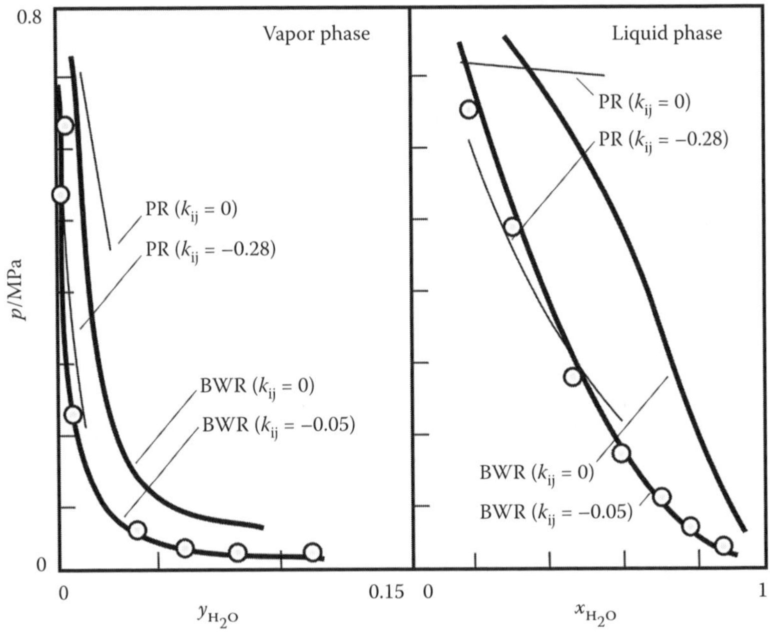

Figure 7.5 $p(y_{H_2O})_T$ and $p(x_{H_2O})_T$ sections for the vapor and liquid phases, respectively, for water + ammonia at $T = 293.15$ K. ———: Peng-Robinson equation of state with $k_{ij} = 0$ and $k_{ij} = -0.28$; ———: Benedict, Webb, and Rubin equation of state with $k_{ij} = 0$ and $k_{ij} = -0.05$; ○: measured values.

for the preferred binary interaction parameter for each equation of state and $k_{ij} = 0$ to illustrate further the importance of this parameter. Although in the vapor phase the two approaches might look similar, in the liquid phase the differences are significantly greater. It is clearly evident that if the correct value of the binary interaction parameter is not employed, the VLE can not be predicted.

7.5.6 How Do I Construct a VLE Composition Diagram?

The use of Equation 7.17 to construct a composition diagram of a mixture will be given with, for example, (ethanol + benzene) at a temperature of 333 K. In this example, the activity coefficients were obtained from the Wilson model, the vapor pressure from Antoine's equation for each substance (Equation 4.20) and the fugacity coefficient from the virial equation of state. All required parameters and the computer program used for these calculations were obtained from Assael et al. (Assael et al. 1996); the program can be obtained without cost from anonymous ftp at ftp://transp.cheng.auth.gr/. The estimates obtained from these calculations are shown in Figure 7.6 and are in excellent agreement with

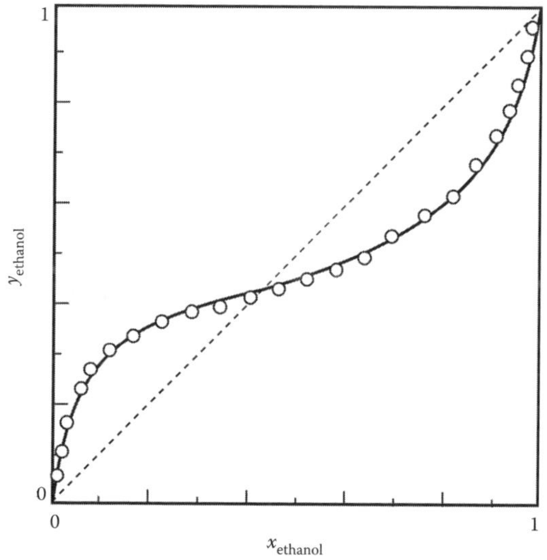

Figure 7.6 The gas and liquid mole fractions x_{ethanol} and y_{ethanol} for (ethanol + benzene) at temperatures of 333 K. ———, calculated with parameters and computer programs reported by Assael et al. (1996); ○, Han et al. 2007; and - - -, solely to illustrates $y = x$ that is often included by chemical engineers. According to IUPAC nomenclature, the axis labels should be written as $y(C_2H_5OH)$ and $x(C_2H_5OH)$ for the ordinate and abscissa, respectively, rather than the form shown that is typically adopted by chemical engineers.

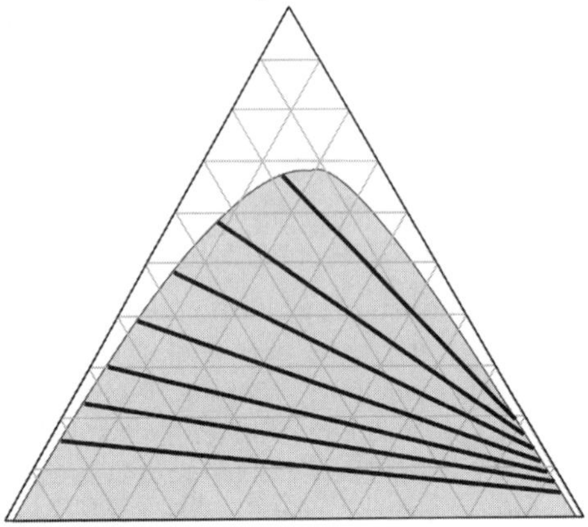

Figure 7.7 Schematic of an LLE composition diagram for (propan-2-one + methylbenzene + water).

the measured values and demonstrate that the activity-coefficient model is the preferred method for VLE calculations.

7.5.7 How Do I Construct a LLE Composition Diagram?

Equation 7.17 can also be applied to the estimation of (liquid + liquid) equilibria (often given the acronym LLE). As an example, we have estimated the LLE for (propan-2-one + methylbenzene + water), at $T = 283.15$ K and $p = 0.1$ MPa with the activity coefficient obtained from the UNIQUAC model and all other required parameters obtained from Assael et al. (Assael et al. 1996) with the results shown in Figure 7.7.

7.6 HOW DO I CALCULATE TRANSPORT PROPERTIES?

The transport properties of fluids are often expressed (Assael et al. 1996) as the sum of three contributions, a zero-density contribution, which depends only on temperature (essentially the value at the limit of zero density), a critical enhancement term, and an excess contribution that describes the density dependence away from the critical region.

The zero-density contribution is well understood and is readily obtained (Assael et al. 1996). The critical enhancement is also understood and can be calculated in most cases (Assael et al. 1996). The excess contribution, however, is more difficult to obtain. For dense gases and liquids away from the critical region, methods based on the Enskog theory for hard spheres give an excellent representation of experimental data (Assael et al. 1992a, 1992b). A more generalized approach can be obtained by adopting a scheme based on the principle of corresponding states. Although this formalism lacks a rigorous theoretical background, the addition of so-called "shape factors" permits a description of the liquid and vapor phases for pure fluids and their mixtures with sufficient certainty for the purpose of engineering. A corresponding-states approach has very successfully been applied to hydrocarbons (Huber 1998) and to refrigerants (Gallagher et al. 1999). To illustrate the use of the corresponding states in this regard, we calculate the viscosity of liquid ($0.5\,C_8H_{18} + 0.5\,C_{12}H_{26}$),* at $T = 323.22$ K and pressures of (0.1 and 96.1) MPa. Two methods were used for these calculations: (1) based on the principle of corresponding-states as provided within the computer package SUPERTRAPP (Huber 1998), and (2) a scheme based on hard spheres encoded in the computer package TRANSP (Assael and Dymond 1999). The values obtained from these calculations are listed in Table 7.4 together with measured values (Assael et al. 1991) that have an expanded uncertainty of about ±1 %. The values listed in Table 7.4 also include the differences between the measured and estimated viscosity, which is never more than about 2.2 % that is about twice the estimated expanded uncertainty of the measurements and would be considered excellent agreement. Unfortunately, this is a best case and estimates with similar differences from measured values cannot be obtained for all other fluid mixtures. The program SUPERTRAPP covers the whole liquid and vapor phases for a large number of hydrocarbons and their mixtures, the application of TRANSP is limited to the liquid phase and to a small number of components and mixtures.

TABLE 7.4 THE MEASURED VISCOSITY η(EXPT) OF AN EQUIMOLAR (0.5 OCTANE + 0.5 DODECANE) AT A TEMPERATURE $T = 323.22$ K AS A FUNCTION OF PRESSURE p ALONG WITH THE ESTIMATED VALUES η(CALC) DETERMINED FROM TWO ALGORITHMS, ONE KNOWN BY THE ACRONYM SUPERTRAPP (HUBER 1998) THE OTHER TRANSP (ASSAEL AND DYMOND 1999), AND DIFFERENCE $\Delta\eta = \eta$(CALC) − η(EXPT). THE η(EXPT) WERE REPORTED BY ASSAEL ET AL. (1991)

p/MPa	η(expt)/ µPa · s	η(calc)/ µPa · s	$100 \cdot \Delta\eta/\eta$	η(calc)/ µPa · s	$100 \cdot \Delta\eta/\eta$
		SUPERTRAPP		TRANSP	
0.1	623	635	1.9	637	2.2
96.1	1482	1501	1.3	1482	0

* C_8H_{18} is octane and $C_{12}H_{26}$ is dodecane.

For the prediction of transport properties we mention one final source of the theoretically based scheme reported by Vesovic and Wakeham (1989; Royal et al. 2005), which provides estimates of the viscosity and thermal conductivity of gases and liquid mixtures with densities on the order of 100 kg·m^{-3} from the pure component values.

For temperatures of the order of 1000 K even the measured thermal conductivity obtained from a variety of methods and sources exhibit differences. For example, the thermal conductivity of KCl(l) and NaCl(l) reported by different workers differ, as Figure 7.8 shows, by >±100 %. The uncertainties of the different measurement techniques used were cited by each of the authors to be of the order of ±1 %. When the user is faced with measurements of the same property that differ by ±100 %, discriminating values that are plausibly more reliable than the others in view of the cited uncertainties of the measurements requires either considerable knowledge of the measurement technique or of the procedure used. In this particular case, chance selection through

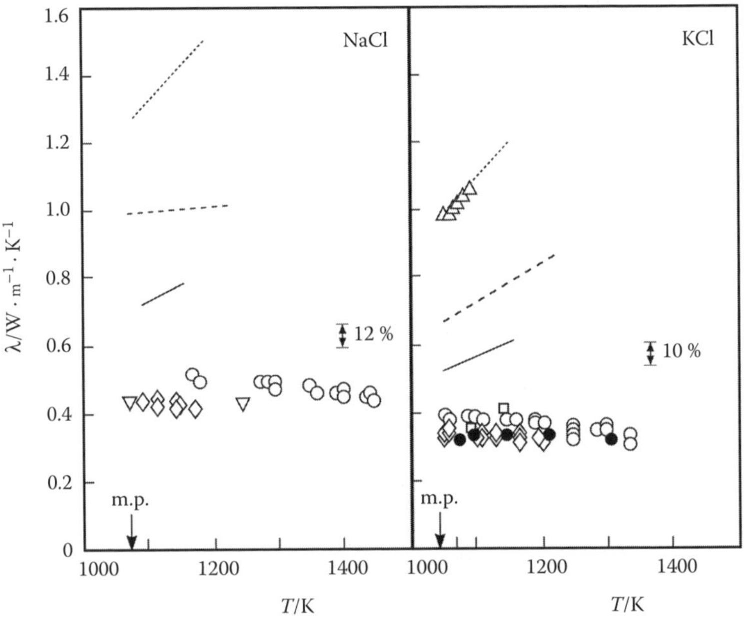

Figure 7.8 Thermal conductivity λ of molten KCl(l) and NaCl(l) as a function of temperature T reported in the archival literature. ·········, Bystrai et al. (1974); - - - -, Fedorov and Machuev (1970); ——; Smirnov et al. (1987); ▽, Golyshev et al. (1983); ○, Nagasaka et al. (1992); ◇, Harada (1992, personal communication); □, McDonald and Davis (1971); △, Polyakov and Gildebrandt (1974).

a blind-folded scientist with a pin is probably as good a means of selection as any other; which demonstrates there is much still to be done.

7.7 REFERENCES

Assael M.J., and Wakeham W.A., 1992, "Vibrating-wire viscometry on liquids at high pressure," *Fluid Phase Equilib.*, **75**:269–285.

Assael M.J., and Dymond J.H., 1999, *TRANSP V.2.0: A computer package for the calculation of high-pressure liquid phase transport properties.* Available from the authors.

Assael M.J., Dymond J.H., and Patterson P.M., 1992a, "Correlation and prediction of dense fluid transport coefficients—V. Aromatic hydrocarbons," *Int. J. Thermophys.* **13**:895–905.

Assael M.J., Dymond J.H., Papadaki M., and Patterson P.M., 1992b, "Correlation and prediction of dense fluid transport coefficients—I. n-Alkanes," *Int. J. Thermophys.* **13**:269–281.

Assael M.J., Nieto de Castro C.A., and Wakeham W.A., 1978, *The estimation of physical properties of fluids. Part II. The economic advantages of accurate transport property data, Proceedings of CHEMPOR,* pp. 16.1–16.9.

Assael M.J., Papadaki M., Richardson S.M., Oliveira C., and Wakeham W.A., 1991, "Vibrating-wire viscometry on liquid hydrocarbons at high pressure," *High Temp. – High Press.*, **23**:561.

Assael M.J., Trusler J.P.M., and Tsolakis T.F., 1996, *Thermophysical Properties of Fluids. An Introduction to Their Prediction.* Imperial College Press, London.

Bird R.B., Stewart W.E., and Lightfoot E.N., 1960, *Transport Phenomena,* John Wiley & Sons Inc., New York.

Bystrai G.P., Desyatnik V.N., and Zlokazov V.A., 1974, "Thermal conductivity of molten uranium tetrachloride mixed with sodium chloride and potassium chloride," *Atom. Energ.* **36**:517–518.

Clarke A.G., and Smith E.B., 1968, "Low-temperature viscosities of argon, krypton and xenon," *J. Chem. Phys.* **48**:3988–3991.

Dix M., Drummond I.W., Lesemann M., Peralta-Martinez V., Wakeham W.A., Assael M.J., Karagiannidis L., and van den Berg H.R., 1998, *Proceedings of 5th Asian Thermophysical Properties Conference,* Seoul, pp. 133–136.

Dymond J.H., Marsh K.N., Wilhoit R.C., and Wong, K.C., 2002, *Virial Coefficients of Pure Gases and Mixtures Group IV Physical Chemistry Volume 21* in *Subvolume A Virial Coefficients of Pure Gases.* Landolt-Börnstein *Numerical Data and Functional Relationships in Science and Technology.* Martienssen W. (chief), eds. Frenkel M., and Marsh K.N., Springer-Verlag, New York.

Dymond J.H., Marsh K.N., and Wilhoit R.C., 2003, *Virial Coefficients of Pure Gases and Mixtures Group IV Physical Chemistry Vol. 21 Subvolume B Virial Coefficients of Mixtures.* Landolt-Börnstein *Numerical Data and Functional Relationships in Science and Technology.* Martienssen W. (chief), eds. Frenkel M., and Marsh K.N., Springer-Verlag, New York.

Experimental Thermodynamics, Volume I, Calorimetry of Non-Reacting Systems, 1968, eds. McCullough J.P., and Scott D.W., for IUPAC, Butterworths, London.

Experimental Thermodynamics, Volume II, Experimental Thermodynamics of Non-Reacting Fluids, 1975, eds. Le Neindre B., and Vodar B., for IUPAC, Butterworths, London.

Experimental Thermodynamics, Volume III, Measurement of the Transport Properties of Fluids, 1991, eds. Wakeham W.A., Nagashima A., and Sengers J.V., for IUPAC, Blackwell Scientific Publications, Oxford.

Experimental Thermodynamics, Volume IV, Solution Calorimetry, 1994, eds. Marsh K.N., and O'Hare P.A.G., for IUPAC, Blackwell Scientific Publications, Oxford.

Experimental Thermodynamics, Volume V, Equations of State for Fluids and Fluid Mixtures, Parts I and II, 2000, eds. Sengers J.V., Kayser R.F., Peters C.J., and White H.J., Jr., for IUPAC, Elsevier, Amsterdam.

Experimental Thermodynamics, Volume VI, Measurement of the Thermodynamic Properties of Single Phases, 2003, eds. Goodwin A.R.H., Marsh K.N., and Wakeham W.A., for IUPAC, Elsevier, Amsterdam.

Experimental Thermodynamics, Volume VII, Measurement of the Thermodynamic Properties of Multiple Phases, 2005, eds. Weir R.D., and de Loos T.W., for IUPAC, Elsevier, Amsterdam.

Experimental Thermodynamics, Volume VIII, Applied Thermodynamics of Fluids, 2010, eds. Goodwin A.R.H., Sengers J.V., and Peters C.J., for IUPAC, RSC Publishing, Cambridge.

Fedorov V.I., and Machuev V.I., 1970, "Thermal conductivity of fused salts," *Teplofiz. Vys. Temp.* **8**:912–914.

Fredenslund A., and Mollerup J., 1974, "Measurement and prediction of equilibrium ratios for C_2H_6+CO_2 system," *J. Chem. Soc. Faraday Trans. I* **70**:1653–1660.

Fröba A., and Leipertz A., 2003, "Accurate determination of liquid viscosity and surface tension using surface light scattering (SLS): Toluene under saturation conditions between 260 and 380 K," *Int. J. Thermophys.* **24**, 895–920.

Gallagher J., McLinden M., Morrison G., and Huber M., 1999, *REFPROP V.5.0: A Computer Package for the Calculation of the Thermodynamic Properties of Refrigerants and Refrigerant Mixtures*, Available from the National Institute of Standards and Technology, Gaithersburg, MD.

Gallieto G., and Boned C., 2008, "Dynamic viscosity estimation of hydrogen sulfide using a predictive scheme based on molecular dynamics," *Fluid Phase Equilib.* **269**:19–26.

Golyshev V.D., Gonik M.A., Petrov V.A., and Putilin Yu.M., 1983, *Teplofiz. Vys. Temp.* **21**:899.

Han K.-J., Hwang I.-C., Park S.-J., and Park I.-H., 2007, "Isothermal vapor-liquid equilibrium at 333.15 K, density, and refractive index at 298.15 K for the ternary mixture of dibutyl ether plus ethanol plus benzene and binary subsystems," *J. Chem. Eng. Data* **52**:1018–1024.

Harvey A., Peskin A.P., and Klein S.A., 2008, *NIST/ASME STEAM, Version 2.2, Formulation for General and Scientific Use*, Available by National Institute of Standards and Technology, Gaithersburg, MD.

Huber M., 1998, *SUPERTRAPP V.2.0: A Computer Package for the Calculation of the Transport Properties of Nonpolar Fluids and their Mixtures*, Available by National Institute of Standards and Technology, Gaithersburg, MD.

Jensen K.F., 2001, "Microreaction engineering—is small better?," *Chem. Eng. Sci.* **56**:293–303.

Kern D.Q., 1950, *Process Heat Transfer*, McGraw-Hill International, Tokyo.

Kobayashi K., and Nagashima A., 1985, "Measurements of the viscosity of sea water under high pressure," *High Temp.-High Press.* **17**:131–140.

Lee B.I., and Kesler M.G., 1975, "Generalized thermodynamic correlation based on 3-parameter corresponding states," *A.I.Ch.E. J.* **21**:510–527.

Lemmon E.W., McLinden M.O., and Huber M.L., 2004, *NIST Standard Reference Database 23 version 7.1 (REFPROP)*, National Institute of Standards and Technology.

Lemmon E.W., McLinden M.O., Huber M.L., 2009, *NIST Standard Reference Database 23 version 8.0 (REFPROP)*, National Institute of Standards and Technology.

Lemmon E.W., McLinden M.O., and Huber M.L., *REFerence fluid PROPerties program 23*, Physical and Chemical Properties Division, National Institute of Standards and Technology Boulder, Colorado.

Lenoir J.M., Hayworth K.E., and Hipkin H.G., 1971, "Enthalpies of benzene and mixtures of benzene with n-octane," *J. Chem. Eng. Data* **16**:280–284.

Maitland G.C., Rigby M., Smith E.B., and Wakeham W.A., 1981, *Intermolecular Forces. Their Origin and Determination.* Clarendon Press, Oxford.

McDonald J., and Davis H.T., 1971, "Determination of the thermal conductivities of several molten alkali halides by means of a sheathed hot-wire technique," *Phys. Chem. Liq.* **2**:119–134.

Millat J., Dymond J.H., Nieto de Castro C.A., eds., 1995, *Transport Properties of Fluids. Their Correlation, Prediction and Determination*, Cambridge University Press, London.

Mostert R., van den Berg H.R., and van der Gulik P.S., 1989, "A guarded parallel-plate instrument for measuring the thermal conductivity of fluids in the critical region," *Rev. Sci. Instrum.* **60**:3466–3474.

Nagasaka Y., "Effect of atmosphere on surface tension and viscosity of molten LiNbO3 measured by the surface laser-light scattering method," *Proceedings of 16th European Conference on Thermophysical Properties*, London, 1–6 September, 2002.

Nagasaka Y., Nakazawa N., and Nagashima A., 1992, "Experimental determination of the thermal diffusivity of molten alkali-halides by the forced Rayleigh-scattering method. 1. Molten LiCl, NaCl, KCl, RbCl, and CsCl," *Int. J. Thermophys.* **13**:555–574.

Oye H.A., and Torklep K., 1979, "Absolute oscillating cylinder (or cup) viscometer for high temperatures," *J. Phys. E: Sci. Instrum.* **12**:875–885.

Peng D.Y., and Robinson D.B., 1976, "A new two-constant equation of state," *Ind. Eng. Chem. Fundam.* **15**:59–64.

Plöcker U., Knapp H., and Prausnitz J.M., 1978, "Calculation of high-pressure vapor-liquid equilibria from a corresponding-states correlation with emphasis on asymetric mixtures," *Ind. Eng. Chem. Proc. Des. Dev.* **17**:324–332.

Polyakov P.V., and Gildebrandt E.M., 1974, *Teplofiz. Vys. Temp.* **12**:1313.

Rayleigh M.A., 1970. *Argon*, in *The Royal Institution Library of Science,* Volume 4. eds. Bragg L., and Porter G., Applied Science Publishers, London.

Royal D., Vesovic V., Trusler, J.P.M., and Wakeham W.A., 2005, "Predicting the viscosity of liquid refrigerant blends: Comparison with experimental data," *Int. J. Ref. Rig.* **28**:311–319.

Sato H., Higashi Y., Okada M., Takaishi Y., Kafawa N., and Fukushima M., 1994, *JAR Thermodynamic Tables Volume 1: HFCs and HCFCs.* Japanese Association of Refrigeration, Tokyo.

Smirnov M.V., Khokholov V.A., and Filatov E.S., 1987, *Electrochim. Acta.* **32**:1019.

Starling K.E., and Han M.S., 1972, "Thermodata refined for LPG. 15. Industrial applications," *Hydrocarb. Process.* **51**:129–132.

Thermodynamic Research Center (TRC), (1942–2007), *Thermodynamic Tables Hydrocarbons,* ed. Frenkel M., National Institute of Standards and Technology Boulder, CO, Standard Reference Data Program Publication Series NSRDS-NIST-75, Gaithersburg, MD.

Thermodynamic Research Center (TRC), (1955–2007), *Thermodynamic Tables Non-Hydrocarbons,* ed. Frenkel M., National Institute of Standards and Technology Boulder, CO, Standard Reference Data Program Publication Series NSRDS-NIST-74, Gaithersburg, MD.

Tufeu R.,1971, "Etude experimental en fonction de la temperature et de la pression de la conductivite thermique de l' ensemble des gaz rares et des melanges helium-argon," PhD Thesis, Paris University.

Vesovic V., and Wakeham W.A., 1989, "Prediction of the viscosity of fluid mixtures over wide ranges of temperature and pressure," *Chem. Eng. Sci.* **44**:2181–2189.

Vogel E., 1972, "Construction of an all-quartz oscillating-disk viscometer and measurements on nitrogen and argon," *Wiss. Zeit. Rostock.* **21**:169–179.

Will S., and Leipertz A., 2001, "Thermophysical properties of fluids from dynamic light scattering," *Int. J. Thermophys.* **22**:317–338.

Will S., Fröba A., and Leipertz A., 1998, "Thermal diffusivity and sound velocity of toluene over a wide temperature range," *Int. J. Thermophys.* **19**:403–414

Wu G.Z.A., and Stiel L.I., 1985, "A generalized equation of state for the thermodynamic properties of polar fluids," *AIChE J.* **31**:1632–1644.

Index

Note: Page numbers followed by "n" denote footnotes.

D

F

G

O

P

Q

R

S

T

U

Z